Semiconducting and Metallic Polymers

Alan J. Heeger

Department of Physics, University of California, Santa Barbara

N. Serdar Sariciftci

Linz Institute for Organic Solar Cells (LIOS), Department of Physical Chemistry, Johannes Kepler University, Linz, Austria

Ebinazar B. Namdas

Centre for Organic Photonics and Electronics (COPE), School of Mathematics and Physics, University of Queensland, Brisbane, Australia

OXFORD
UNIVERSITY PRESS

OXFORD
UNIVERSITY PRESS
Great Clarendon Street, Oxford OX2 6DP

Oxford University Press is a department of the University of Oxford.
It furthers the University's objective of excellence in research, scholarship,
and education by publishing worldwide in

Oxford New York

Auckland Cape Town Dar es Salaam Hong Kong Karachi
Kuala Lumpur Madrid Melbourne Mexico City Nairobi
New Delhi Shanghai Taipei Toronto

With offices in

Argentina Austria Brazil Chile Czech Republic France Greece
Guatemala Hungary Italy Japan Poland Portugal Singapore
South Korea Switzerland Thailand Turkey Ukraine Vietnam

Oxford is a registered trade mark of Oxford University Press
in the UK and in certain other countries

Published in the United States
by Oxford University Press Inc., New York

First published 2010

British Library Cataloguing in Publication Data
Data available

Library of Congress Cataloging in Publication Data
Data available

Typeset by SPI Publisher Services, Pondicherry, India
Printed in Great Britain
on acid-free paper by
CPI Antony Rowe, Chippenham, Wiltshire

ISBN 978–0–19–852864–7

1 3 5 7 9 10 8 6 4 2

Preface

Conducting (conjugated) polymers have a unique set of properties: The electronic properties of metals and semiconductors *and* the processing advantages and mechanical properties of polymers. It is the "*and*" that makes conducting polymers special materials. There are, after all, many excellent metals; for example, copper, nickel, silver and gold, to name but a few. Similarly, there are many excellent semiconductors. Indeed, the sophisticated modern electronics industry uses silicon as the semiconductor of choice. It would be difficult to improve on the quality of copper as a metal or silicon as a semiconductor.

Why, then has there been interest in semiconducting and metallic polymers? The first answer is simple: they did not exist. Now, 35 years after their discovery, metallic conducting polymers have been demonstrated in the laboratory with electrical conductivities approaching that of copper and with mechanical strengths exceeding that of steel, a remarkable achievement. A wide variety of electrical and optical devices have been demonstrated using semiconducting polymers. For example, light-emitting devices have been made which are as bright as fluorescent lamps at applied voltages of only a few volts; photovoltaic solar energy conversion using conjugated polymer composites is in industrial production; and conjugated polymer transistors, circuits and chips are included in products.

A more general answer, however, goes back to the "*and*"; polymers are much easier to process than metals or crystalline semiconductors. The possibility of processing electronic materials at low temperatures offers the promise of low-cost manufacturing. Indeed, semiconducting and metallic polymers can be thought of as electronic "inks". The advances in printing technology (ink-jet printing, off-set printing, etc.) combined with the science and technology of conducting polymers will revolutionize the way in which electronic devices are manufactured. In addition, semiconducting and metallic polymers can be used in applications which require special mechanical properties, such as flexibility.

Thus, the study of semiconducting and metallic polymers has been important for two reasons:

(i) Since they did not exist, an entirely new field of science and technology has been created.
(ii) They offer a unique combination of properties not available from other known materials.

The first expresses a scientific and intellectual challenge; the second expresses a promise for utility in a wide range of applications.

This book summarizes a selection of topics which address the scientific and intellectual challenges and the potential for applications of such materials. It is a rapidly growing field and therefore the selection of topics in this book and their coverage therein cannot be truly "complete". The authors have used their experience, their judgement and their direct involvement in the research and development of this field to choose the content of the chapters of this book.

The field of semiconducting and conducting polymers is one of the most attractive areas of interdisciplinary materials science and technology. With necessary input from physics, chemistry, polymer science, fluid dynamics, electrical and electronic engineering, as well as optical sciences, working in this field requires an unusually wide range of interdisciplinary knowledge. Organic light-emitting diodes, organic solar cells, organic transistors, organic photodetectors and organic biosensors are entering the market place. Therefore, our goal was to present the science and technology in a coherent manner that is widely accessible to young scientists and engineers as well as the people in industry who have the responsibility to commercialize the next generation products.

Alan J. Heeger
Niyazi Serdar Sariciftci
Ebinazar B. Namdas
December 2009

Contents

1 Electronic structure of π-conjugated polymer chains 1

1.1 Carbon in the sp^2p_z configuration 1

1.2 Traditional polymers (based upon sp^3 hybridized carbon) and conducting polymers (based upon sp^2p_z hybridized carbon) 3

1.3 The electronic structure of *trans*-polyacetylene $(CH)_x$: The simplest conducting polymer 4

1.4 Band structure of the uniform *trans*-$(CH)_x$ chain in the tight-binding approximation 5

1.5 Band structure of the dimerized *trans*-$(CH)_x$ chain in the tight-binding approximation 8

1.6 Experimental verification of the band structure of *trans*-$(CH)_x$ 10

1.7 Energy band structure of conjugated polymers: General comments 12

1.8 Importance of disorder and the electron–electron Coulomb repulsion 12

1.9 Metal or antiferromagnetic insulator: The Mott transition 14

Further reading 16

2 Doping of conducting polymers 18

2.1 Introduction 18

2.2 Chemical doping by charge transfer 21

2.3 Electrochemical doping 22

2.4 Electrochemical charge injection: Measurement of E_g 24

2.5 Polymer battery 26

2.6 Polyaniline: Doping by acid–base chemistry 27

2.7 Photoinduced "doping" 28

2.8 Charge injection at a metal–semiconducting polymer (MS) interface 29

Further reading 29

3 Novel properties generate new opportunities 31

3.1 Introduction 31

3.2 Semiconducting and metallic polymers: Materials that are soluble in and processable from common solvents 31

3.2.1 Side-chain functionalization 33

3.2.2 Precursor route 34

3.2.3 Counter-ion induced processability of polyaniline 35

3.2.4 Dispersions of metallic polymers in water 36

3.3 Blends of conducting polymers with insulating polymers: Percolation at low concentrations 37

3.4 Transparent metallic conductors 40
Further reading 43

4 Disorder induced metal–insulator (M-I) transition in conducting polymers 44
4.1 Introduction to the consequences of strong disorder 44
4.2 Anderson localization 46
4.3 Variable range hopping 48
4.4 Critical regime of the M-I transition for "metallic" polymers 51
4.5 Electrical conductivity in the critical regime 52
4.6 Magnetic field induced cross-over from metal to insulator 53
4.7 Zabrodskii plots: Characterization of the metal–insulator transition as a function of magnetic field and pressure 54
4.8 The metal–insulator transition as observed through infrared measurements 56
Further reading 65

5 Metallic state of conducting polymers 66
5.1 Introduction 66
5.1.1 Highest occupied band is partially filled with a well-defined Fermi energy 66
5.1.2 Optical properties consistent with the physics of metals 67
5.1.3 Transport properties indicative of "free" electrons 67
5.2 Magnetic susceptibility of doped metallic polymers 67
5.3 Specific heat 69
5.4 Thermopower 69
5.5 Metallic reflection 70
5.6 Transport measurements 71
5.7 Striving toward higher conductivities in the metallic state 75
5.8 Estimating the intrinsic electrical conductivity of metallic polymers 78
Further reading 80

6 Nonlinear excitations in conjugated polymers: Solitons, polarons and bipolarons 81
6.1 Polyacetylene: Semiconductor with degenerate ground state 81
6.2 Solitons in *trans*-polyacetylene 83
6.3 Nondegenerate ground state polymers 85
6.4 Polarons and bipolarons in conjugated polymers with nondegenerate ground state 87
Further reading 90

7 Solitons, polarons and bipolarons: Experiments 91
7.1 Doping induced optical absorption in polyacetylene 91
7.2 ESR measurements of the spin susceptibility of doped polyacetylene 91
7.3 Infrared active vibrational (IRAV) modes 94
7.4 Photoinduced absorption in *trans*-polyacetylene 97
7.5 Photoinduced ESR in *trans*-polyacetylene: Photogeneration of spinless solitons 100

7.6 Time resolved photoinduced absorption in *trans*-$(CH)_x$; solitons in < 250 fs 101
7.7 Doping induced absorption in conjugated polymers with nondegenerate ground state 101
7.8 ESR measurements of the spin susceptibility in conjugated polymers with nondegenerate ground state 104
7.9 IRAV modes observed in semiconducting polymers with nondegenerate ground state 105
7.10 Photoinduced absorption in polythiophene (nondegenerate ground state) 105
7.11 Time resolved photoinduced absorption: Nondegenerate ground state 106
Further reading 107

8 Conjugated polymers as semiconductors 108
8.1 Solution cast thin film diodes 108
8.2 Semiconductors in which the Fermi energy can be controlled and shifted across the energy gap 111
8.3 Optical properties of semiconducting polymers 113
8.4 Optical properties of oriented and aligned semiconducting polymer chains 115
8.4.1 Vibronic structure 117
8.5 Structural disorder in semiconducting polymers 122
8.6 Photoconductivity measurement of the onset of the interband absorption 123
8.7 Photon energy required for the photogeneration of charged polarons 125
8.8 Measurement of the exciton binding energy 126
8.9 Recombination of charge carriers (monomolecular and bimolecular) 127
8.10 Transport of charge carriers in disordered semiconducting polymers 128
8.11 Summary 130
Further reading 131

9 Polymer based light-emitting diodes (PLEDs) and displays fabricated from arrays of PLEDs 132
9.1 LEDs fabricated from semiconducting polymers 132
9.2 Accurate measurement of OLED/PLED device parameters 135
9.3 Fowler–Nordheim tunneling in semiconducting polymer MSM diodes 138
9.3.1 Single carrier devices 140
9.3.2 PLED operating voltage and efficiency 145
9.3.3 Limits of the MIM model 145
9.3.4 Approaches to improved carrier injection: Hole-blocking layers, electron–acceptors and hole–acceptors 146
9.4 Thickness dependence of electroluminescence efficiency 148
9.5 Limits on the electroluminescence efficiency 149
9.5.1 Phosphorescent LEDs 149
9.5.2 Fluorescent LED (singlet–triplet ratio) 150
9.6 Pixelated displays 151
9.7 White light emission 153

9.8 Lifetime of solution processed light-emitting diodes 157
9.9 Conclusion 157
Further reading 157

10 Light-emitting electrochemical cells (LECs) 158
10.1 Introduction 158
10.2 LEC device architecture 158
10.3 LEC operating mechanism 159
10.4 LECs in the surface geometry configuration 163
10.5 Direct confirmation of electrochemical junction formation in the LEC 164
10.6 Control of phase separation: Conjugated polymer plus solid electrolyte 166
10.7 LECs versus LEDs: Advantages and disadvantages 167
10.8 Junction formation in the "sandwich" configuration 169
10.9 Frozen junction LECs 169
10.10 LECs fabricated with ionic luminescent polymers and with ionic liquid additives 172
10.11 LECs with long operating lifetimes: Ionic liquid additives 174
10.12 LECs: Summary 175
Further reading 176

11 Semiconducting polymers as laser materials 177
11.1 Optical gain in semiconducting polymers 177
11.2 Photoinduced absorption as a loss mechanism 181
11.3 Amplified stimulated emission (ASE) from films of conjugated, semiconducting polymers 184
11.4 Direct measurements of gain, loss and ASE spectra in planar waveguides of semiconducting polymers 186
11.5 Lowering the lasing threshold with energy transfer (to increase the Stokes shift) 189
11.6 Laser resonant structures and true laser emission 191
11.7 Progress towards electrically pump lasing 195
11.8 Laser lookalikes and cases of "mistaken identity" 198
11.9 Concluding comments 199
Further reading 200

12 Photoinduced charge transfer from semiconducting polymers to acceptors 201
12.1 Introduction 201
12.2 Photophysics of conjugated polymer/acceptor composites 202
12.3 Photoinduced electron transfer from semiconducting polymers to C_{60} 204
12.4 Light induced electron spin resonance (LESR) 205
12.5 Ultrafast photoinduced charge transfer 206
12.6 Sensitization of photoconductivity 211
Further reading 213

13 Photovoltaic cells and photodetectors 214
13.1 Dual function polymer diodes 214
13.2 Donor–acceptor bilayer heterojunctions 217
13.3 Self-assembly of BHJ nanomaterials by spontaneous phase separation 219
13.4 Nanomorphology of BHJ solar cells 219
13.5 Optimizing the nanomorphology by thermal annealing 221
13.6 Characterization of the BHJ nanomorphology 222
13.7 The origin of the open circuit voltage, V_{oc} 227
13.8 Dynamics of the BHJ cell 230
13.9 Optimizing the performance of new materials: Processing additives 232
13.10 Design rules toward bulk heterojunction solar cells with > 10% PCE 235
13.11 Titanium oxide as multifunctional components in BHJ solar cells 235
13.12 BHJ tandem solar cells 238
13.13 BHJ solar cells with internal quantum efficiency approaching 100% 240
13.14 Recombination in PCDTBT: An example 244
13.15 Photodetectors 246
Further reading 249

14 Polymer field effect transistors (FETs) 250
14.1 Introduction 250
14.2 Operating principles of OFETs 251
14.3 Different architectures used in organic FETs 254
14.4 Contact resistance and errors in the mobility measurements 256
14.5 Polarons as the gate induced carriers in OFETs 257
14.6 Charge transport at high carrier density 259
14.7 Organic dielectrics as gate insulators in OFETs 262
14.8 OFETs for both n-channel and p-channel circuit operation 265
14.9 Light-emitting field effect transistors (LEFETs) 267
14.10 High brightness LEFETs 270
14.10.1 Bilayer LEFETs 271
14.10.2 Phosphorescent LEFETs 272
Further reading 272

Index 273

1

Electronic structure of π-conjugated polymer chains

1.1 Carbon in the sp^2p_z configuration

Semiconducting and metallic "organic" polymers are based on linear carbon chains. The carbon atom (C_6) has six electrons outside the nucleus, of which four are valence electrons; i.e. four (the 2s and the 2p electrons) take part in chemical bonds. In free space, or where the potential is spherically symmetric, the 1s and 2s orbitals of the carbon atom are filled, and the 2p orbitals are occupied with two electrons, resulting in the atomic configuration:

$$1s^2 2s^2 2p^2$$

In order to lower the total energy and form spatially directed chemical bonds, in molecules and crystals, carbon can form two primary hybrid structures; tetrahedrally directed covalent bonds as in diamond and saturated polymers, and hexagonally directed covalent bonds as in graphite and conjugated polymers.

To form directed tetrahedral bonds and directed hexagonal bonds:

Step 1: Promote one electron from 2s to 2p. The atomic configuration is then therefore:

$$1s^2 2s 2p^3$$

This change in occupation costs energy, ΔE (~180 kcal/mole > 7.8eV), since the ground state, $1s^2 2s^2 2p^2$, has the lowest energy for the carbon atom in free space. This energy and more will be gained through the formation of covalent bonds (otherwise the bonds will not form!).

The s and p wavefunctions, s, p_x, p_y, and p_z, are defined in terms of an orthogonal coordinate system (the s-orbital is spherically symmetric while p_x, p_y, and p_z are directed along the x, y and z-axes, respectively). The spatial functions, $f_s(r)$ and $f_p(r)$ describe the amplitude as a function of r and each is normalized to unity.

$$\left.\begin{aligned} s &= 1 f_s(r) \\ p_x &= \sqrt{3}\sin\theta\cos\phi f_p(r) \\ p_y &= \sqrt{3}\sin\theta\sin\phi f_p(r) \\ p_z &= \sqrt{3}\cos\theta f_p(r) \end{aligned}\right\} \text{normalized to } \frac{1}{\sqrt{4\pi}}$$

Step 2: Take linear combinations of s and p_x, p_y, p_z to form spatially directed orbitals.
sp^3 hybridization – for bonding orbitals with tetrahedral symmetry;
sp^2p_z hybridization – for bonding orbitals with hexagonal symmetry (with the p_z- orbital available to form a π-bond).

Step 3: Form covalent bonds using the directed orbitals; the energy gained in forming the covalent bonds must be significantly greater than ΔE (see Step 1). The "correct" linear combinations are as follows:

For tetrahedral symmetry:

$$\left.\begin{aligned} t_{111} &= \frac{1}{2}(s + p_x + p_y + p_z) \\ t_{1\bar{1}\bar{1}} &= \frac{1}{2}(s + p_x - p_y - p_z) \\ t_{\bar{1}1\bar{1}} &= \frac{1}{2}(s - p_x + p_y - p_z) \\ t_{\bar{1}\bar{1}1} &= \frac{1}{2}(s - p_x - p_y - p_z) \end{aligned}\right\} sp^3 \text{ hybridization}$$

For hexagonal symmetry:

$$t_z = p_z$$

$$\left.\begin{aligned} t_0 &= \sqrt{\frac{2}{3}}\left(\frac{\sqrt{2}}{2}s + \frac{\sqrt{3}}{2}p_x - \frac{1}{2}p_y\right) = \sqrt{\frac{2}{3}}\left[\frac{\sqrt{2}}{2}s + \sqrt{3}\sin\theta\cos(\phi + 30^\circ)\right] \\ t_{120} &= \sqrt{\frac{2}{3}}\left(\frac{\sqrt{2}}{2}s + p_y\right) = \sqrt{\frac{2}{3}}\left(\frac{\sqrt{2}}{2}s + \sqrt{3}\sin\theta\sin\phi\right) \\ t_{240} &= \sqrt{\frac{2}{3}}\left(\frac{\sqrt{2}}{2}s - \frac{\sqrt{3}}{2}p_x - \frac{1}{2}p_y\right) = \sqrt{\frac{2}{3}}\left[\frac{\sqrt{2}}{2}s + \sqrt{3}\sin\theta\cos(\phi + 150^\circ)\right] \end{aligned}\right\} sp^2 \text{ hybridization}$$

Here the p_z orbital is directed perpendicular to the plane defined by the other sigma orbitals (t_0, t_{120}, t_{240}) as shown in Fig. 1.1.

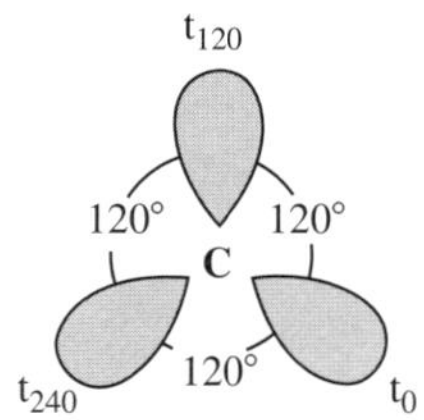

Fig. 1.1 Planar backbone created by the sp^2 hybridized sigma bonds of carbon. The p_z electrons are perpendicular to this plane.

1.2 Traditional polymers (based upon sp^3 hybridized carbon) and conducting polymers (based upon sp^2p_z hybridized carbon)

The tetrahedral configuration of an sp^3 hybridized carbon results in "traditional" plastics such as polyethylene where the valence of carbon is saturated. As sketched in Fig. 1.2, in polyethylene, there are two carbon–carbon and two carbon hydrogen covalent bonds (σ-bonds).

Saturated polymers are flexible and transparent. Since they have low melting points, they can be extruded and injection molded. Traditional saturated polymers are used in higher quantities than any other materials – more by volume and more by weight – in a wide range of applications, including fabrics, engineering plastics and, more generally, "plastic everything".

The wide range of utility of traditional saturated polymers can be seen in terms of the simplest and most fundamental member of this class of materials, polyethylene (Fig. 1.2). Polyethylene can be processed into amorphous materials (low-cost plastics) or materials fabricated from the same chemical structure can be made to be very strong by orienting the chain-extended macromolecules via gel processing. If sufficiently well aligned, the limiting tensile strength can approach that of the carbon–carbon covalent bond. In fact, highly oriented polyethylene is used in the fabrication of high strength skis and bullet-proof vests.

Traditional polymers are electrical insulators. Since all of the valence electrons are bound in covalent bonds, there are no mobile electrons to participate in electronic transport. Conjugated polymers are formed from sp^2p_z hybridized carbon. Polyacetylene is the simplest conjugated polymer (see Fig. 1.3). The three in-plane σ-orbitals create the "backbone"; two of them bonded to the neighboring carbons and the third σ-orbital bonded to a hydrogen atom. The fourth electron resides in the p_z-orbital and, because of its orthogonality to the plane defined by the three σ-bonds, it is in a first approximation independent of them. Throughout

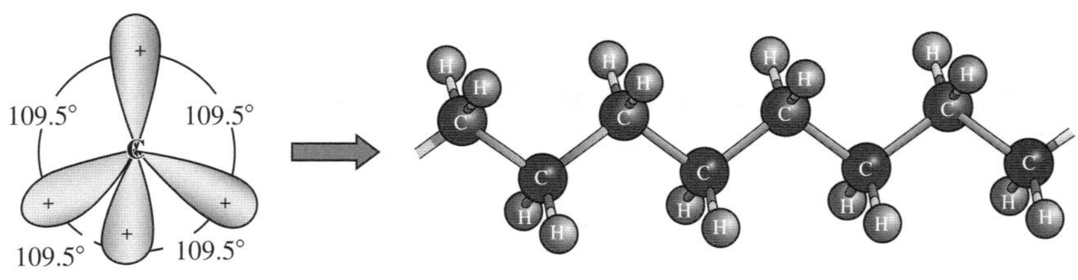

Fig. 1.2 Schematic diagram of sp^3 hybridized carbon and the structure of polyethylene.

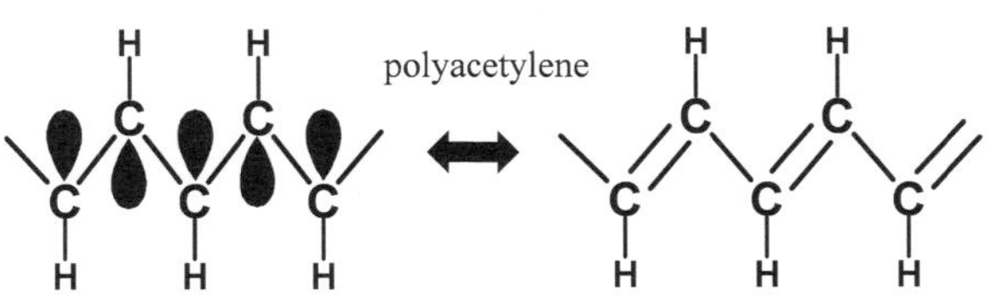

Fig. 1.3 Schematic diagram of sp^2p_z hybridized carbon and the structure of polyacetylene.

this book, the p_z-electron and the $p_z$orbital will be designated as the π-electron and the π-orbital, respectively.

This one-electron picture of the p_z-electron being decoupled from the backone σ-orbitals gives these polymers special electronic properties.

In the sp^2p_z configuration of Fig. 1.3, the electrons in the sp^2-orbitals form covalent bonds and are therefore lower in energy than the nonbonded electron in the p_z-orbital. Thus, the sp^2p_z configuration of each of the carbon atoms in the polyacetylene chain is analogous to a "pseudo-lithium" atom; i.e. a filled core (sp^2) with one outer electron (the π-electron). The π-electron on a given carbon atom is attracted to the nuclei of the neighboring carbon atoms. As a result, there is a tendency for the π-electron to delocalize along the chain. If the outer π-electron were able to delocalize over many, many sites along the chain (as do the electrons in lithium metal), then polyacteylene would be a metal! More generally, as discussed in the following sections, polymer chains with carbon atoms in the sp^2p_z configuration, "conjugated" polymers, are semiconductors that can be doped (see Chapter 2) to become conducting and even metallic materials.

1.3 The electronic structure of *trans*-polyacetylene $(CH)_x$: The simplest conducting polymer

In general, the electronic structure of π-conjugated polymers consists of two parts: the energy bands that originate from the bonding and anti-bonding energy levels associated with the σ-bonds formed from the sp^2 wavefunctions and the π-band that originates from the delocalized p_z wavefunctions. In the one-electron approximation (i.e. neglecting the electron–electron Coulomb interaction) the number of sub-π-bands is determined by the number of carbon atoms in the unit cell (the repeat unit).

The polymer chain structure is held together by the σ-bonds. In sp^3 hybridized carbon chains such as polyethylene, any excitation of the σ-bonded electrons will promote electrons

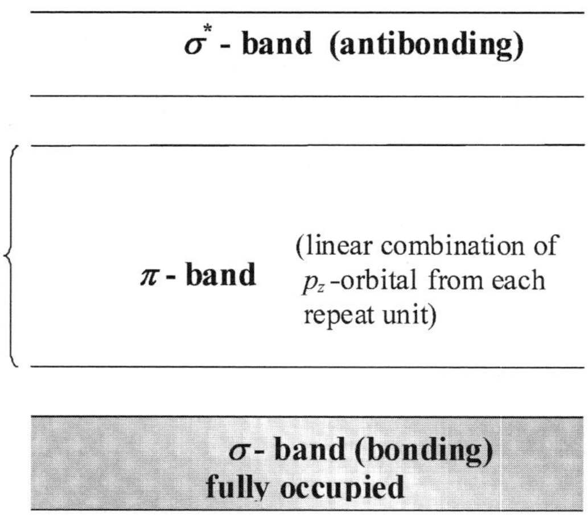

Fig. 1.4 Schematic band diagram for π-conjugated polymers.

into an antibonding σ*-band and cause structural instability. When electrons are excited into the σ*-antibonding orbitals, degradation usually occurs because the electrons in the σ-bonding orbitals are essential for holding the polymer chain together. For this reason saturated polymer can be designed as good photoresist materials. Consequently, saturated polymers tend not to have interesting electronic or optical properties. For π-conjugated polymers, however, excitations within the π-band do not typically lead to bond breaking. As we will see, π-conjugated polymers are photoconductors rather than "photoresists". Although formally orthogonal, the σ- and π-bands can overlap in energy; i.e. the top of the σ-band can be at higher energy than the bottom of the π-band, etc.

1.4 Band structure of the uniform *trans*-(CH)$_x$ chain in the tight-binding approximation

For the uniform bond length structure of polyacetylene shown on the left side of Fig. 1.3, there is a single π-band which can accommodate $2N$ electrons where N is the number of atoms in the chain (the factor of 2 arises from the electron spin; each state in the band can accommodate two electrons, one with spin up and one with spin down). Since each carbon atom contributes one p_z-electron, the π-band is half-filled. Uniform bond-length $(CH)_N$ would be a metal (see Fig. 1.5).

Shortly after the initial discovery of doping and the metal–insulator transition in polyacetylene, a theoretical description of the electronic structure was proposed. The construction of the remarkably successful Su–Schrieffer–Heeger (SSH) Hamiltonian was based on two assumptions:

(a) the π-electronic structure can be treated in the tight-binding approximation with a transfer (hopping) interaction, t_o; and
(b) the chain of carbon atoms is coupled to the local electron density through the length of the chemical bonds.

The first assumption defines the lowest order transfer energy, t_0, which is responsible for the delocalization of the π-electrons along the chain. The second assumption provides the first-order correction to t_o (as a function of the bond length) as well as two intuitive terms related to the lattice degrees of freedom: a harmonic "spring constant" term which represents the increase in potential energy that results from displacement from the uniform bond lengths in $(CH)_x$ and a kinetic energy term for the nuclear motion. These two lattice terms, coupled to the electronic

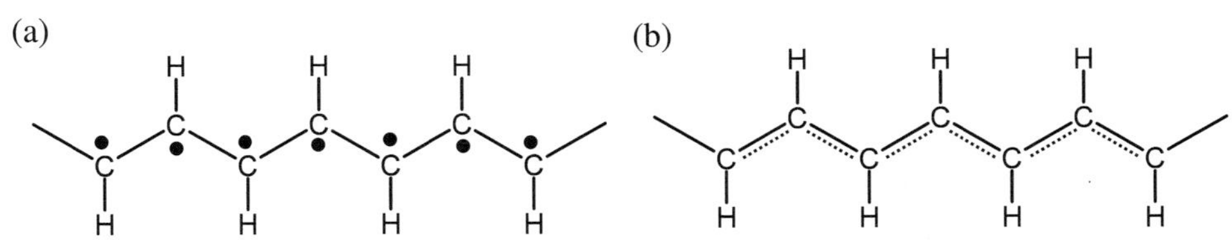

Fig. 1.5 Structure (a) and band structure (b) of uniform bond-length $(CH)_N$. The structure diagram indicates one electron per carbon atom (top), but that electron is delocalized over the chain (bottom).

structure via the bond-length dependent transfer term, are the origin of the nonlinear excitations of the polymer chain, solitons, polarons and bipolarons.

The SSH Hamiltonian is written as the sum of three terms:

$$H_{SSH} = \sum_{n,\sigma} [-t_0 + \alpha(u_{n+1} - u_n)](c^+_{n+1,\sigma}c_{n,\sigma} + c^+_{n,\sigma}c_{n+1,\sigma}) + \sum_n \frac{p_n^2}{2m} + \frac{1}{2}K\sum_n (u_{n+1} - u_n)^2 \quad (1.1)$$

where $c^+_{n,\sigma}$ and $c_{n,\ \sigma}$ are the fermion creation and annihilation operators for site n and spin σ, p_n are the nuclear momenta, u_n are the displacements from equilibrium, m is the carbon mass, and K is an effective spring constant of the σ-bond. The first term describes the electronic structure, and the second and third terms describe the vibrations of the nuclei with respect to their equilibrium positions.

In the tight–binding approximation of the π-electron band structure, one assumes that the wavefunctions are linear combinations of p_z-orbitals, that there is a transfer or hopping interaction with magnitude, $-t_o$ (often denoted as β in molecular orbital theory) that transfers an electron on site i to near neighbors sites $i \pm 1$, and that all contributions to the wavefunction from other atomic orbitals (such as the sp^2 atomic orbtials) can be neglected. The neglect of contributions from the sp^2 orbtials requires that $t_o/\Delta E_{\pi\text{-}\sigma} << 1$, where $\Delta E_{\pi\text{-}\sigma}$ is the π-σ energy difference (the difference between the energy of the nonbonded p_z-orbital and the energies of the σ-bond and the σ*-bond) an assumption which is consistent with the basic σ-π orthogonality and with the general electronic structure shown in Fig. 1.4.

The transfer energy arises from the attractive interaction between an electron on a carbon atom at site i along the chain and the nucleus of the carbon atom at site $i \pm 1$. Thus, the magnitude of $-t_o$ is roughly proportional to the ionization potential of the carbon atom (I_p). Since I_p arises from the attraction of the electron to the nucleus on the same site, $-t_o$ is approximately —$2I_p$ times the overlap of the wavefunction on neighboring sites

$$t_0 \approx 2I_p S \quad (1.2)$$

where $S = \langle i \mid i \pm 1\rangle$ represents the overlap of the π-electron wavefunctions on site i and site $i \pm 1$. The factor of 2 arises from the fact that I_p has two contributions, potential energy and kinteitc energy (the kinetic energy term is positive and approximately half as large as the attractive potential energy). For polyacetylene, electron energy loss measurements of the band dispersion yield $t_o \approx 3$ eV. Since $I_p \approx 5$ eV, the implied overlap of the wavefunctions is relatively large, $\langle i \mid i \pm 1\rangle \approx 0.3$. Thus, there is a strong tendency toward delocalization of the electrons along the conjugated chain of carbon atoms.

The magnitude of the transfer interaction is strongly dependent on the distance between successive carbon atoms;

$$\mathrm{S} \sim \exp - [(\mathrm{a} \pm \delta)/\mathrm{r}_\pi] = \mathrm{S}_0 \pm \alpha\delta + \text{terms of order } \delta^2 \quad (1.3)$$

where r_π is the effective "radius" of the p_z wavefunction, δ is the magnitude of the change in bond length ($a \rightarrow a \pm \delta$), and α is a constant. To include this coupling of the electronic structure to the molecular structure in the Hamiltonian, SSH generalized the transfer term to include the bond-length dependent hopping interaction from site n to $n + 1$:

$$t_{n,n+1} = t_o + \alpha(u_{n+1} - u_n) \tag{1.4}$$

where u_n is the displacement from equilibrium of the nth carbon atom. This effectively couples the electronic states to the molecular geometry, giving the electron–phonon (el-ph) interaction term where α is the el-ph coupling constant. The precise form of $t_{n,n+1}$, in which the dependence of the hopping energy on the C–C distance is linearized for small deviations about t_o, is the first term in a Taylor expansion. The bond-length dependence of the transfer energy is an essential feature of the description of conjugated polymers since the ground state bond alternation makes the hopping interaction across double bonds larger than the corresponding hopping interaction across single bonds.

Consider first only the tight-binding terms which describe the electronic structure of a $(CH)_N$ chain with uniform bond lengths:

$$H_{TB} = -I_P \Sigma_{n,\sigma} [c^{\dagger}_{n,\sigma} c_{n,\sigma}] + t_0 \Sigma_{n,\sigma} [c^{\dagger}_{n+1,\sigma} c_{n,\sigma} + c^{\dagger}_{n,\sigma} c_{n+1,\sigma}] \tag{1.5}$$

where $c_{n,\sigma}{}^{\dagger}$ and $c_{n,\sigma}$ are the creation and annihilation operators for electrons on site n with spin σ (up or down) and $|< c_{n,\sigma}{}^{\dagger} c_{n,\sigma}>|^2 = 1$, the average number of electrons per site in $(CH)_N$. The tight-binding terms are readily diagonalized by Fourier transformation to k-space:

$$c^{\dagger}_{n,\sigma} = (1/Na)^{1/2} \sum\nolimits_n e^{-ikn} c^{\dagger}_{k,\sigma}, \tag{1.6}$$

$$c_{n,\sigma} = (1/Na)^{1/2} \sum\nolimits_n e^{+ikn} c_{k,\sigma} \tag{1.7}$$

where $k = 0, \pi/Na, 2\pi/Na, \ldots, \pi/a$, $| <c_{k,\sigma^{\dagger}} c_{n,\sigma}>|^2 = 1$, and N is the number of atoms in the chain. The diagonalized Hamiltonian is given by

$$H = \sum\nolimits_{k,\sigma} [-I_P + (1 - 2\cos ka)]\, c^{\dagger}_{k,\sigma} c_{k,\sigma} \tag{1.8}$$

and the π-electron energies are given by

$$E_{\pi} = <H> = -I_P + (1 - 2\cos ka) \tag{1.9}$$

where energy is measured with respect to the vacuum.

The electronic band structure of the uniform chain is sketched in Fig. 1.6. In the tight-binding approximation, the π-electron band structure of the uniform chain is a single band centered at $-I_P$ with a total bandwidth of W, where $W = 4t_o$. The average energy of the

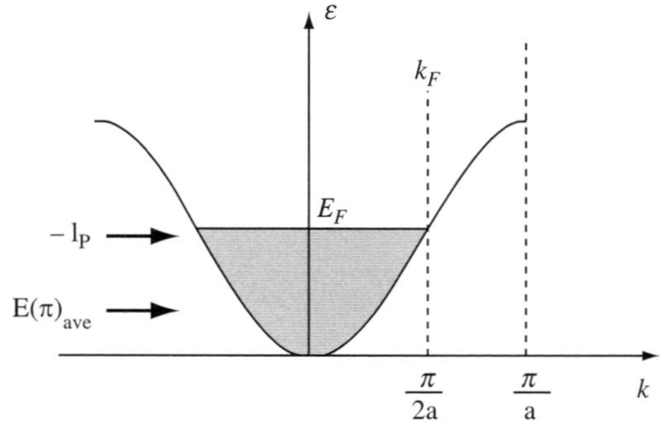

Fig. 1.6 Tight-binding band structure of the uniform chain, $(CH)_N$.

π-electrons, $E(\pi)_{ave}$ (the average energy of the occupied states in Fig. 1.6) in the metallic uniform $(CH)_N$ chain is well below the center of the π-band.

1.5 Band structure of the dimerized *trans*-$(CH)_x$ chain in the tight-binding approximation

Although the π-electrons in polyacteylene are delocalized along the chain, polyacetylene is not a metal. The polymerization of polyacetylene from the monomer acetylene yields a dimerized (bond alternating) structure. The resulting polymer is insoluble and intractable. Consequently, the molecular weight cannot be directly determined. Thus, because the Staudinger index, N, is unknown, polyacetylene is typically designated as $(CH)_x$.

As sketched in Fig. 1.7, the molecular structure of "real" polyacetylene has alternating single and double bonds which are, respectively, longer and shorter than the equilibrium value of the bond length in uniform $(CH)_N$. In this structure, the π-electrons on neighboring carbon atoms form weaker and stronger π-π bonds resulting in the bond alternating structure. The bond alternation sketched in Fig 1.3 has been determined from analysis of X-ray diffraction data and from analysis of nuclear magnetic resonance data; the shorter bond length is 1.35 Å and the longer bond length is 1.45 Å.

Such a bond alternating structure doubles the unit cell (compare the structures on the left and on the right of Fig. 1.3), thereby opening a gap in the electronic structure. As a result, because of the bond alternating structure, polyacetylene is a semiconductor.[1]

In the dimerized form, the SSH Hamiltonian reduces to

$$H = \sum_{m,\sigma} [(-t_0) + \alpha(u_{m+1} - u_m)] \times [c^{+}_{m+1,\sigma} c_{m,\sigma} + c^{+}_{m,\sigma} c_{m+1,\sigma}] \tag{1.10}$$

where $u_m = \langle u_m \rangle = (-1)^m u$. Thus, $t_m = t_0 + (-1)^m 2\alpha u = t_0 + (-1)^m \Delta$, and

$$H = \sum_{m,\sigma} [(-t_0) \pm \Delta] \times [c^{+}_{m+1,\sigma} c_{m,\sigma} + c^{+}_{m,\sigma} c_{m+1,\sigma}] \tag{1.11}$$

Diagonalization of the Su–Schrieffer–Heeger Hamiltonian for dimerized *trans*-$(CH)_x$ yields the band structure of the dimerized chain as sketched in Fig. 1.8; the energy versus wave-number (k) is given by

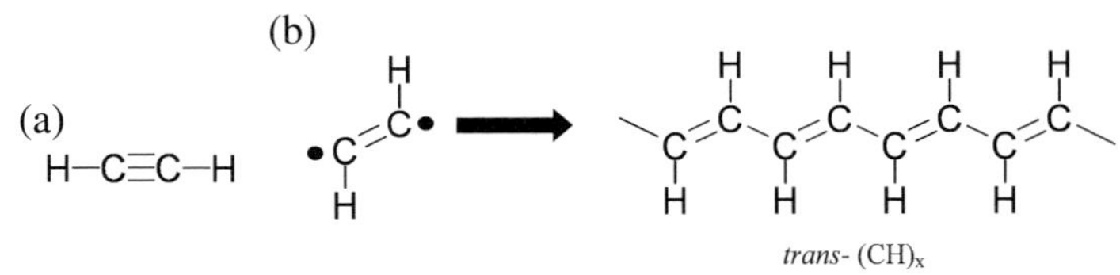

Fig. 1.7 (a) Molecular structure of acetylene; (b) polymerization of acetylene into *trans*-polyacetylene, *trans*-$(CH)_x$.

[1] See for example the discussion of the dimerized chain in C. Kittel, *Introduction to Solid State Physics*, John Wiley, New York.

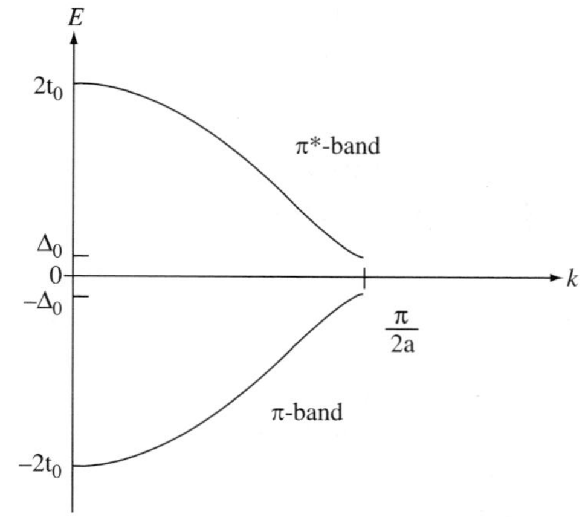

Fig. 1.8 Tight-binding band structure, $E(k)$ versus k, of the dimerized *trans*-$(CH)_x$. chain with molecular structure shown in Fig. 1.7. *Trans*-$(CH)_x$ is a semiconductor with energy gap $E_g = 2\Delta_o$.

$$E_k = \pm\sqrt{\epsilon_k^2 + \Delta^2} \tag{1.12}$$

where $\epsilon_k = 2t_0 \cos ka$.

In the terminology of traditional condensed matter (semiconductor) physics, the π-band corresponds to the valence band, and the π*-band corresponds to the conduction band. The density of states, $\rho_o(E)$, as a function of energy is shown in Fig. 1.9. All states in the π-band are filled, and all states in the π*-band are empty.

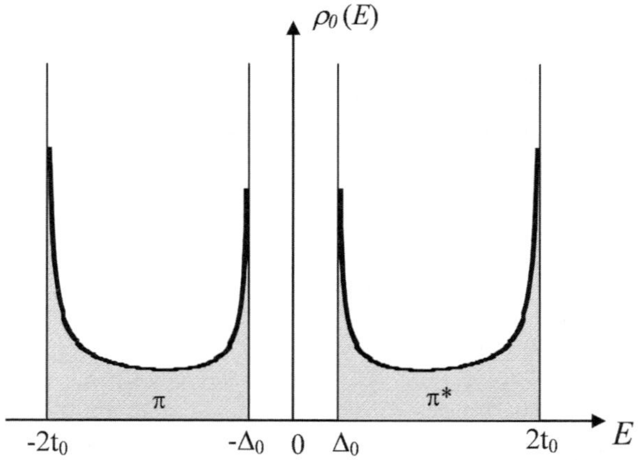

Fig. 1.9 The density of states as a function of energy, $\rho_o^{\dim}(E)$, for the dimerized *trans*-$(CH)_x$. chain with molecular structure shown in Fig. 1.7. All states in the π-band are filled and all states in the π*-band are empty.

The bond alternation in polyacetylene is an example of the well-known "Peierls instability" of one-dimensional (1D) metals. As first shown by Rudolf Peierls, a 1D metal such as the uniform CH chain sketched in Fig. 1.3 is always unstable to the formation of a lattice distortion with periodicity Λ where $2\pi/\Lambda = 2k_F$. The fundamental reason why such 1D metallic structures are unstable against any infinitesimal distortion of the lattice can be understood using the example of the uniform $(CH)_N$ chain. With one π-electron contibuted to the band by each of the N carbon atoms in the chain, the average π-electron energy, $E(\pi)_{ave}$, is indicated in Fig. 1.6. Consider a distortion with periodicity $2a$; i.e. the ith bond length goes from $a \rightarrow (a + \delta)$ and the $(i + 1)$st bond length goes from $a \rightarrow (a - \delta)$, etc., where δ is the magnitude of the bond length distortion. In the presence of such bond distortions, an energy gap opens precisely at the Fermi energy with magnitude proportional to the magnitude of the distortion, as shown in Fig. 1.8. Upon opening the energy gap at the Fermi energy, the total energy of the occupied electronic states decreases as the gap pushes the occupied states to lower energy (see Figs 1.8 and 1.9). Since the magnitude of the energy gap is proportional to the magnitude of the distortion, the average energy of the π-electrons decreases by an amount proportional to δ. This decrease in energy is partially compensated by the increase in the total bond energy (lattice energy) that arises from the distortion away from uniform bond lengths (proportaional to δ^2). Thus, for a 1D metal, the decrease in $E_{ave}(\pi)$ is always greater than the increase in lattice energy. The gap at the Fermi energy that opens as a result of this instability is often referred to as "Peierls gap".

The energy balance described in the previous paragraph is not generally true for three dimensional metals. For 3D systems, a distortion of the lattice away from uniform bond lengths will typically open a gap on only a small portion of the Fermi surface. In a 3D metal, the lattice energy increase is typically greater than the decrease in electronic energy, and the metallic state is stable.

As described in Chapter 2, semiconducting polymers can be doped to high carrier densities, sufficiently high to reach the metallic state. If these doped polymers were truly one dimensional, the Peierls theorem would imply that a chain distortion would be spontaneously generated, and the semiconducting state restored. The metallic state of doped conjugated polymers is stabilized by interchain interactions sufficiently strong that the systems are anisotropic three-dimensional metals (quasi-1D metals).

1.6 Experimental verification of the band structure of *trans*-$(CH)_x$

Electron energy loss spectroscopy was utilized to probe the excitation spectrum of *trans*-polyacetylene. When a beam of electrons with incoming energy E_o and momentum $\hbar q$ passes through a thin film of $(CH)_x$, the energetic incident electrons can "kick" an electron from an initial state (i) in the π-band with wavenumber k_i to a final state (f) in the π*-band with wavenumber k_f. The experiment is sketched in Fig. 1.10.

Electron energy loss experiments measure the change in momentum and the change in energy of an electron beam passing through a thin film: $\Delta E_{loss} = E_{\pi^*}(k_f) - E_{\pi}(k_i)$ where $k_f - k_i = q$. The momentum change is determined by the scattering angle (see Fig. 1.10). Thus, the electron energy loss experiment enables the reconstruction of the $E(\kappa)$ versus k curves of the band structure. As shown in Fig. 1.11, electron energy loss measurements indicate a bandwidth of approximately 12 eV with a band gap of approximately 1.8 eV.

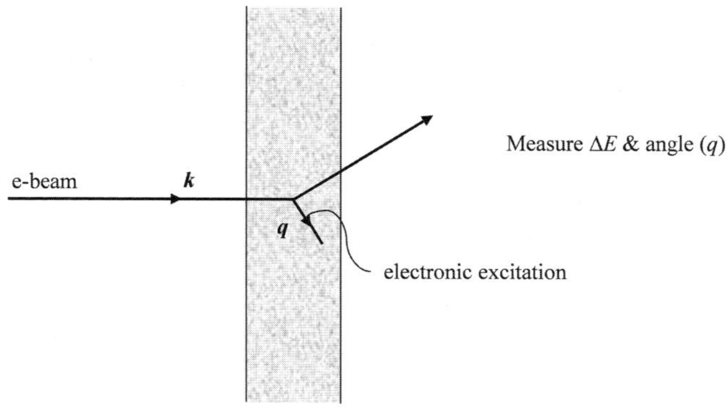

Fig. 1.10 An electron beam with energy E_o and momentum $\hbar q$ pass through a thin film of semiconducting polyacetylene. Measurement of the energy loss and the exit angle of the scattered electrons determines the excitation energy for promoting an electron from the π-band with $k = k_i$ to the π*-band with $k = k_f$ where $k_f — k_i = q$.

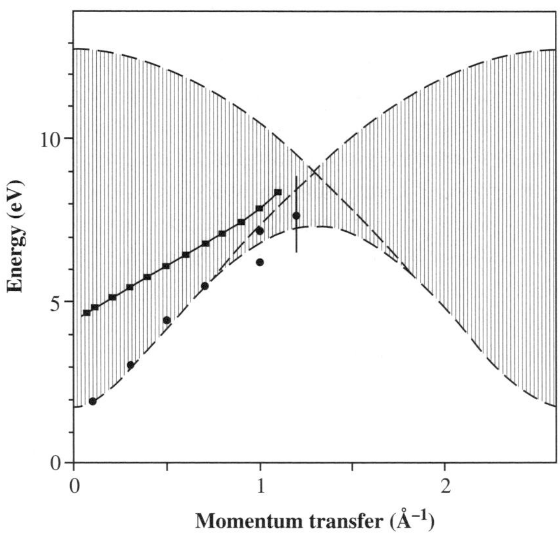

Fig. 1.11 Excitation of π-electrons in *trans*-$(CH)_x$ with momentum transfer q parallel to the chain axis. The solid circles represent the maxima in the π-π* interband transitions. The hatched area indicates the region of possible interband transitions calculated in the tight-binding approximation. The transitions near the upper dashed line (starting at 12.8 eV and $q = 0$ and ending at 1.8 eV and $q = 0.26$) are suppressed by the matrix element. The solid squares represent the maxima in $1/\mathrm{Im}\varepsilon(\omega)$; i.e. the π-plasmon spectrum. [Reproduced from J. Fink and G. Leising, *Phys. Rev. B*, **34**, 5320 (1986) with permission of The American Physical Society.]

The data shown in Fig. 1.11 are in good agreement with the predictions of the simple picture of the dimerized polyacetylene chain (see Fig. 1.8) with the following parameters:

$$E_g = 2(t_2 - t_1) = 2\Delta = 1.8\,\text{eV}$$
$$W = 4t_0 = 12.8\,\text{eV}$$

These data provide the most direct measurements of the electronic structure of *trans*-polyacetylene and the remarkably simple band picture shown in Fig. 1.8.

1.7 Energy band structure of conjugated polymers: General comments

Conjugated polymers are defined as bond-alternating linear chains of carbon atoms in the sp^2p_z configuration. The molecular structures of some of the more important members of this class are shown in Fig. 1.12. The electronic structure diagram sketched in Fig. 1.4, with a broad π-band in the energy range between the bonding π-band and the antibonding σ*-band is qualitatively correct for the entire class of conjugated polymers.

Specifically, however, the number of π-bands in the electronic structure of a conjugated polymer is determined by the number of atoms in the repeat unit. If there are n atoms in the main chain of the repeat unit, the π-band is divided into n subbands. Thus, for poly(paraphenylene), there are six carbon atoms in the repeat unit and six π-subbands (corresponding to the six molecular orbitals of benzene).

Poly(phenylene vinylene), PPV, is an alternating copolymer of the repeat units of polyacteylene and poly(paraphenylene). PPV and its soluble derivatives have emerged as the prototypical luminescent semiconducting polymers. Since there are eight carbon atoms in the repeat unit, the π-band is split into eight subbands. Each subband can hold two electrons per atom (spin up and spin down). Thus, the four lowest subbands are filled (the π–subbands) and the four highest subbands are empty (the π*-subbands). Consequently, PPV is not metallic since there are no partially filled bands. The energy difference between the highest occupied π-subband and the lowest unoccupied π*-subband is the π-π* energy gap of the polymer. Since the magnitude of the π-π* energy gap is approximately 2.5 eV, PPV is a semiconductor. The π–electron band structure of PPV is sketched in Fig. 1.13.

1.8 Importance of disorder and the electron–electron Coulomb repulsion

Although this very elementary description of conjugated polymers in terms of the concepts of energy band theory explains in qualitative terms how conjugation can lead to semiconducting and metallic properties, it is incomplete for three important reasons:

(1) Typically, these macromolecular systems are disordered in the solid state as a result of chain entanglement, etc. The disorder causes localization of electronic states, band tailing, and broadening of the optical transitions. Because of the disorder, even heavily doped polymers can be insulators rather than metals (see Chapter 4). In principle, however, disorder can be controlled. Chain extended and chain aligned samples can

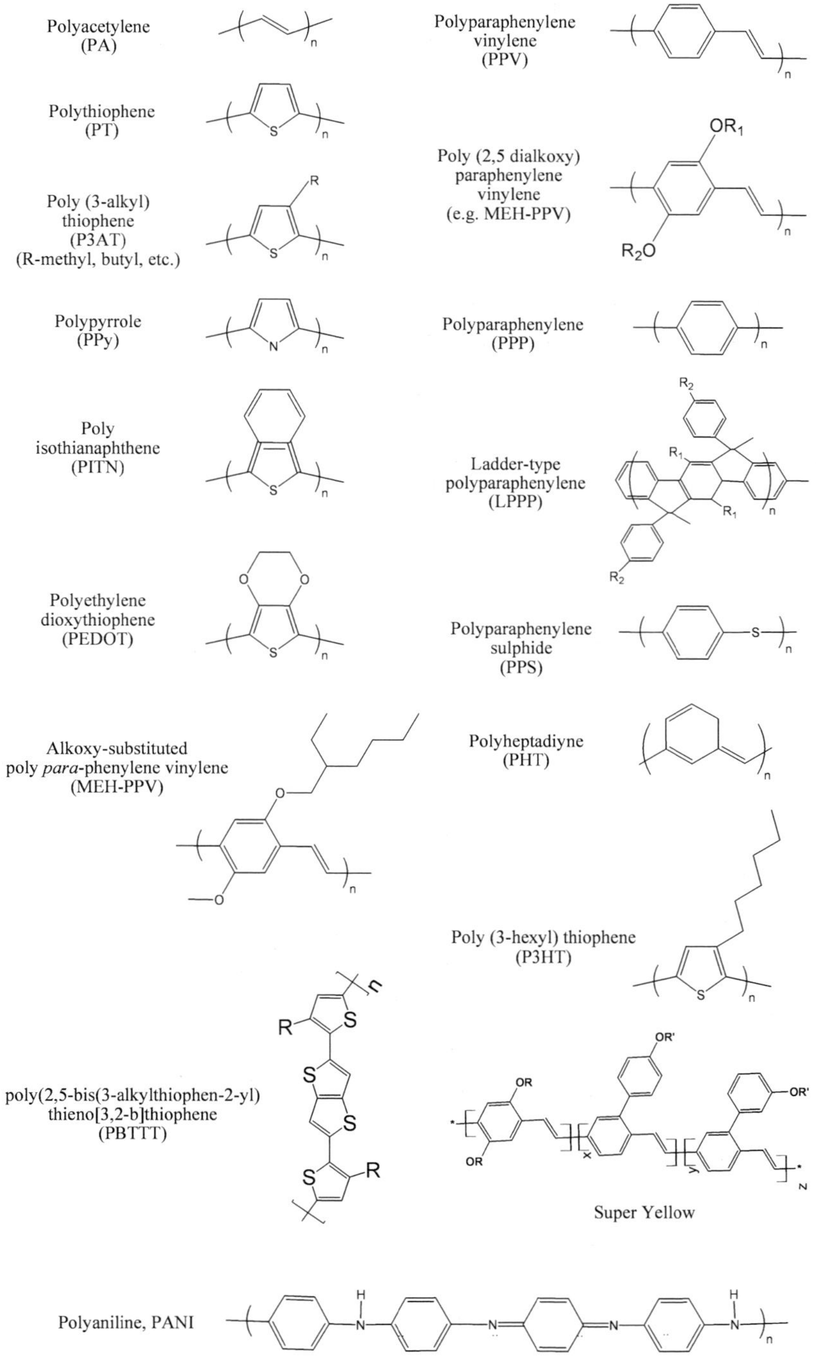

Fig. 1.12 Molecular structures of a few of the more important conjugated polymers.

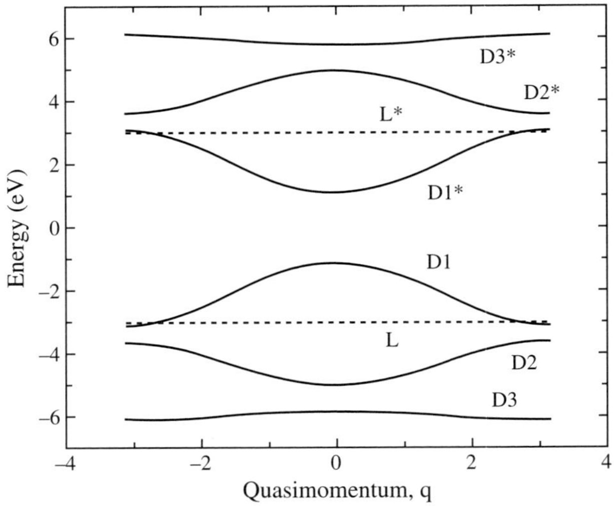

Fig. 1.13 π-electron band structure of PPV. [Reproduced from N. Kirova *et al.*, *Synthetic Metals*, **100**, 29 (1999), Elsevier Science.]

be prepared with properties that approach those of idealized ordered materials. Indeed, by utilizing the method of gel-processing of blends of conjugated polymers in polyethylene, a high degree of structural order has been attained. Thus, as a starting point, it is useful to consider idealized samples in which the macromolecular chains are chain extended and chain aligned.

(2) The two-fold coordination associated with linear conjugated polymers and the proportionality of bond length to bond strength (or bond order) in conjugated systems, leads to a fundamental interconnection between their electronic and chemical structures. As a result of this interconnection, the elementary excitations in semiconducting polymers (electrons and holes in classic inorganic semiconductors) involve lattice (geometrical) relaxation around the electrons and holes. The lattice relaxation in the excited state leads to self-localization and the formation of solitons, polarons and bipolarons.

(3) The band description ignores correlations between electrons that result from the electron–electron repulsive interaction. An introduction to the possible effects introduced by strong electron–electron interactions is given in the following section.

1.9 Metal or antiferromagnetic insulator: The Mott transition

If all the bond lengths in polyacetylene were equal (as in Fig. 1.3), there are two limiting cases.

(1) If the electron–electron repulsive interaction associated with two electrons in the p_z-orbital (magnitude denoted as U) is well-screened and small compared to the transfer interaction $2t_o$ (the factor of 2 arises since there are two nearest neighbors in a linear

chain), this simple polymer would be a metal; i.e. for $U < 2t_o$, polyacetylene with uniform bond lengths would be a metal.

(2) If the electron–electron repulsive interaction, U, is large compared to $2t_o$, this simple polymer would be an antiferromagnetic insulator; i.e. for $U >> 2t_o$, polyacetylene with uniform bond lengths would be an antiferromagnetic insulator. The transition from metal to insulator (the "metal–insulator transition") driven by the electron–electron Coulomb repulsion was discussed in great detail in the 1970s with fundamentally important contributions by Neville Mott (hence the metal–insulator transition is commonly known as the Mott transition). Without going into the detail of the theory of the Mott transition, one can understand the limiting cases in relatively simple terms.

In the metallic form, the energy level associated with the π-orbital (E_π) would broaden into the π-band with a total bandwidth, $W = 4t_o$, spanning the energy range from $E_\pi - 2t_o$ to $E_\pi + 2t_o$ Since that band would be half filled by the single π-electron contributed to the band by each of the carbon atoms in the chain, the average energy would be approximately $E_\pi - t_o$; i.e. in a metal the average kinetic energy is lowered by approximately t_o (see Fig. 1.6).

$$\Delta E_\pi(\text{kinetic}) \approx -t_o \tag{1.13}$$

However, because the electronic occupation is not correlated, there is significant probability of double occupancy. Thus, if $n_\uparrow$ is the number of electrons on the carbon atom with spin up and $n_\downarrow$ is the number of electrons on the same carbon atom with spin down, then the strength of the Coulomb repulsion will be $U\langle n_\uparrow\rangle\langle n_\downarrow\rangle$ where $\langle\ \rangle$ denotes an average value. Because

$$\langle n_\uparrow\rangle + \langle n_\downarrow\rangle = 1 \tag{1.14}$$

and $\langle n_\uparrow\rangle = \langle n_\downarrow\rangle$ in a nonmagnetic metal,

$$\langle n_\uparrow\rangle = \langle n_\downarrow\rangle = 1/2 \tag{1.15}$$

Thus,

$$\Delta E_\pi(\text{Coulomb}) \approx U/4 \tag{1.16}$$

and

$$\Delta E_\pi(\text{total}) = \Delta E_\pi(\text{kinetic}) + \Delta E_\pi(\text{Coulomb}) \approx -t_o + U/4 \tag{1.17}$$

The metallic state will have lower energy only if

$$U/4 < t_o \tag{1.18}$$

or

$$U < 4t_o \tag{1.19}$$

$$U < W$$

where W is the π-electron bandwidth. On the other hand, if $U >> W$, the energy is lowest if the electrons localize with one electron per site with spin up or spin down; i.e either with $\langle n_\uparrow\rangle = 1$ and $\langle n_\downarrow\rangle = 0$ or with $\langle n_\uparrow\rangle = 0$ and $\langle n_\downarrow\rangle = 1$. In this case, there is an energy gap with magnitude of approximately U to the lowest energy charged excitation which involves transfer of an electron to a neighboring occupied site.

With one electron per site, the energy is lowered still more if the spins form an antiferromagnetic (AF) ground state ... ↑↓↑↓↑↓↑↓↑↓↑↓↑↓↑↓↑↓↑↓ ... When the neighboring spins are antiparallel, the spin on site i can partially delocalize by hopping to site $(i+1)$ and back via the transfer interaction (t_o) at a cost of energy of U in the intermediate state. Thus, for the antiferromagnetic ground state, the energy per site is lowered in second order perturbation theory by

$$\Delta E_{\pi}(\text{AF}) = -2(t_o)^2/U \equiv J_{\text{AF}}$$

per site where again the factor of 2 comes from the two nearest neighbors; J_{AF} is defined as the antiferromagnetic exchange interaction ("super exchange").

The ground state of polyacetylene is a semiconductor with an energy gap. When doped p-type (by oxidation) or n-type (by reduction), polyacetylene is readily converted into a highly conducting state with electrical conductivity comparable to that of elemental metals. In addition to the magnitude of the electrical conductivity, one can distinguish the metallic state from the antiferromagnetic insulator by the magnitude of the magnetic susceptibility.

The Pauli spin susceptibility of a metal is given by the following expression:

$$\chi_P \propto g\mu_B^2/E_F$$

For metallic polyacetylene,

$$\chi_P \propto g\mu_B^2/2t_0.$$

where μ_B is the Bohr magnetron and $g \approx 2$ $(2\pi\gamma/h = g\mu_B$ is the gyromagnetic ratio and h is Planck's constant). By contrast, the magnetic susceptibility of an antiferromagnet is given by the following expression

$$\chi_{AF} \propto g\mu_B^2/J_{AF}$$

Note that $2t_o >> J_{\text{AF}}$ since $U >> t_o$ in the insulating antiferromagnetic state. Thus, quite generally, $\chi_{\text{AF}} >> \chi_{\text{P}}$. However, measurements of the magnetic susceptibility of highly conducting, doped polyacetylene demonstrate that the spin susceptibility is consistent with the Pauli spin susceptibility expected for a metal.

We conclude, therefore, that for polyacetylene and related π-conjugated polymers, $U< 2t_o$; i.e polyacetylene with uniform carbon–carbon bond lengths would be on the metallic side of the metal–insulator transition. The conclusion that $U< 2t_o$ is consistent with the electron energy loss experiments which are in good agreement with the predictions of the simple tight-binding model.

Further reading

J. M. Andre, J. Delhalle and J. L. Bredas, *Quantum Chemistry Aided Design of Organic Polymers*, World Scientific, Singapore, 1991.

A. J. Heeger, S. Kivelson, J. R. Schrieffer and W. P. Su, *Rev. Mod. Phys.*, **60**, 781 (1988).

E. Heilbronner and H. Bock, *The HMO Model and its Applications*, John Wiley, London, 1976.

R. E. Peierls, *Quantum Theory of Solids*, Oxford University Press, Oxford, 1955.
L. Salem, *Molecular Orbital Theory of Conjugated Systems*, Benjamin, New York, 1966.
T. A. Skotheim (ed.), *Handbook of Conducting Polymers*, Marcel Dekker, New York, 1986.
T. A. Skotheim and J. R. Reynolds (eds.), *Handbook of Conducting Polymers*, 3rd edition, CRC Press, Taylor and Francis, Boca Raton, 2007.

2
Doping of conducting polymers

2.1 Introduction

Conjugated polymers in their undoped, pristine state are semiconductors/insulators with energy gaps that are too large for thermal excitation of a significant number of charge carriers ($e^{\bar{E}_g/kT}$). Therefore, undoped conjugated polymers such as polythiophene and polyacetylene have very few carriers and, therefore, electrical conductivities of only 10^{-10}–10^{-8} S/cm (Fig. 2.1). Upon doping conjugated polymers, there is a rapid increase in electrical conductivity by several orders of magnitude up to values of around 10^{-1} S/cm even at a very low level of doping such as <1%. Subsequent doping to higher levels results in saturation of the conductivity at values in the range 10^2–10^5 S/cm (Fig. 2.2). Heavily doped, stretch-oriented polyacetylene has the highest reported electrical conductivity with confirmed values of approximately 80,000 S/cm.

Charge injection onto conjugated, semiconducting macromolecular chains, denoted as "doping", is quite different from doping in conventional inorganic semiconductors such as silicon and germanium. In inorganic crystalline semiconductors, doping occurs through substutional insertion of atoms which have higher or lower valence than the host semiconductor

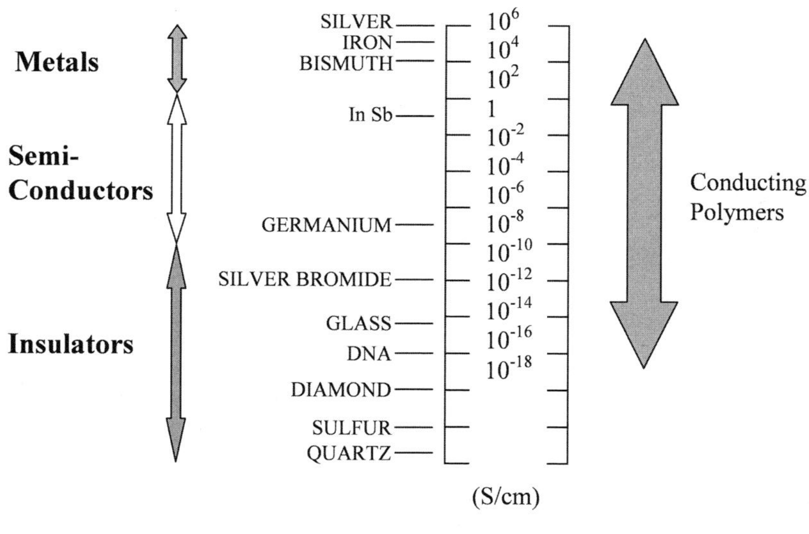

Fig. 2.1 Full range of electrical conductivity from insulators to metals.

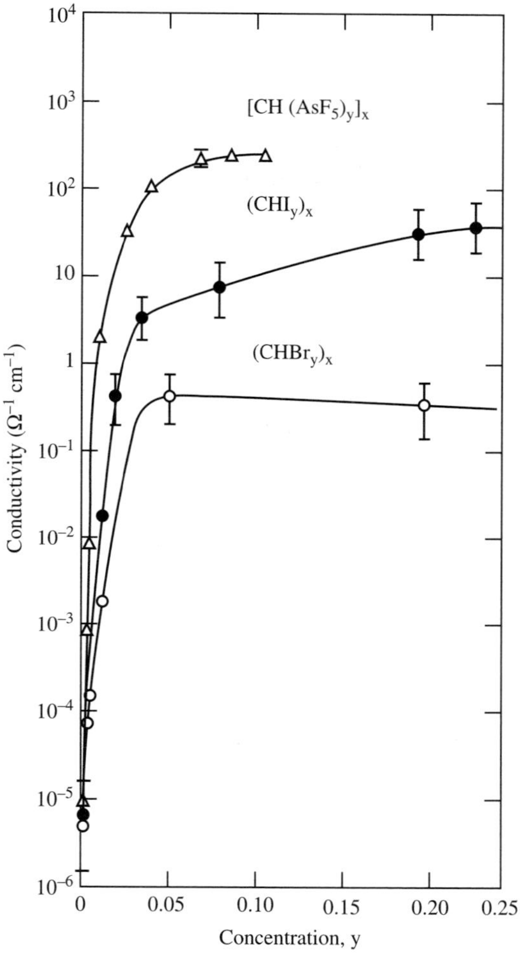

Fig. 2.2 Electrical conductivity (room temperature) of polyacetylene as a function of dopant concentration. [Reproduced from C. K. Chiang *et al.*, *Appl. Phys. Lett.*, **33**, 18 (1978) with permission of The American Physical Society.] Stretch-oriented samples exhibit much higher conductivity values.

for n-type or p-type doping, respectively. Thus, a phosphorous atom in the silicon lattice contributes an extra electron to the semiconductor band structure (phosphorous has a valence of five compared to four for silicon) and leads to n-type doping. Overall charge neutrality is maintained; the extra electron charge is compensated by the positive charge on the nucleus of the impurity atom. Similarly, p-type doping would involve the insertion of atoms from column three of the periodic table, such as, for example gallium.

For semiconducting polymers, doping occurs via oxidation or reduction of the π-electron system for p-type or n-type doping, respectively. Since there can be no net uncompensated charge, in semiconducting polymers a counter-ion is inserted into the structure. Because of the strong intrachain covalent bonds in semiconducting polymers, doping is reversible and does not interrupt the polymer chain structure. The advantage of this "anisotropic chemistry" is clear. Within the conjugated polymer chain the strength of the σ-bonds approaches that of diamond. Between the chains, however, the forces are much weaker (e.g. van der Waals forces and hydrogen bonding). This anisotropic chemical bonding enables the counter-ions to diffuse into and out of the structure between the polymer chains. Therefore, it is relatively easy to intercalate ions between the semiconducting polymer chains in response to the oxidation or reduction of the backbone. As summarized below in Fig. 2.3, reversible charge injection by "doping" can be accomplished in a number of ways.

Each of the methods of charge-injection doping indicated in Fig. 2.3 leads to unique and important phenomena. For chemical and/or electrochemical doping, the induced electrical conductivity is permanent, until the carriers are either chemically compensated or until the carriers are purposely removed by "undoping" (for example by reversing the electrochemical reactions). In the case of photoexcitation, the photoconductivity is transient and lasts only until the excitations are either trapped or decay back to the ground state. For charge injection at an interface, electrons reside in the π*-band and/or holes reside in the π-band only as long as the bias voltage is applied.

Doping also serves to (reversibly) shift the electrochemical potential of the polymer; thereby making possible polymer batteries and electrochemically active polymer electrodes, electrochromic phenomena ("smart" windows), and polymer p-i-n junctions (light-emitting electrochemical cells).

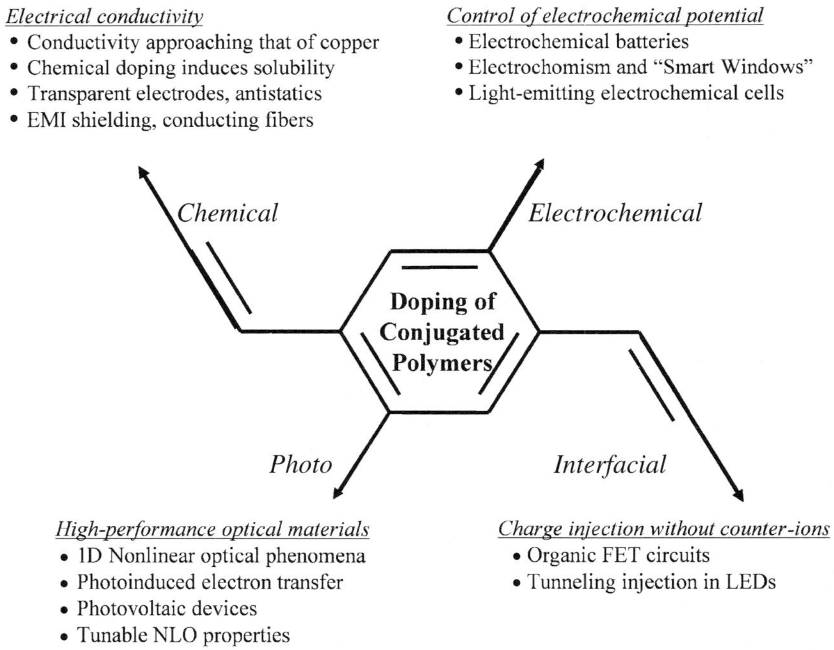

Fig. 2.3 Doping methods of conjugated polymers.

Because of the self-localization associated with the formation of solitons, polarons and bipolarons, charge injection leads to the formation of localized structural distortions and electronic states in the energy gap. In the case of "photodoping", the redistribution of oscillator strength associated with the subgap infrared absorption and the corresponding bleaching of the interband (π-π^*) transition provide a route to nonlinear optical (NLO) response. Real occupation of low energy excited states and virtual occupation of higher energy excited states (in the context of perturbation theory) lead to, respectively, resonant and nonresonant NLO response.

2.2 Chemical doping by charge transfer

During gas phase doping, an oxidizing or reducing agent is brought in contact with the conjugated polymer for p-type or n-type doping, respectively. For example, when polyacetylene is exposed to iodine vapor, doping occurs through the following chemical reaction:

$$3/2xy(I_2) + (CH)_x \rightarrow [(CH)^{+y}(I_3{}^-)_y]_x$$

where $[(CH)^{+y}(I_3{}^-)_y]_x$ denotes an oxidized (p-type doped) polyacetylene chain with y positive charges per average repeat unit and with the same number of $I_3{}^-$ ions inserted to compensate the charge. In this example, the I_2 serves as the oxidizing agent.

Alternatively, for solution phase doping, one can use the oxidant NO^+ in the salt $(NO)PF_6$. When polyacetylene films are dipped into a solution containing $(NO)PF_6$, the components react spontaneously (the NO species is the oxidizing agent):

$$2(NO^+PF_6{}^-)_{y/2} + (CH)_x \rightarrow (N_2O_2)_{gas} + [(CH)^{+y}(PF_6{}^-)_y]_x$$

This reaction is especially elegant because the N_2O_2 escapes as a gas without contamination of the polymer sample while the $PF_6{}^-$ is the counter-ion.

For n-type doping, a reducing agent must be used. An example of an n-type doping reaction in solution is the following:

$$y(Na^+)(\text{naphthalene}^-) + (CH)_x \rightarrow (\text{naphthalene})_{\text{neutral}} + [Na_y{}^+(CH)^{-y}]_x$$

where the naphthalene$^-$ anion is the reducing agent, and Na^+ is the counter-ion; this reaction is carried out in solution in tetrahydrofuran, THF.

Both oxidative and reductive doping result in mobile charges in the π-electronic structure of the conjugated polymer chain. Since the mobile charges are compensated by counter-ions which diffuse into the material between polymer chains, doped conjugated polymers are salts. The structure of Na-doped metallic polyacetylene is shown in Fig. 2.4. The general structure of parallel semiconducting chains with counter-ions intercalated between chains has distinct advantages:

(1) The counter-ions do not disrupt the chains.
(2) The charges on counter-ions are separated from the chain, analogous to modulation doping in inorganic semiconductors.

The former implies that the electronic strucure of the semiconductor remains intact after doping; the latter implies that local disorder along the stacks of counter-ions will not cause strong scattering of the electrons or holes on the semiconducting chains. Both of these features

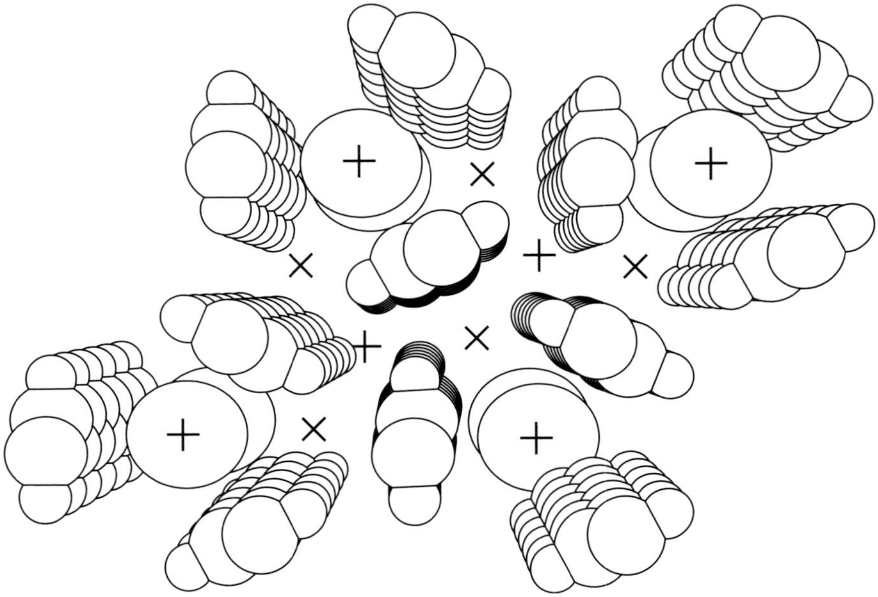

Fig. 2.4 Structure of metallic sodium dopd polyacetylene. [Reproduced from M. Winokur *et al.*, *Phys. Rev. Lett.*, **58**, 2329 (1987) with permission of The American Physical Society.]

imply that relatively high carrier mobilities might be expected if one can achieve chain extension, chain alignment and structural order.

2.3 Electrochemical doping

For electrochemical doping, the conjugated polymer is attached to one of the electrodes in an electrochemical cell. For simplicity in Fig. 2.5, a two electrode configuration with a working electrode and a counter-electrode is displayed. In electrochemical doping, the semiconducting

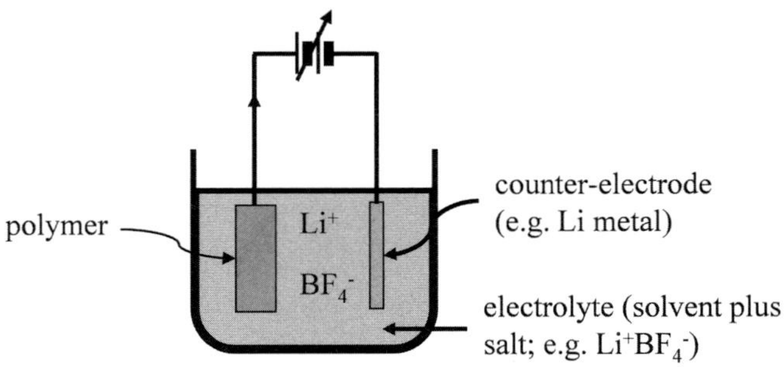

Fig. 2.5 Schematic diagram of an electrochemical cell used for doping a semiconducting polymer.

polymer is oxidized or reduced by the electrode and the counter-ion moves from the electrolyte into the polymer, again diffusing into the structure between chains.

In a pure semiconductor, the chemical potential is located at the center of the energy gap. If there are no states in the gap and if the chemical potential of the counter-electrode is locked at a specific value (as is the case when the counter-electrode is a metal), the chemical potential of the semiconductor can be readily moved across the gap by the application of an external voltage to the electrochemical cell.

The neutral point occurs when the applied voltage is equal to the difference in chemical potential between the metallic counter-electrode (Li in Figs 2.5 and 2.6) and the chemical potential of the neutral semiconducting polymer. Since the ionization potential (the energy of top of the π-band relative to the vacuum) and the energy gap of the polymer are known, and since the ionization potential of the counter-electrode is known, the neutral point can be estimated with reasonable accuracy. In polyacetylene versus Li, the neutral point is at an applied voltage of approximately 2.4 V (polymer positive with respect to the Li counter-electrode as shown in Figs 2.5 and 2.6). At higher voltages, the chemical potential moves down in energy toward the top of the π-band. When the chemical potential reaches the top of the π-band, the semiconductor will be oxidized (electrons removed from the π-band) and p-type doped with associated insertion of BF_4^- ions (see Fig. 2.6). Larger applied voltages lead to higher doping levels. For n-type doping, the external voltage must be decreased with respect to the neutral point. At lower voltages, the chemical potential moves up in energy toward the bottom of the π*-band. When the chemical potential reaches the bottom of the π*-band, the semiconductor will be reduced (electrons injected into the π*-band) and n-type doped with associated insertion of the Li^+ ions.

For p-type doping in the $LiBF_4$ electrolyte (for example in propylene carbonate) and with Li as the counter-electrode, the electrochemical reaction is described as follows:

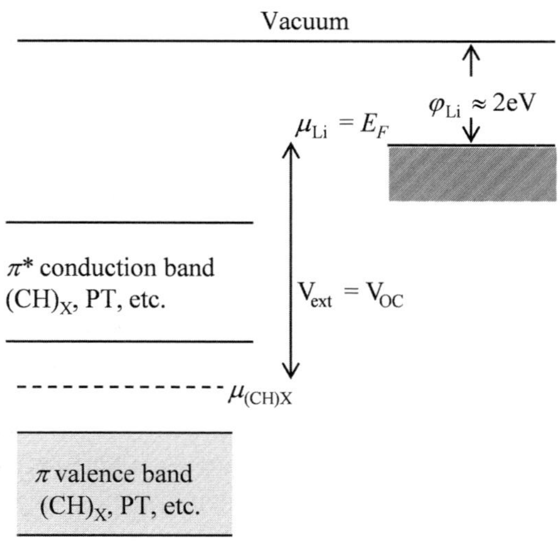

Fig. 2.6 Band diagram of the electrochemical doping of a semiconducting polymer; ϕ_{Li} is ionization potential of Li electrode and $\mu_{(CH)_x}$ is chemical potential of the conjugated polymer.

$$(\pi - \text{polymer})_n + [\text{Li}^+(\text{BF}_{4^-})]_{\text{electrolyte}} \rightarrow [(\pi - \text{polymer})^{+y}(\text{BF}_{4^-})_y]_n + \text{Li(electrode)}$$

For n-type doping with the same electrolyte and counter-electrode,

$$(\pi - \text{polymer})_n + \text{Li(electrode)} \rightarrow [(\text{Li}^+)_y(\pi - \text{polymer})^{-y}]_n + [\text{Li}^+(\text{BF}_{4^-})]_{\text{electrolyte}}$$

After oxidation or reduction, charge is stored in the electronic structure of the conjugated π-electron system of the polymer leading to mobile charge carriers.

Electrochemical doping at constant potential (potentiostatic doping) has the important advantage of directly controlling the chemical potential. As shown in Fig. 2.6, by varying V_{ext}, one shifts the chemical potential of the polymer with respect to the counter-electrode in the two electrode cell (or with respect to the reference electrode in the three electrode cell). If, after doping to a specific level, the conjugated polymer working electrode is discharged (leading to the neutral undoped polymer) there are two possibilities for the intercalated ions:

(1) The counter-ions diffuse out of the polymer returning it to the original undoped state.
(2) Alternatively, oppositely charged ions from the electrolyte move into the polymer and neutralize the charges of the pre-existing intercalated ions.

Which of these two alternatives occurs in an electrochemical doping/undoping cycle depends on the respective ions in the polymer matrix. The process can be monitored experimentally using a quartz microbalance covered with the conjugated polymer and used as the working electrode. For a quartz oscillator, the fundamental resonance frequency will shift as a function of the mass of the resonator. Thus, a reduction of the mass of the electrode indicates that the counter-ions are being driven out (case 1). In contrast, an increase of the mass of the quartz electrode indicates the intercalation of ions of opposite charge which neutralize the counter-ions (case 2). Quartz microbalance studies have demonstrated both of the two alternatives listed above for different counter-ions.

2.4 Electrochemical charge injection: Measurement of E_g

From the above discussion, n-type doping occurs with onset at a voltage, $V_n{}^o$, when V_N is the open circuit voltage of the neutral conjugated polymer versus the reference electrode (or, for two electrode configuration, the counter-electrode)

$$V_n^0 = V_N - \frac{\Delta}{e}.$$

The onset of p-type doping occurs at $V_p{}^o$ where

$$V_p^0 = V_N + \frac{\Delta}{e}$$

Thus, by measuring the onset of p-type and n-type doping in a single electrochemical cell, one can determine the energy gap of the semiconducting polymer; see Fig. 2.7.

$$E_g = 2\Delta = (V_p^0 - V_n^0)e$$

Electrochemical voltage spectroscopy has been used to monitor the doping process under quasi-equilibrium conditions. After first setting the external voltage at V_N, the voltage is

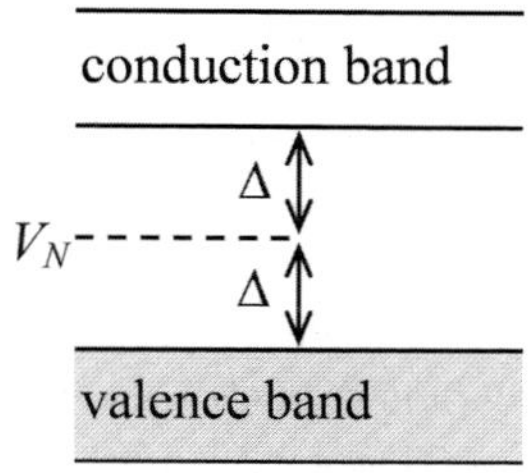

Fig. 2.7 Band diagram showing the neutral point in the center of the gap.

increased (or decreased) in small steps. After each such step, the current through the cell is integrated as the cell approaches electrochemical equilibrium at the new voltage, and the charge that has been injected is recorded. In this way for each ΔV, the corresponding ΔQ is determined. The plot of $\Delta Q/\Delta V$ versus V can be thought of as an "infinitesimally slow" cyclic voltammogram. This process is continued until the full V versus Q curve is obtained. Data obtained in this way for *trans*-polyacetlyene are shown in Fig. 2.8; for *trans*-$(CH)_x$ versus Li, $V_N \approx 2.4$ volts. These results indicate that the onset of p-type doping occurs at 3.05 V and the onset of n-type doping occurs at 1.75 V (both voltages referenced with respect to Li). Thus, the energy gap as determined by electrochemical redox is $E_g \approx 1.4$ eV for *trans*-polyacetylene; in agreement with the energy gap as determined from optical absorption (see Fig. 7.2), but somewhat smaller than obtained from the electron energy loss experiments ($E_g = 1.8$ eV).

In general, for conjugated polymers, the energy gap as determined by electrochemical oxidation and reduction is in reasonably good agreement with the optical gap; i.e. the π-π^* gap as determined by optical absorption. Note, however, that there are entropy contributions (not included) since the number of degrees of freedom changes when the counter-ions move from the electrolyte into the doped polymer.

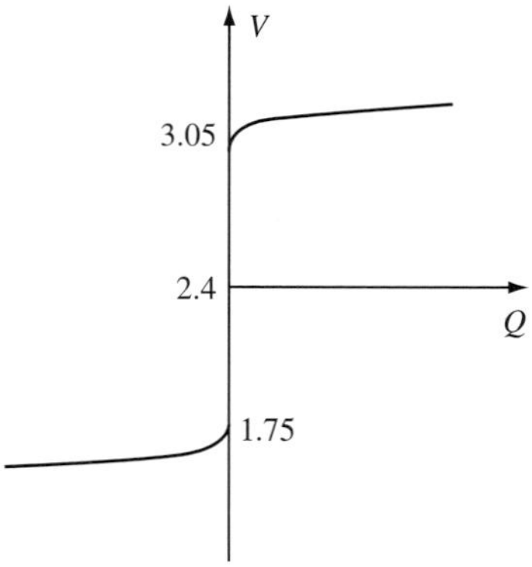

Fig. 2.8 Voltage (V) as a function of the charge injected for *trans*-polyacetylene as obtained by electrochemical voltage spectroscopy. The onset of p-type doping occurs at 3.05 V and the onset of n-type doping occurs at 1.75 V.

2.5 Polymer battery

In a polymer battery the working electrode consists of an electroactive conjugated polymer. In principle, since conjugated polymers can be doped both n-type and p-type, both electrodes can be made from the same material, resulting in the all-polymer battery shown in Fig. 2.9. Electrical charge is stored in the doped polymer; the battery operates through reversible electrochemical doping of the conjugated polymer. The difference in the electrochemical potentials of the n-type and p-type doped polymer (see Fig. 2.8) provides the electromotive force for generating currents in the external circuit.

For an all-polymer battery with both electrodes made of the same electroactive conjugated polymer, the electrochemical reactions can be described as follows:

Anode reaction:

$$(\pi - \text{polymer}) + [\text{Li}^+(\text{BF}_4{}^-)]_{\text{solv}} \rightarrow [(\pi - \text{polymer})^{x+}(\text{BF}_4{}^-)_x] + x\text{Li}^+ + xe^-$$

Cathode reaction:

$$(\pi - \text{polymer}) + xe^- + [\text{Li}^+(\text{BF}_4{}^-)]_{\text{solv}} \rightarrow [(\pi - \text{polymer})^{x-}(\text{Li}^+)_x] + x(\text{BF}_4{}^-)$$

Although this simple, lightweight battery concept attracted industrial interest, the energy density is not large enough for large scale commercial applications. The maximum doping level is relatively low (approximately 16%) in polyacetylene with similar numbers per carbon in other polymers. However, the easy intercalation and de-intercalation of the counter-ions enables conducting polymer battery systems to be able to handle high current densities in charge/discharge cycles. The basic concept of the all polymer battery has re-emerged in the light-emitting electrochemical cell (see Chapter 10).

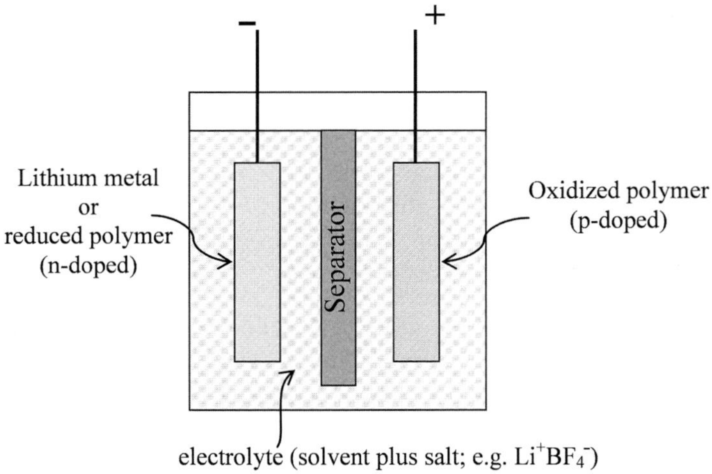

Fig. 2.9 Schematic diagram of the polymer battery.

2.6 Polyaniline: Doping by acid–base chemistry

Polyaniline represents a special case in conjugated polymers of "protonic-acid doping". In polyaniline, the transition from insulator to metal occurs through a protonation induced change in the π-electron structure with no addition of electrons to or withdrawal of electrons from the π-electron system. Protonation by acid–base chemistry leads to an internal redox reaction and the conversion from semiconductor (the emeraldine base) to metal (the emeraldine salt).

Polyaniline is synthesized in the emeraldine salt form, as shown in Fig 2.10 where A^- denotes the counter-ion, e.g. Cl^-, HSO_4^-, etc. Upon washing the emeraldine salt form with a strong base, the polymer converts into the insulating emeraldine base form (blue in color). The chemical structure of the semiconducting emeraldine base form of polyaniline is that of an alternating copolymer as shown in Fig. 2.11.

The structure of the emeraldine base can be denoted as [(1A)(2A)]n, where

$$(1A) = (B—NH—B—NH)$$
$$(2A) = (B—N = Q = N—).$$

In this shorthand notation, B denotes a C_6H_4 phenyl ring in the benzenoid form, and Q denotes a C_5H_4 phenyl ring in the quinoid form. The emeraldine base can be fully reduced to leucoemeraldine, (1A)*n*, a large band-gap insulator with π-π* transition peaked at about 3.8 eV.

Schematically, the emeraldine base form has two resonance forms. In the lower energy form, one of the rings in the repeat unit is in quinoid form (see Fig. 2.11). In the higher energy form (Fig. 2.12) all of the rings are in the benzenoid configuration. In Fig. 2.12, a localized unpaired spin appears on each of the imine nitrogens. This alternative, higher energy, structure for the emeraldine base form can be written as follows:

This structure can be stabilized by adding protons to these imine nitrogens in acidic media with associated conversion to the emeraldine salt. The emeraldine salt has an unpaired spin on every second nitrogen along the polyaniline backbone, i.e. a polaron lattice is formed. This polaron lattice structure has a unit cell consisting of only two aniline units in contrast to the base form which has four aniline units. Thus, the proton induced spin unpairing mechanism leads to a structural change with one unpaired spin per repeat unit, *but with no change in the number of electrons*:

$$[(B—NH—B—\underset{A^-}{NH^+})^\bullet]_N$$

where the dot denotes an unpaired spin in every repeat unit and A^- denotes the counter-ion. Thus, the emeraldine salt is formally a half-filled band system and, potentially, a metal. Again, as with polyacetylene, if the mean electron–electron repulsive energy is larger than the bandwidth, the structure with one unpaired electron per repeat unit would be an antiferromag-

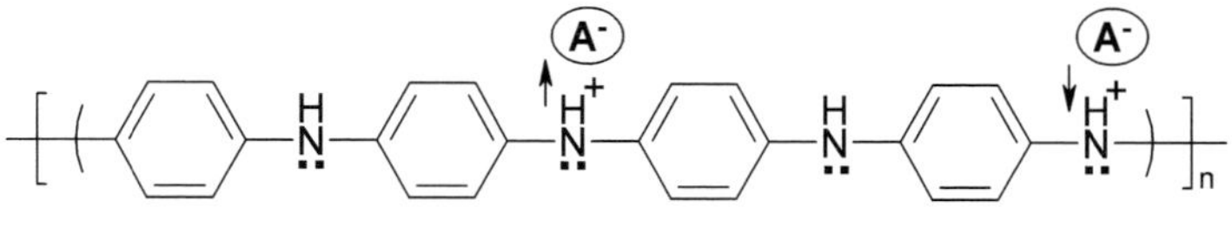

Fig. 2.10 Emeraldine salt form of polyaniline.

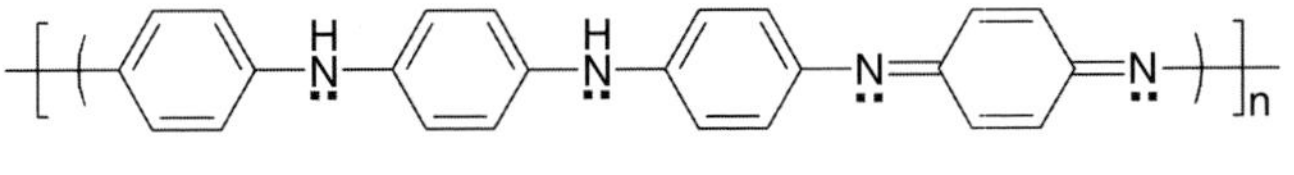

Fig. 2.11 Emeraldine base form of polyaniline.

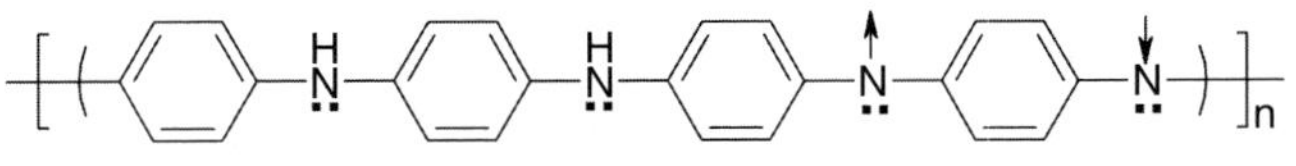

Fig. 2.12 Alternative structure: higher energy form of the emeraldine base.

netic Mott insulator (see Chapter 1). The experimental results demonstrate, however, that the emeraldine salt has the electronic structure of a metal, although as shown in Chapter 4 disorder can cause a transition from metal to Fermi "glass" insulator.

The transition from insulating emeraldine base to metallic emeraldine salt is particularly interesting, because it occurs without any change in the number of π-electrons. Thus, formally the insulator-to-metal transition in polyaniline occurs without doping, but rather through a protonation induced first-order quantum phase transition in the electronic structure.

2.7 Photoinduced "doping"

During photoexcitation of a semiconducting polymer, one excites an electron from the filled π-band to the empty π*-band. From the point of view of chemistry, this can be visualized as the local oxidation of the conjugated polymer backbone with local reduction of the same chain at a different place along the backbone, i.e.

$$(\pi - \text{polymer})_n + h\nu \rightarrow [\{\pi - \text{polymer}\}^{+y} + \{\pi - \text{polymer}\}^{-y}]_n$$

where y is the number of electron–hole pairs (dependent upon the pump rate in competition with the recombination rate). Such photoexcited charges on a single chain will in general respond to their mutual Coulomb attraction and form a neutral bound state called an "exciton". The branching ratio between free carriers and bound excitons, and the closely related issue of the magnitude of the exciton binding energy) will be discussed in more detail in Chapter 8.

Photoexcitation can lead to mobile charges, photoinduced positive and negative polarons, and thereby to photoinduced conductivity ("photoconductivity"). The photoinduced positive and negative polarons can be created on a single chain or they can appear on separate polymer chains by interchain electron transfer. More rapid recombination can be expected if the positive and negative polarons are on the same chain. Since the interchain transfer integral is a sensitive function of the interchain packing density, the details of the chain packing and the local structure will determine the quantum yield of mobile charge generation in semiconducting polymers.

If a neutral exciton is created near (or diffuses to) an interface such as a semiconducting polymer/electrode metal interface or a donor/acceptor interface, then the local potential discontinuity at such an interface can create a sufficiently large local electric field to ionize the neutral exciton into a positive/negative polaron pair. One of the most effective ways to achieve charge separation is through the use of donor/acceptor bulk heterojunction materials. In such a system, the donor phase is intimately mixed with an acceptor phase (either covalently bonded or phase separated on an nanometer length scale), so that the exciton is always close to a charge separating interface. For semiconducting polymers as electron donors and fullerenes as electron acceptors, the quantum yield for charge separation can approach 100% (see Chapter 12).

2.8 Charge injection at a metal–semiconducting polymer (MS) interface

Electrons and holes can be injected from metallic contacts into the π and π* bands, respectively:

(a) Hole injection into an otherwise filled π-band

$$(\pi - \text{polymer})_n - y(\text{e}^-) \rightarrow [(\pi - \text{polymer})^{+y}].$$

(b) Electron injection into an empty π*-band

$$(\pi^* - \text{polymer})_n + y(\text{e}^-) \rightarrow [(\pi^* - \text{polymer})^{-y}]_{n.}$$

In the case of charge injection at an MS interface, the polymer is oxidized or reduced (electrons are added to the π* band or removed from the π-band). The polymer is not doped in the sense of chemical or electrochemical doping, however, for there are no counter-ions. This distinction becomes particularly clear when comparing charge injection in polymer light-emitting diodes (where there are no ions) and polymer light-emitting electrochemical cells (where electrochemical doping with associated redistribution of ions provides the mechanism for charge injection).

Using carrier injection at the MS interface, diodes with excellent current–voltage characteristics can be fabricated by casting the active polymer directly from solution. Dual carrier injection in metal/polymer/metal structures provides the basis for polymer LEDs. In polymer LEDs, electrons and holes are injected from the cathode and anode, respectively, into the undoped semiconducting polymer; light is emitted when the injected electrons and holes meet in the bulk of the polymer and recombine radiatively. By charge-injection doping at an MIS interface, the polymer semiconductor can be used as the active element in thin film field effect transistors (FETs). These charge injection devices are discussed in greater detail in Chapters 9, 10 and 14.

Thus, as summarized in Fig. 2.3, doping is a common feature of conducting polymers; doping leads to a remarkably wide range of electronic phenomena.

Further reading

G. Hadziioannou and P. F. van Hutten (eds.), *Semiconducting Polymers*, Wiley-VCH, Weinheim, 2000.

G. Hadziioannou and G. G. Mallarias (eds.), *Semiconducting Polymers* 2nd edition, Wiley-VCH, Weinheim, 2007.

H. S. Nalwa (ed.), *Handbook of Organic and Conductive Molecules and Polymers*, John Wiley, Chichester, 1997.

T. A. Skotheim (ed.), *Handbook of Conducting Polymers*, Marcel Dekker, New York, 1986.

T. A. Skotheim and J. R. Reynolds (eds.), *Handbook of Conducting Polymers*, 3rd edition, CRC Press, Taylor and Francis, Boca Raton, 2007.

3

Novel properties generate new opportunities

3.1 Introduction

Conducting polymers exhibit novel properties; properties not typically available in metals and semiconductors. As a result, a number of applications have emerged that are enabled by this class of materials. We focus on three specific "novel properties" that are characteristic features of conducting polymers:

(a) semiconducting and metallic polymers that are soluble in and processable from common solvents;
(b) blends of conducting polymers with insulating polymers; percolation at low concentrations;
(c) transparent conductors.

Each of these novel properties is related to a fundamental feature of the chemistry and/or physics of the class of conducting polymers. Although this list is certainly not exhaustive; it illustrates an important point: Novel properties generate both new science and new opportunities for technology.

The "novel properties" listed above are enabling features for a number of applications including polymer light-emitting diodes (LEDs), polymer light-emitting electrochemical cells (LECs), conducting polymers as electrochromic materials, polymer photodetectors and polymer photovoltaic cells. Each of these application areas is discussed in detail in later chapters.

3.2 Semiconducting and metallic polymers: Materials that are soluble in and processable from common solvents

In the early years following the discovery that semiconducting polymers could be doped to the metallic state, there was serious doubt as to whether processing methods could be developed. There were two issues:

1. Without the alkyl- and alkoxy-side-chains, the interchain electron transfer interactions of conjugated polymers are relatively strong compared with interchain interactions typical of saturated polymers. Thus, the early conducting polymers were insoluble and infusible.

2. There was concern that addition of such alkyl- and alkoxy-side-chains would increase the interchain spacing in the solid state and thereby inhibit the interchain electron hopping that is required for macroscopic electrical conductivity.

This early uncertainty regarding processability of conducting polymers is no longer an issue; as shown in Fig. 3.1(a), semiconducting and metallic polymers can be prepared as solutions or dispersions in common organic solvents and in water. Thin, uniform films of these materials can be spin-cast or dip-coated onto substrates. The ability to fabricate optical quality thin films by spin-casting from solution has proven to be a key step in the development of plastic electronic devices such as diodes, photodiodes, LEDs, LECs, optocouplers, and thin film transistors. In fact, as suggested by Fig. 3.1(a), when properly formulated such solutions can be used as "inks" and deposited/patterned onto surfaces with printing technology see Fig. 3.1(b) (for example, ink-jet printing). These unique processing opportunities have enabled the concept of printed "plastic" electronics.

As a result of this remarkable progress, conducting polymers offer a unique combination of properties: the electrical and optical properties of metals and semiconductors in combination with the processing advantages and mechanical properties of polymers.

There are four basic approaches toward processing conducting polymers:

1. side-chain functionalization, principally used for processing semiconducting polymers from solution in organic solvents or from water;
2. precursor route chemistry, principally used for processing polyacetylene and PPV into thin films;
3. counter-ion induced processing, principally used for processing polyaniline in the metallic form from organic solvents;
4. aqueous colloidal dispersions created by template synthesis, principally used for processing polyaniline and poly(ethylenedioxythiophene), PEDOT.

(a) (b)

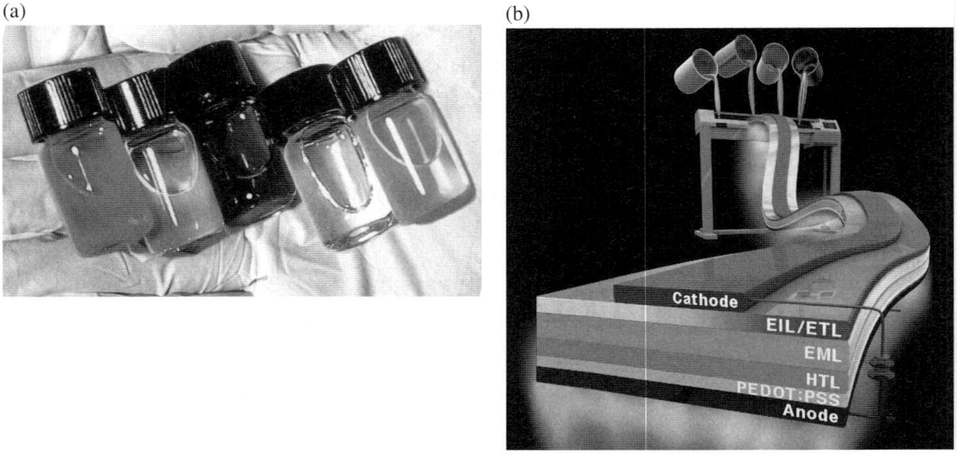

Fig. 3.1 (a) Vials containing semiconducting and metallic polymers in solution. The colored solutions are semiconducting polymers with different band gaps (the colorless solution contains a blue-emitting polymer with absorption in the ultraviolet). The dark solution in the center contains conducting poly (ethylenedioxythiophene). (b) The organic semiconductors in solution form can be used as "inks" and printed onto surfaces with any printing technology. [Reproduced from *Convergence*, the *Magazine of Engineering and the Sciences* at UC Santa Barbara, 9, 2007.] (This figure is reproduced in color in the color plate section.)

3.2.1 Side-chain functionalization

The addition of moderately long side-chains onto the monomer units resulted in derivatives of polythiophene, the poly(3-alkylthiophenes), or P3ATs; see Fig. 3.2(a). Since the side-chains decrease the interchain coupling and increase the entropy, these derivatives can be processed either from solution or from the melt. Similarly, side-chain functionalization of poly(phenylene vinylene), PPV (see Fig. 3.2b) and related structures has progressed to the point where a variety of semiconducting polymers and copolymers are available with energy gaps that span the visible spectrum (see Fig. 3.1).

Water solubility is achieved by terminating the side-chains with polar groups such as $(CH_2)_n SO_3^- M^+$ as shown for example in Fig. 3.3. When a polymer such as the PPV derivative in Fig. 3.3 is dissolved in water, there is an equilibrium

$$SO_3^- Li^+ \leftrightarrow SO_3^- + Li^+$$

such that the polymer in solution can be described as an anionic polyelectrolyte.

By terminating the side-chains with, for example, $NH_3{}^+I^-$, one can make water soluble cationic chains.

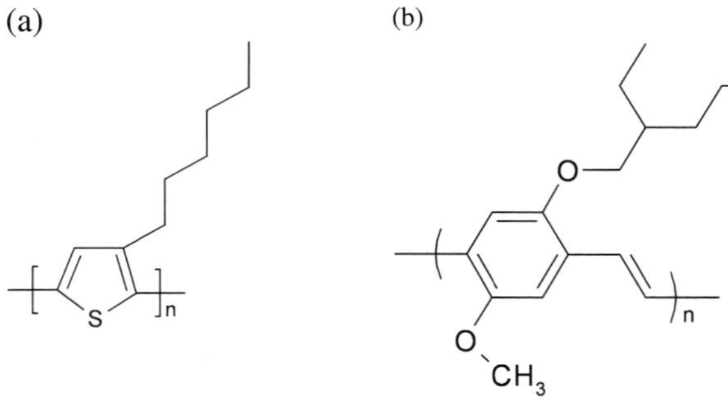

Fig. 3.2 Molecular structure of (a) poly(3-hexylthiophene), P3HT, and (b) poly(2-methoxy,5-(2′-ethyl-hexoxy)-*p*-phenylene vinylene) (MEH-PPV); side-chains, both branched and linear, enable solubility in common organic solvents. Note that the parent polymers, polythiophene and poly(phenylene vinylene) are insoluble.

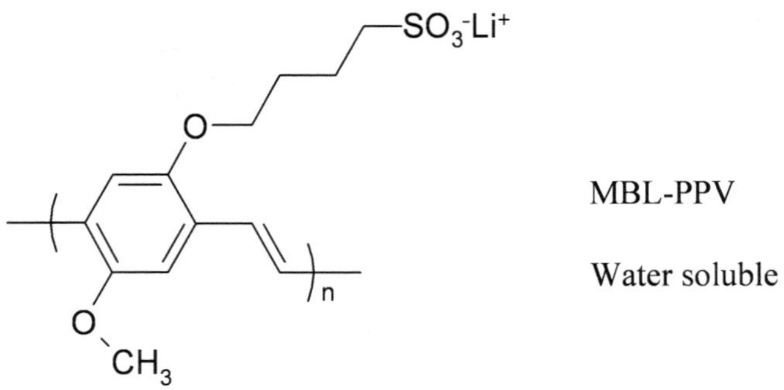

Fig. 3.3 Molecular structure of a water soluble PPV derivative, poly(2-methoxy,5-(2′-butoxy solfonate)-*p*-phenylene vinylene), MBL-PPV.

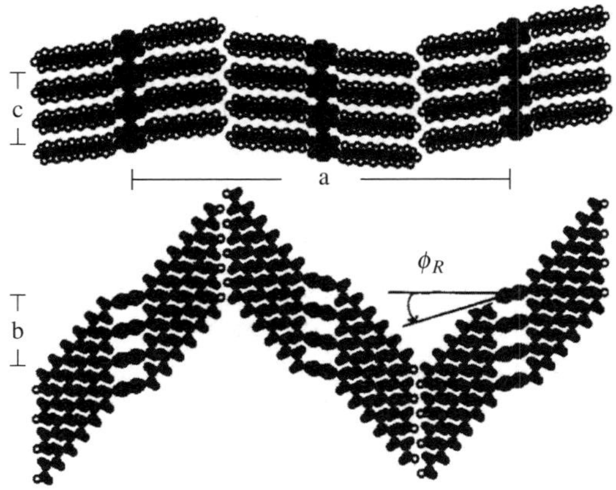

Fig. 3.4 X-ray structure of P3HT show stacked lamellar structure with good π-π packing and good overlap between the π-electron wavefunctions on neighboring chains in a quasi-two-dimensional array. Lattice parameter a=33.6Å, b=7.66 Å, c=7.7 Å and ϕ_R=5. [Reproduced from T. J. Prosa *et al.*, *Macromolecules*, **25**, 4364 (1992) with permission of The American Chemical Society.]

Although side-chain functionalization reduces the interchain electron transfer interactions, the polymer chains typically stack face-to-face in the solid state with good π-π packing and good overlap between the π-electron wavefunctions on neighboring chains in a quasi-two-dimensional array. As an example, the solid state structure of P3HT is shown in Fig. 3.4. Note the stacking of the P3HT backbones and the interdigitation of the side-chains. In such a structure, one expects the highest carrier mobility to be along the chains, but with facile interchain transport between chains within the two-dimensional plane. Thus, although the side-chain functionalization does reduce the interchain interactions, cast films of these materials can exhibit high performance electrical properties. For example, when used as the active semiconductor in field effect transistors, regio-regular P3HT films cast from solution exhibit hole mobilities in excess of 0.1 cm^2/Vs; i.e. comparable to the carrier mobility in amorphous silicon.

3.2.2 Precursor route

The versatile precursor synthesis route to conjugated polymers involves the preparation of a processable precursor (saturated) polymer and the subsequent conversion of that precursor to the conjugated polymer. After conversion to the conjugated form, the resulting polymer is insoluble and intractable, but thermally stable. PPV, for example, is stable at temperatures in excess of 300 °C. The precursor route synthesis was introduced as an alternative synthesis route to polyacetylene. Some years earlier, however, PPV was synthesized by conversion of a soluble and processable precursor polymer as shown in the following scheme:

Step 1: Synthesize water soluble precursor polymer (not conjugated)

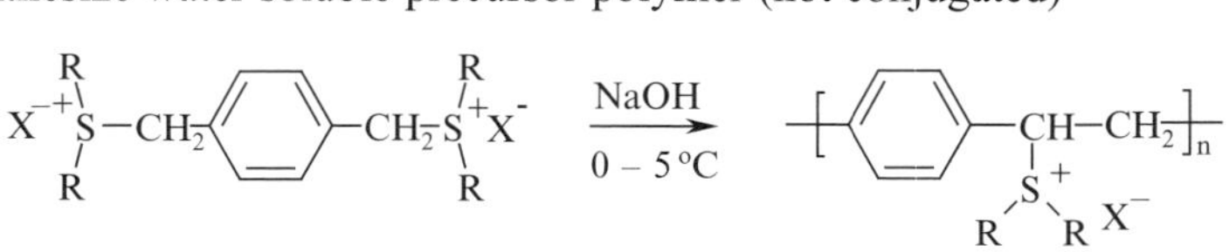

R=CH_3, C_2H_5
X=Cl, Br

Step 2: Convert the precursor (by heating) to the fully conjugated polymer

3.2.3 Counter-ion induced processability of polyaniline

Polyaniline in its emeraldine base form is intractable and insoluble.

The counter-ion induced processability of "metallic" polyaniline utilizes bifunctional counter-ions such as the ionic dodecylbenzenesulfonate group to render the polymer soluble in organic solvents. A variety of sulfonic acids (e.g camphor sulfonic acid, CSA) can be used.

The charge on the SO_3^- head group of the sulfonic acid forms an ionic bond with the positive charge (proton) on the PANI chain, while the hydrocarbon tail "likes" organic solvents (see Fig. 3.5). The use of such bifunctional counter-ions to enable the processing of PANI in the conducting form resulted in materials with improved homogeneity, a higher degree of crystallinity, and with correspondingly improved electrical conductivities. For details on the effects of disorder, see Chapter 4.

The final step (not shown in Fig. 3.5) involves the rearrangement to form the metallic emeraldine salt with the two-ring repeat unit as shown in Fig 3.6.

The solubility of "metallic" PANI in organic solvents has enabled the fabrication of conducting blends of PANI with a variety of insulating host polymers. These polymer blends

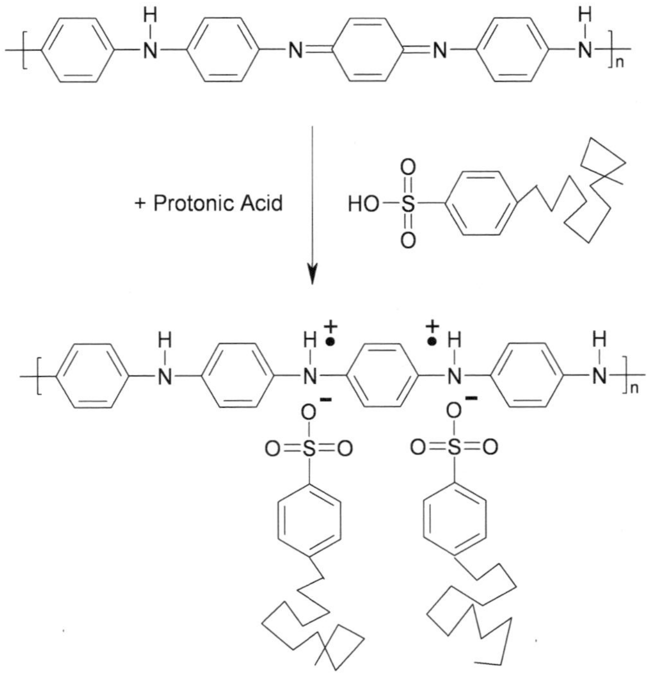

Fig. 3.5 The emeraldine base form of PANI can be converted to the metallic emeraldine salt form by protonation via reaction with dodecylbenzenesulfonic acid.

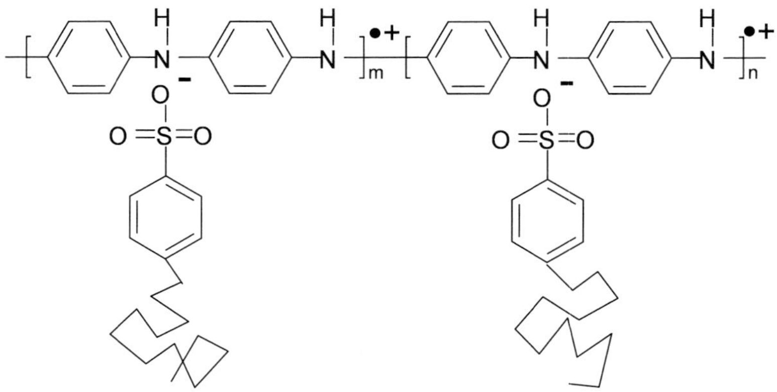

Fig. 3.6 Molecular structure of the metallic state of PANI-DBSA with a two-ring repeat unit. There is one "unpaired" electron per repeat unit. In the metallic state, the unpaired electron is delocalized along the metallic chain.

exhibit a remarkably low percolation threshold as a result of the spontaneous formation of interpenetrating network structures (see Section 3.3).

3.2.4 Dispersions of metallic polymers in water

Template-guided synthesis of conducting polymers was developed to enable the processing of conducting polymers using water as the "solvent". The molecular template, polyacids such as polystyrene sulfonic acid, binds the monomer, for example aniline or ethylene-dioxythiphene (EDOT), to form molecular complexes which are dispersed in water as colloidal particles (see Fig 3.7). The attractive force between template and monomer can be chemical and/or physical. Upon polymerization, the aniline or EDOT monomers form polyaniline or poly(ethylenedioxythiophene), PEDOT, and remain attached to the template to form the template-(conducting polymer) complex.

By judicious choice of the template molecule and the polymerization conditions, stable submicron size colloidal particles can be formed during polymerization. The stabilization against coagulation arises from the Coulomb repulsion between particles, which results from the surface charge provided by the extra sulfonic acid groups in polystyrene sulfonic acid. Stable dispersions of polyaniline-polystyrene sulfonic acid complexes can be made with particle sizes in the nanoscale regime. Transparent films with a specific resistivity of 1–10 Ω-cm can be cast from such dispersions.

Bayer commercialized poly(ethylenedioxythiophene)-polystyrene sulfonate (PEDOT-PSS), under the trade name "Baytron". The PEDOT-PSS is prepared as a stable dispersion in water. Films of PEDOT-PSS are semitransparent and can be spin-cast with a surface resistance of approximately 500 Ω/square and with 75% transmission.

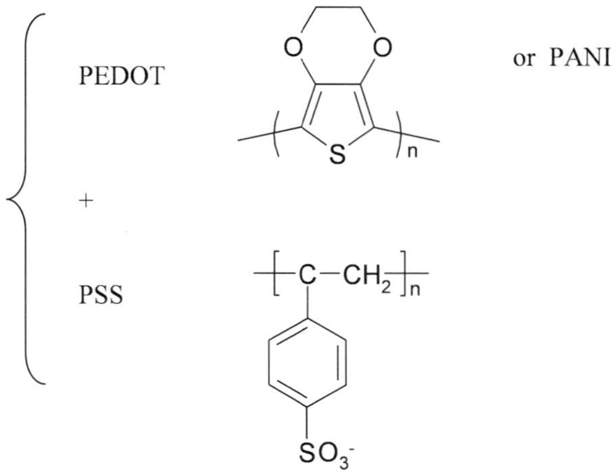

Fig. 3.7 Dispersions of PANI and PEDOT are processable from water. Polycations such as PSS serve as the counter-ions.

3.3 Blends of conducting polymers with insulating polymers: Percolation at low concentrations

As a result of their solubility in common organic solvents, metallic polymers can be co-dissolved with a variety of insulating polymers in the same solvent. When cast from solution, phase separation typically occurs. Under such circumstances, the electrical conductivity of the polymer blend is determined by the "connectivity" of the conducting pathways. The percolation threshold can be defined through the example of the "vandalized grid". In the schematic picture of Fig. 3.8, two electrodes are connected through a metallic grid. Subsequent cutting of the elements of the grid will be monitored through measurements of the total resistance between the two electrodes.

If none of the grid elements are cut (the fraction of uncut bonds is 1), the resistance is small because there are many parallel highly conducting pathways. Cutting the conductingelements will reduce the number of connected pathways between the two electrodes, and the resistance will increase. When a critical fraction of network elements is cut, the current will fall to zero since there will be no interconnected pathways between the two electrodes. This critical fraction is defined as the "percolation threshold". In the two-dimensional experiment sketched in Fig. 3.8, the percolation threshold is at 50%.

For a summary of percolation theory and its applications, the reader is referred to *The Physics of Amorphous Solids* by R. Zallen (John Wiley, New York, 1983). In three dimensions, the critical volume fraction needed for continuous conducting pathways is approximately 16%.

As demonstrated in Fig. 3.9, for simple blends of conducting polymers in an insulating polymer matrix, the percolation threshold is observed at approximately 16% as expected from percolation theory, implying that the conducting polymer is embedded into the insulating matrix as globular conducting units.

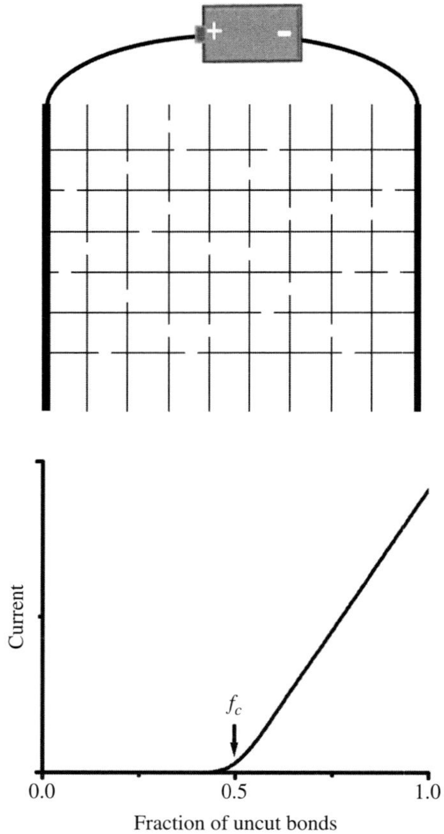

Fig. 3.8 The "vandalized grid" with percolation threshold at approximately 50%.

Since the conducting polymer component in such blends is relatively expensive, it would be advantageous to have conducting blends with threshold well below 16%. Can this be done? The answer is yes. The percolation threshold can be lowered by orders of magnitude through controlled phase separation and the formation of interpenetrating networks. The concept is sketched in Fig. 3.10 (on the right).

The interpenetrating network concept was implemented using PANI/CSA in poly(methylmethacrylate), PMMA. The films prepared for these measurements were cast from meta-cresol. The onset of conductivity is at approximately 0.3% (see Fig. 3.11), approximately 50 times below the percolation threshold for globular objects. The ultra-low percolation threshold indicated in Fig. 3.11 is indicative of the formation of an extended interpenetrating network as sketched schematically in Fig. 3.10.

The transmission electron microscopy (TEM) image of this network is shown in Fig. 3.12. Although the details of the self-assembly of these tenuous networks is not well understood, it is clear that the surface energy for the PANI/CSA in PMMA must be negative, for otherwise the conducting component would form globular structures with minimal surface area. The importance of the network formation is clear; because of the tenuous conducting networks, an electrical conductivity of 0.5 S/cm can be achieved at a volume fraction of only 0.5%.

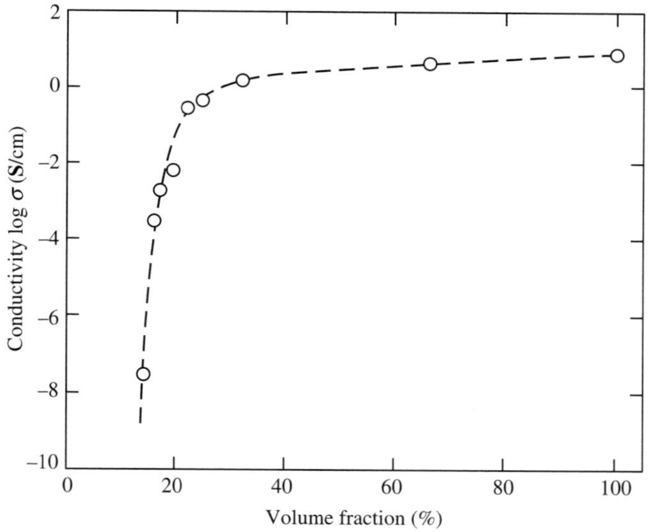

Fig. 3.9 Dependence of the room temperature d.c. conductivity of the doped composite comprising P3HBzT in polystyrene as a function of volume fraction (%) of P3HBzT. [Reproduced from S. Hotta *et al.*, *Synthetic Metals*, **22**, 79 (1987) with permission of Elsevier.]

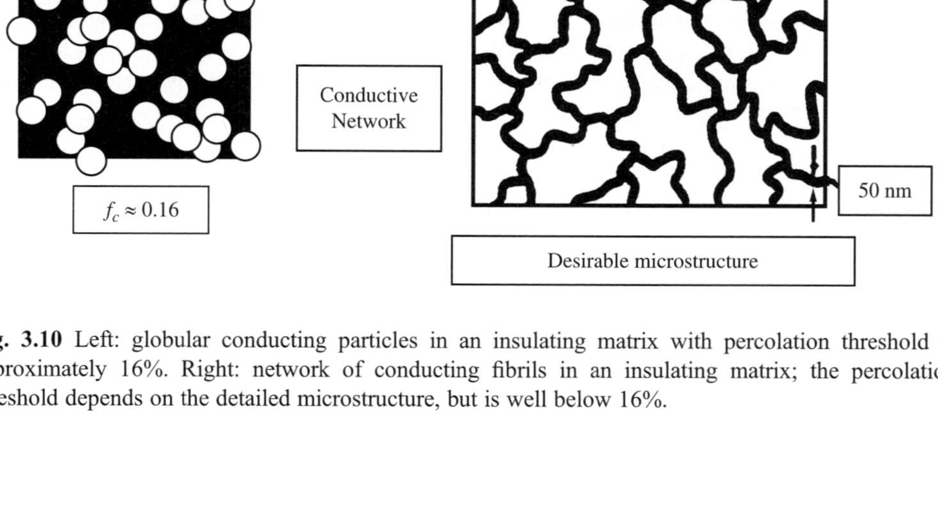

Fig. 3.10 Left: globular conducting particles in an insulating matrix with percolation threshold at approximately 16%. Right: network of conducting fibrils in an insulating matrix; the percolation threshold depends on the detailed microstructure, but is well below 16%.

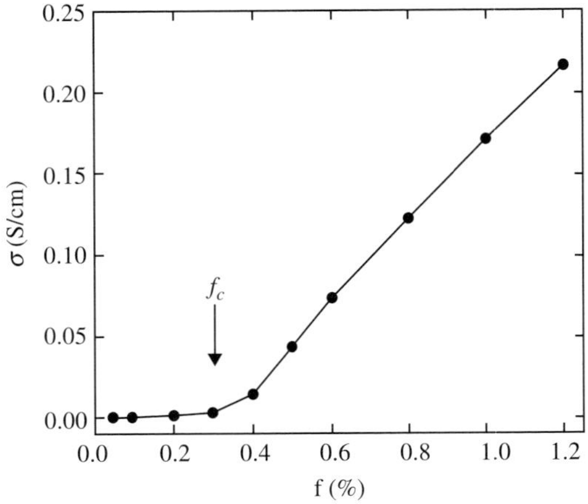

Fig. 3.11 Concentration dependence of the conductivity of PANI-CSA in polymethylmethacrylate. The percolation threshold is at approximately 0.3%.

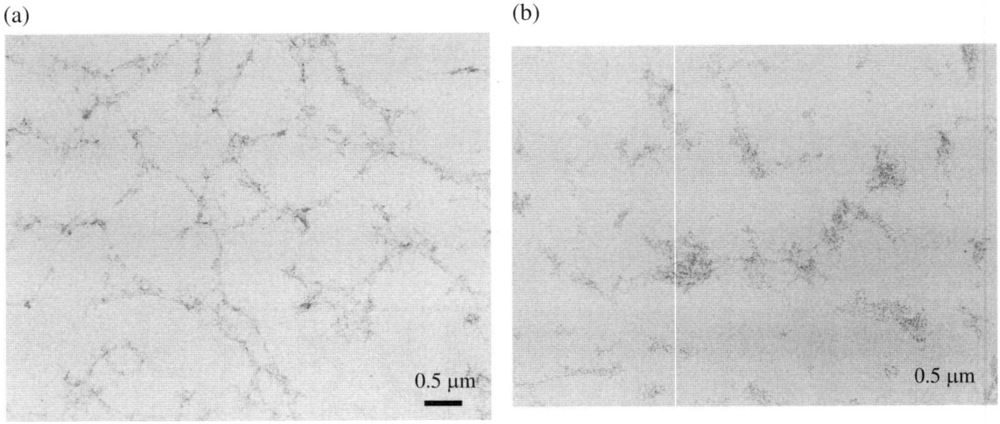

Fig. 3.12 TEM micrographs of blends of PANI/CSA in PMMA. The dark fibrils are the conducting polyaniline fibrils. Note that for the 0.5% blends the number of connected paths is significantly lower than for the 1% blends. [Reproduced from M. Reghu *et al.*, *Phys. Rev. B*, **50**, 13931 (1994) with permission of The American Chemical Society.]

3.4 Transparent metallic conductors

The high reflectance of a metal originates from the contribution to the dielectric function by the mobile free electrons. In general, the dielectric function for a free electron metal can be written in terms of its real and imaginary parts; $\varepsilon(\omega) = \varepsilon_1(\omega) + i\varepsilon_2(\omega)$:

$$\varepsilon_1(\omega) = \varepsilon_\infty - \omega_p^2\tau^2/[1 + \omega^2\tau^2] \quad (3.1a)$$
$$\varepsilon_2(\omega) = (4\pi/\omega)\sigma(\omega) \quad (3.1b)$$

where

$$\sigma(\omega) = \sigma_o/[1 + \omega^2\tau^2] \quad (3.1c)$$
$$\sigma_o = (1/4\pi)\omega_p^2\tau \quad (3.1d)$$

and

$$\omega_p^2 = 4\pi Ne^2/m^*.$$

In Eqn 3.1, τ is the mean time between scattering events that relax the electrons back toward equilibrium, and ε_∞ is the high frequency dielectric constant ($\varepsilon_\infty = 1.0 + \varepsilon_{core}$ where ε_{core} is the contribution from the ion cores). Equation 3.1c is the standard expression for the frequency dependent conductivity of a metal, $\sigma(\omega) = \sigma_o/(1 + \omega^2\tau^2)$, where $\sigma_o = Ne^2\tau/m^*$, N is the number of electrons per unit volume and m^* is the electron effective mass.

Recalling that $\varepsilon = (n+ik)^2$ where n and k are the real and imaginary parts of the index of refraction, in terms of n and k, the reflectance, $R(\omega)$, is given by

$$R(\omega) = [(n - 1)^2 + k^2]/[(n + 1)^2 + k^2] \quad (3.2)$$

For a good metal, the mean free path and the scattering time are long, $\tau \to \infty$, so that $\omega_p\tau >> 1$ and $\omega\tau >> 1$ for $\omega \sim \omega_p$. In this limit, $\sigma(\omega) \sim 1/\tau$ implying that the imaginary part is small. Thus,

$$\varepsilon_{tot}(\omega) = \varepsilon_1(\omega) = \varepsilon_\infty - \omega_p^2/\omega^2.$$

The reflectance for $\tau \to \infty$ is sketched as the solid curve in Fig. 3.13. For $\omega > \omega_p$, the only contribution to the index is that of the atomic cores. Thus, as shown in Fig. 3.13, the reflectance is relatively small for $\omega > \omega_p$, $R = [(\varepsilon_\infty)^{1/2} - 1]^2/[(\varepsilon_\infty)^{1/2} + 1]^2$. For example, with $\varepsilon_\infty \approx 4$, $R \approx 11\%$. For $\omega_p/\omega = (\varepsilon_\infty - 1)^{1/2}$, $n = 1$ and $R = 0$. For $\omega < \omega_p/(\varepsilon_\infty)^{1/2}$, $\varepsilon_1 < 0$ and n is imaginary implying that the light wave is exponentially damped inside the metal; i.e. $R = 1$.

For finite τ, the $\sigma(\omega)$ contribution must be included in the calculation. The dashed curve in Fig. 3.13 shows $R(\omega)$ for finite τ. Although the turn-on of the reflectance for $\omega < \omega_p$ is more rounded for finite τ, the general shape is not changed.

Thus, metals reflect light at frequencies below the plasma frequency, ω_p. At frequencies above the plasma frequency, metals are transparent. For conventional metals (Na, Cu, Ag, etc.), N is of order 10^{23} per unit volume. As a result, the plasma frequency is in the ultraviolet, $h\nu \approx 4$–5 eV. Therefore, conventional metals appear shiny and "metallic-looking" in the spectral range to which the human eye is sensitive.

Metallic polymers have a lower density of electrons than conventional metals. Both the length of the repeat unit along the chain and the interchain spacing are relatively large compared to the interatomic distances in conventional metals. For metallic polymers, N is typically of order 2–5 $\times$ 10^{21} cm^{-3}. Thus, for metallic polymers, the plasma frequency is approximately $\hbar\omega_p \approx 1$ eV; i.e. at wavelengths in the infrared. Metallic polymers exhibit high reflectance (and thus look "shiny") in the infrared, but they are semitransparent in the visible part of the spectrum. The reflectance of high quality, metallic polyaniline (PANI:CSA) is shown in Chapter 5, Fig. 5.8.

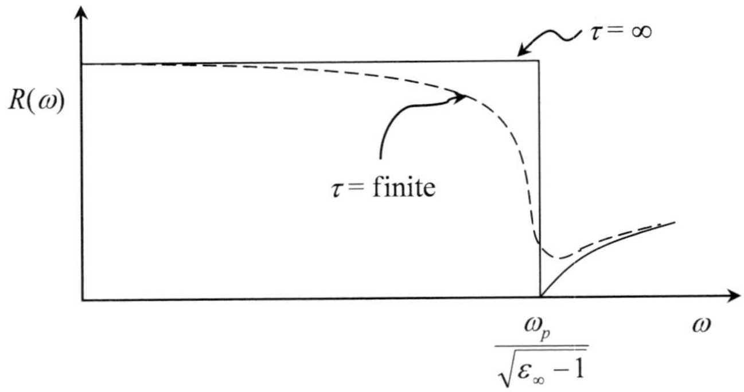

Fig. 3.13 Reflectance of a metal; $R(\omega) \rightarrow 1$ below the plasma frequency.

The frequency dependent conductivity is plotted versus ω in Fig. 3.14 (see Eqn 3.1). When $\omega_p\tau \geq 1$, the absorption that results from $\sigma(\omega)$ is confined to frequencies in the infrared at frequencies below ω_p. When both the reflectance and the absorption are in the infrared, a thin film of the conducting material is transparent in the visible portion of the spectrum.

The residual absorption at frequencies above the plasma frequency arises from two sources:

1. If the scattering time is sufficiently short, the high frequency "tail" of $\sigma(\omega)$ can extend into the visible part of the spectrum. Thus, $\sigma(\omega)$ provides a contribution to the imaginary part of the index, k, and thereby to absorption with absorption coefficient
$$\alpha = 2(\omega/c)k.$$
2. Interband (π-π^*) transitions between the partially filled conduction band and the lowest energy unoccupied band.

Both of these contributions can, in principle, be small. The interband contribution is specific to the molecular structure. For polyaniline, for example, this interband absorption gives rise to the green color. For PEDOT:PSS, there is no interband absorption in the visible portion of the spectrum.

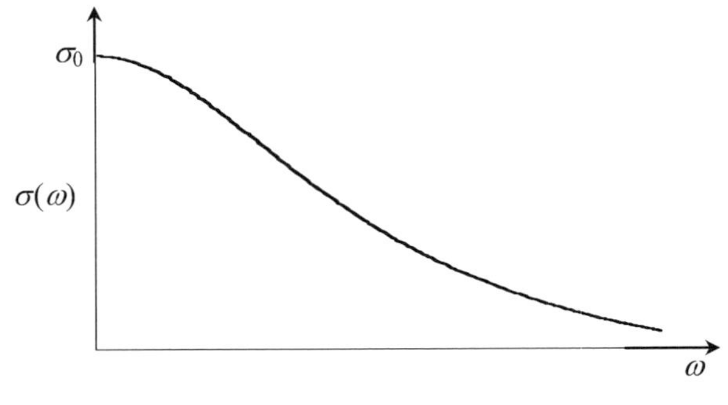

Fig. 3.14 The frequency dependent conductivity, $\sigma(\omega)$, of a metal.

The scattering time can be made longer (and the high frequency tail reduced in magnitude) by improving the quality of the material; e.g. by increasing the molecular weight and the degree of structural order. Increasing the magnitude of the scattering time yields a double benefit; the absorption of visible light decreases (better transparency) and the dc conductivity increases (lower surface resistance).

Optical quality thin films of metallic polymers are useful, therefore, as transparent electrodes. For example, polyaniline, polypyrrole and PEDOT have been used as transparent hole-injecting electrodes in polymer LEDs. The initial demonstration of mechanically flexible polymer LEDs utilized PANI as the anode. More generally, transparent conducting films can be used for a variety of purposes; for example, as antistatic coatings on CRT screens, as electrodes in liquid crystal displays, or as the electrode in the fabricating electrochromic windows.

Further reading

A. R. Blythe and D. Bloor, *Electrical Properties of Polymers*, Cambridge University Press Cambridge, 2005.

J. L. Bredas (ed.), *Conjugated Oligomers and Polymers and Dendrimers: From Polyacetylene to DNA*, Proceedings of the Fourth Francqui Colloquium, De Boeck & Larcier, Paris, 1999.

H. J. Mair and S. Roth (eds.), *Elektrisch leitenden Kunststoffe*, Carl Hanser Verlag, München, 1989.

H. S. Nalwa (ed.), *Handbook of Organic and Conductive Molecules and Polymers*, John Wiley, Chichester, 1997.

W. R. Salaneck, D. T. Clark and E. J. Samuelsen (eds.), *Science and Applications of Conducting Polymers*, IOP Publishing, Bristol, 1991.

4

Disorder induced metal–insulator (M-I) transition in conducting polymers

4.1 Introduction to the consequences of strong disorder

In a perfect crystal there is translational symmetry. As a result of this symmetry, the energy bands of a metal or a semiconductor can be described in *k*-space (Bloch's theorem).

In this chapter, we discuss the consequences of strong disorder; i.e. disorder that is sufficiently strong to cause localization of the wavefunctions and the breakdown of band theory.

In a metal, the electrical conductivity results from the displacement of all the electrons in the Fermi sea (i.e. all electrons with $k < k_F$) in response to an applied electric field by an amount Δq, where, according to Newton's second law,

$$\hbar\Delta q = -eE \tag{4.1}$$

In this equation, $\hbar$ is the Planck constant and $\hbar\Delta q$ is the change of momentum of each of the N electrons in the Fermi sea. Scattering of the electrons by impurities, imperfections and lattice vibrations tends to restore the Fermi sea to the undisturbed equilibrium at $\Delta q = 0$. This relaxation toward $\Delta q = 0$ can be expressed as follows:

$$(\mathrm{d}/\mathrm{dt})\Delta q = -\Delta q/\tau - (\mathrm{e}/\hbar)\mathrm{E} \tag{4.2}$$

where τ is the mean scattering time, i.e. the mean time between collisions between the electrons and impurities, defects or thermal vibrations (phonons). Under steady-state conditions,

$$\Delta q = (\mathrm{e}\tau/\hbar)\mathrm{E} \tag{4.3}$$

The mean drift velocity, v_d, of the electrons is defined by the semiclassical relation

$$\mathrm{m}^*\mathrm{v_d} = \hbar\Delta q \tag{4.4}$$

where m^* is the effective mass of the electrons in the highest partially filled band of the metal. Thus,

$$\mathrm{v_d} = (\mathrm{e}\tau/\mathrm{m}^*)\mathrm{E} \tag{4.5}$$

and the total current density is given by

$$\mathrm{j} = \mathrm{Nev_d} = (\mathrm{Ne^2}\tau/\mathrm{m}^*)\mathrm{E} \tag{4.6}$$

From Ohm's law, $j = \sigma E$, we identify the electrical conductivity, σ, of the metal,

$$\sigma = Ne^2\tau/m^* \tag{4.7}$$

Defining the carrier (electron) mobility by the relation $v_d \equiv \mu E$, this simple model of metallic transport yields

$$\mu = e\tau/m^* \tag{4.8a}$$

$$\sigma = Ne\mu. \tag{4.8b}$$

Note that the electrons at or near the Fermi surface have velocity equal to v_F. Typically, in a metal such as copper, $v_F \equiv 10^8$ cm/s whereas the drift velocity (v_d) is much smaller, only a few cm/s.

The mean free path of electrons (l) in a metal is defined as the mean distance between collisions

$$l = v_F.\tau \tag{4.9}$$

From the magnitude of the electrical conductivity of copper at room temperature, one can estimate the mean free path to be approximately 150Å. Because this distance is much greater than the interatomic spacing (the lattice constant of Cu, $a \approx 2$ Å), the momentum, $\hbar k$ is well defined and the electronic wavefunctions in the metal are delocalized.

Suppose, however, that the disorder is increased; more impurities and more defects are added which scatter the electrons so strongly that the mean free path becomes comparable to the lattice constant, $l \approx a$. In the presence of such strong disorder, the electrons scatter almost even before they can move! Therefore, one can define two regimes:

(1) Weak scattering: k is a good quantum number (translational symmetry exists over many structural repeat units),

$$\Delta k/k << 1 \tag{4.10}$$

The Fermi surface is well-defined. The collection of "free" electrons exhibits metallic behavior.

(2) Strong scattering: $\Delta k/k \sim 1$: The Ioffe–Regel inequality

$$k_F l < 1 \tag{4.11}$$

defines the strong scattering limit in which the disorder is so important that the k-states are not well defined. In the Ioffe–Regel limit, where the mean free path is less than the de Broglie wavelength (or the interatomic spacing), multiple scattering will cause the electron wavefunctions to become spatially localized about a site r_i. The localized wavefunctions will decay exponentially with distance,

$$\psi_i \sim \exp(-\alpha|r - r_i|) \tag{4.12}$$

where α^{-1} is the localization length. Since α^{-1} is typically somewhat larger than the interatomic spacing, ψ_i and ψ_j will overlap. Thus, thermally activated hopping from site to site within the exponential fall-off will lead to electrical transport at finite temperature. At zero temperature, however, multiple tunneling steps among localized states will lead to zero conductivity between sites that are separated by a macroscopic length.

Therefore, in the strong scattering limit (localized states), the metal is transformed into an insulator, a Fermi "glass", rather than a Fermi gas. A Fermi glass is described in terms of a finite and continuous density of states, $N(E)$, in which all states are localized. Within the density of states (each localized in the vicinity of a different r_i) of a Fermi glass, all states are filled up to the Fermi energy. Only electrons within $k_B T$ of the Fermi energy can be promoted to empty states. The Fermi glass has no dc electrical conductivity at $T = 0$ even though $N(E_F) \neq 0$:

$$\sigma_{T\to 0} \to 0 \qquad (\text{mobility}\,\mu_{T\to 0} \to 0) \tag{4.13}$$

In a good metal, the mean free path increases at low temperatures as a result of the decrease in the number of phonons. In a Fermi glass, however, the electrical conductivity occurs only via thermally activated hopping between localized states. Therefore, in a Fermi glass, $\sigma(T)$ increases as a function of T.

4.2 Anderson localization

The pioneering work of Mott and Anderson showed that a metal-to-insulator (M-I) transition can result from electron–electron repulsive interactions (the Mott transition) and/or from localization of the electronic wavefunctions in the presence of disorder (the Anderson transition). As described briefly in Chapter 1, Section 1.9, the Mott transition is a first order phase transition from metal to insulator. Correlation effects split the metallic band into filled levels (single occupancy on each site) and empty levels (double occupancy on each site), thereby creating an energy gap at the Fermi energy. The Mott transition has been observed in transition metal oxides and in organic charge transfer salts. The Mott transition occurs even in a perfect lattice as a result of localization caused by strong electron–electron interactions.

The Anderson transition, on the other hand, results from disorder-induced localization. Schematically, the disorder can be described by a one-dimensional model (see Fig 4.1) in which the periodic potential of a metallic lattice is replaced by a potential which is random from site to site.

One can add random disorder by varying the "atomic" potential at each site (site diagonal), or one can add disorder by displacing the position of each "atomic center" by a random amount (see Fig 4.2) as in a frozen liquid. In the context of the tight-binding model (see Chapter 1, Section 1.4), site diagonal disorder adds a random quantity to the ionization potential at each site, whereas random displacement corresponds to adding a random component to the transfer integral.

Consider the simple model in which one adds to the periodic potential of a three-dimensional tight-binding model (bandwidth W) a random potential with magnitude between 0 and $\pm V_o/2$. How does the electronic structure evolve as V_0 increases from zero? In the low disorder limit, all the energy states within the various tight-binding bands are delocalized and

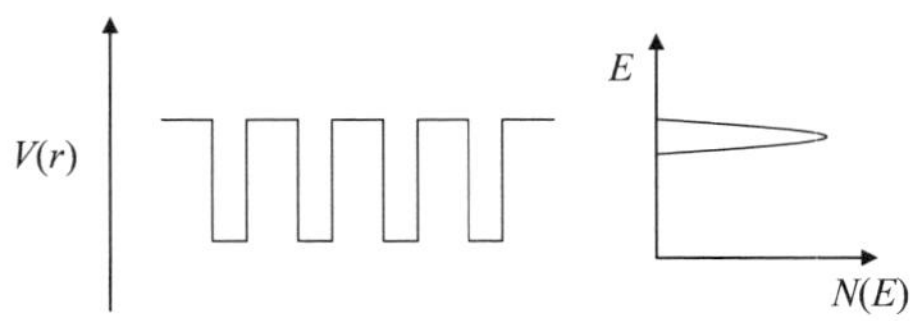

Fig. 4.1 Left: ordered one-dimensional lattice with the same potential at each site. Right: narrow density of delocalized states in the energy band of the ordered lattice.

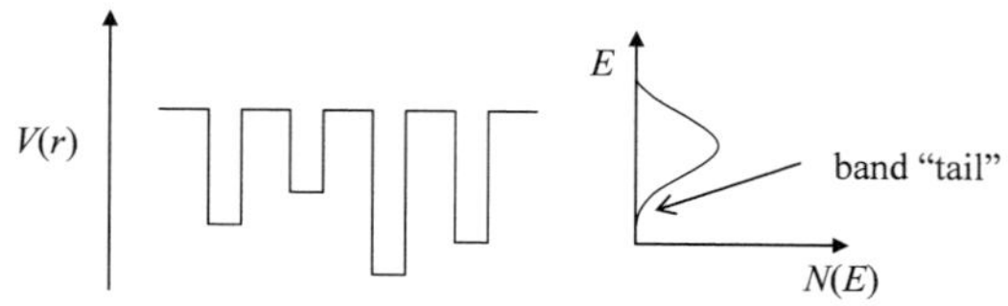

Fig. 4.2 Left: disordered one-dimensional lattices with random potential at each site. Right: broad distribution of localized states (with band "tails") resulting from the disorder.

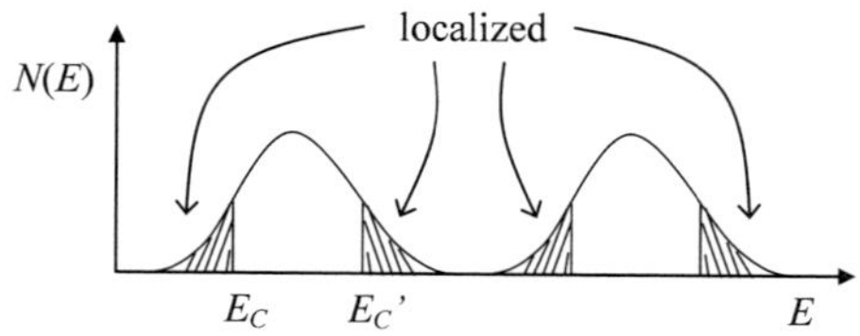

Fig. 4.3 Density of states for two bands in a tight-binding model with disorder. Localized states appear near the edges of the bands; the boundary between localized and delocalized states (E_C and $E_{C'}$ in the drawing) is called the "mobility edge".

the system is a metal or an insulator according to the usual rules of band filling. As V_o is increased, band-tail states appear, and the states near the band edge become localized. As sketched in Fig. 4.3, the boundary between localized and delocalized states is called the "mobility edge".

As the degree of disorder (V_0) increases, E_C and $E_{C'}$ move toward the center of energy band until for $V_o \approx W$ all states within the band are localized.

Suppose the lower band in Fig. 4.3 is filled and the upper band is empty; i.e. the system is a semiconductor. Disorder leads to the formation of localized states within the gap (band-tail states) and to mobility edges within the bands.

Alternatively, suppose the lower band in Fig. 4.3 is partially filled. Without disorder, the system would be a metal. As the disorder is increased, the mobility edges move toward the band center. In the Fermi glass regime (insulator), the Fermi level is located at an energy between the band edge and the mobility edge ($E_F < E_C$), the energy interval in which the states are localized.

In the Fermi glass regime, transport occurs via one of two following mechanisms:

(i) thermally activated variable range hopping;
(ii) thermal excitation with transport through extended states.

In the latter, the conductivity will vary as

$$\sigma = \sigma_{\min} e^{-(E_C - E_F)/kT} \tag{4.14}$$

where ($E_C - E_F$) is the activation energy.

As sketched in Fig. 4.4, if one increases E_F (for example, by doping to increase the number of carriers), the activation energy will decrease and go to zero when E_F crosses E_C.

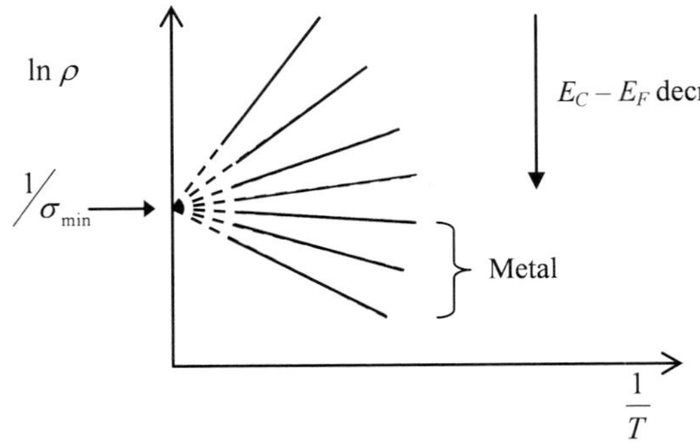

Fig. 4.4 Schematic diagram of lnρ(T) versus 1/T at different values of $(E_C - E_F)$. The cross-over to metallic behavior occurs at $(E_C - E_F) = 0$.

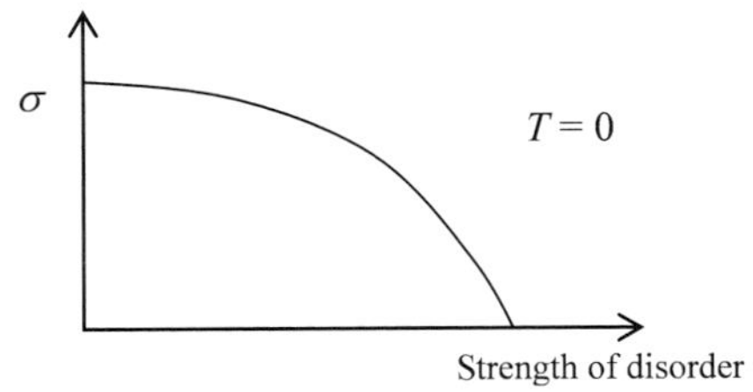

Fig. 4.5 The conductivity at $T = 0$ as a function of the strength of the disorder; the conductivity goes to zero continuously at the M-I boundary.

When $E_F > E_C$, the system will be a metal (a poor metal when $E_F \approx E_C$) with resistivity, $\rho(T)$, that decreases and remains finite as the temperature is lowered toward $T = 0$ K.

More recent calculations show that the Anderson M-I transition is a true phase transition at T=0 (see Fig 4.5), a disorder-induced quantum phase transition from Fermi glass to metal.[1]

4.3 Variable range hopping

Analysis of $\sigma(T)$ in the heavily disordered Fermi glass regime shows that the exponential dependence on $1/T$ predicted by Eqn 4.14, is observed only within a limited temperature range. At low temperatures, plots of ln $\sigma(T)$ versus $1/T$ show curvature. Mott's concept of variable range hopping (VRH) leads to the following remarkable expression for the electrical conductivity at low temperatures in the Fermi glass regime:

$$\sigma_{\mathrm{VRH}}(T) = \sigma_o \exp[-\beta(\mathrm{T_o}/\mathrm{T})^{1/(\mathrm{d}+1)}] \tag{4.15}$$

where the prefactor, σ_o, is defined in Eqn 4.26, β and T_o are constants (see Eqns 4.27 and 4.28) and d is the dimensionality of the system (d = 1, 2 or 3 for one-, two-, or three-dimensional materials).

[1] See E. Abrahams *et al.*, *Phys. Rev. Lett.*, **42**, 673 (1979).

To derive Eqn 4.15, we define the number of electrons per unit volume within k_BT of E_F:

$$n_{\text{eff}} = 2N(E_F)kT \tag{4.16}$$

In the Fermi glass, all the levels at energies less than E_F are doubly occupied. Consider an electron in a localized state centered at site i with energy E_i within approximately k_BT of E_F. Because there is a continuous density of localized states, one can always find a site j with energy E_j arbitrarily close to E_i. Although the state j might be spatially separated from i by a long distance, the probability of finding a site j with energy E_j that is nearly degenerate with E_i is determined by the density of states.

Mott argued that the hopping probability along (opposite to) the external electric field (F) is determined by two competing effects:

$$\nu_{\text{ph}}\exp\left[-2\alpha R - \frac{W \pm eRF}{kT}\right] \tag{4.17}$$

where W is the hopping energy, R is the distance between sites i and j, F is the applied electric field, and ν_{ph} is the number of phonon-assisted hopping "attempts" per unit time. Thus, the current density is given by

$$\begin{aligned} j &= n_{\text{eff}}\, eR(\nu_+ - \nu_-) \\ &= 2eRkTN(E_F)\nu_{\text{ph}}e^{-2\alpha R - \frac{W}{kT}} \sinh\left(\frac{eRF}{kT}\right) \end{aligned} \tag{4.18}$$

For small fields ($eRF/kT <<1$)

$$\sigma = \frac{j}{F} = 2e^2R^2\nu_{\text{ph}}N(E_F)e^{-2\alpha R - \frac{W}{kT}} \tag{4.19}$$

The hopping energy, W, depends on how far the carrier must hop; i.e. the "variable range" of the hopping must be taken into account:

$$W = \frac{1}{N(E_F)\frac{4\pi}{3}R^3} \tag{4.20}$$

Defining the mean energy-weighted hopping distance:

$$\bar{R} = \frac{\int_0^R r(r^2dr)}{\int r^2dr} = \frac{3R}{4} \tag{4.21}$$

the probability of a hop per unit time is given by:

$$\text{Prob. of hop}(\bar{R}) = \nu_{\text{ph}}\exp\left[-2\alpha\bar{R} - \frac{W}{kT}\right] \tag{4.22}$$

Mott argued that the lowest resistance corresponds to the maximum hopping probability; i.e. $\frac{\partial}{\partial R} = 0$ will define, on average, the optimum hopping distance. This maximum occurs when

$$\frac{3}{2}\alpha = \frac{9}{4\pi R^4 N(E_F)kT} \tag{4.23}$$

giving

$$R_{\text{opt}} = \frac{3^{\frac{1}{4}}}{[2\pi\alpha N(E_F)kT]^{\frac{1}{4}}} \tag{4.24}$$

with

$$\bar{R} = \frac{3}{4} R_{opt} \tag{4.25}$$

Substitution into Eqn 4.19 yields

$$\sigma_{\mathrm{VRH}} = 2e^2 N(E_F)\, \bar{R}^2 v_{\mathrm{ph}} e^{-\beta(T_0/T)^{\frac{1}{4}}} \tag{4.26}$$

where

$$T_0 = \frac{\alpha^3}{kN(E_F)} \tag{4.27}$$

and

$$\beta = 2\left(\frac{3}{2\pi}\right)^{\frac{1}{4}} = 1.66 \tag{4.28}$$

Note that T_o is a function of the localization length, $L_{loc} = \alpha^{-1}$ and $N(E_F)$, the two most important parameters. To look for the dependence expressed by Eqn 4.26, one plots $\ln\sigma$ versus $(1/T)^{\frac{1}{4}}$ as shown in Fig. 4.6.

Equation 4.26, which describes variable range hopping in three dimensions, is in excellent agreement with the experimental results from many systems. An example from the field of doped conjugated polymers is shown in Fig. 4.7.

Equation 4.26 assumes that the hopping carriers do not interact; i.e. that screening of the electron–electron Coulomb interaction is efficient. When the Coulomb interaction between the electron which is hopping and the hole left behind is dominant,

$$\ln\rho \propto (T_o'/T)^{1/2}, \tag{4.29}$$

where

$$\mathrm{T_o}' = \beta_I e^2/\epsilon K_B\ L_{loc}, \tag{4.30}$$

e is the electron charge, ε is the dielectric constant, and $\beta_I = 2.8$ (a numerical constant).[2] Thus, in general, in the insulating regime, $\ln\rho$ is proportional to T^{-1}/x where x is determined by the details of the phonon-assisted hopping.

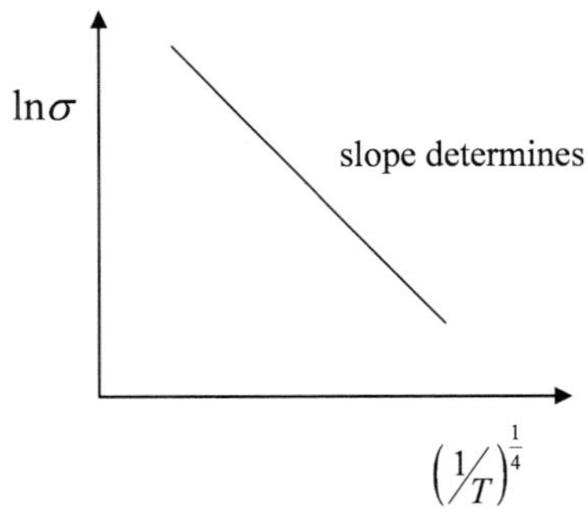

Fig. 4.6 For VRH, a plot of ln σ versus $(1/T)^{\frac{1}{4}}$ yields a straight line with slope T_o.

[2] B. I. Shklovskii and A. L. Efros, in *Electronic Properties of Doped Semiconductors*, p. 212, Springer Verlag, Berlin 1984.

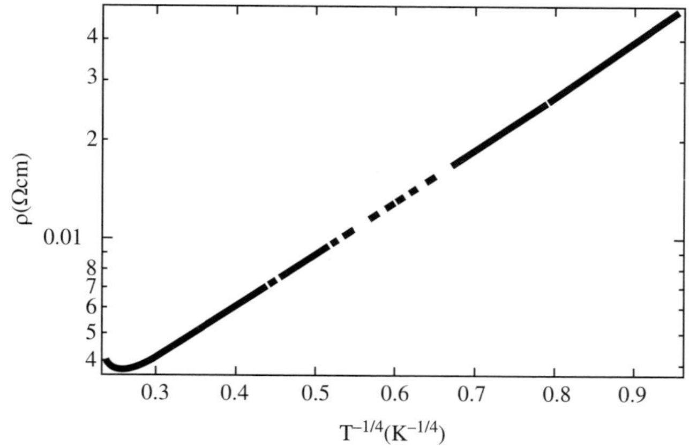

Fig. 4.7 lnρ versus $(1/T)^{\frac{1}{4}}$ for PANI-CSA; the straight line indicates variable range hopping in three dimensions. [Reproduced from R. Menon *et al.*, *Phys. Rev. B*, **48,** 17685 (1993) with permission of The American Physical Society.]

4.4 Critical regime of the M-I transition for "metallic" polymers

The disorder-induced M-I transition is a quantum phase transition. The transition occurs at $T = 0$ as the disorder is increased; on the metallic side of the M-I transition, the conductivity is finite at $T = 0$ whereas on the insulating side of the M-I transition, the conductivity is zero at T=0. Consequently, there is a critical point where the conductivity goes to zero as a power law as $T \to 0$. A schematic phase diagram is shown in Fig. 4.8.

The M-I transition in conducting polymers is particularly interesting; critical behavior has been observed over a relatively wide temperature range in a number of systems, including polyacetylene, polypyrrole, poly(p-phenylene vinylene), and polyaniline.

The resistivity, $\rho(T)$, and the resistivity ratio, $\rho_r = \rho(1.4\text{ K})/\rho(300\text{ K})$, have been successfully used to quantify the relative disorder in different samples and for sorting out the various regimes In general, the resistivity at room temperature increases as the disorder increases. In addition, ρ(T) increases more rapidly upon lowering the temperature so that ρr increases. The resistivity ratio (ρ_r) has proven to be useful as an "effective order parameter" for the M-I transition in conducting polymers. Thus, for a given conducting polymer system (e.g. polyacetylene, polypyrrole, poly(p-phenylene vinylene), or polyaniline) each curve in Fig. 4.8 would represent data obtained from a sample with different resistivity ratio, ρ_r.

In conducting polymers, the critical regime is "tunable" by varying the extent of disorder (i.e. by studying samples with different ρ_r, or by applying external pressure and/or magnetic fields (see Fig. 4.8). The transitions from metallic to critical behavior and from critical to insulating behavior have been induced with a magnetic field, and from insulating to critical and then to metallic behavior with increasing external pressure.

In the metallic regime, the zero temperature conductivity remains finite:

$$\sigma(T) = \sigma_0 + f(T), \tag{4.31}$$

where $f(T)$ is mainly determined by the electron–electron interaction. The magnitude of σ_o depends on the extent of the disorder. Metallic behavior has been demonstrated for conducting

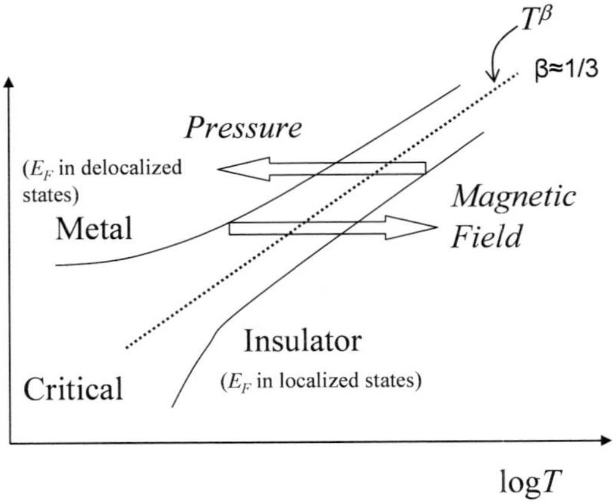

Fig. 4.8 Schematic phase diagram of the disorder induced M-I transition for a metallic polymer.

polymers with $\sigma(T)$ remaining constant as T approaches zero. Well into the metallic regime where the mean free path extends over many repeat units, σo becomes quite large, approaching that of a metal.

In the critical regime, theory predicts that the conductivity should show power law dependence as a function of temperature:

$$\sigma(\mathrm{T}) \approx (4\pi^2 e^2 p_F/h^2)(k_B T/E_F)^{1/\eta} = (e^2/\hbar)k_F(T/T_F)^{\beta} \tag{4.32}$$

where p_F is the Fermi momentum, h is Planck's constant, $k_F = \pi/2c$ is the Fermi momentum, e is the electron charge, k_B is the Boltzmann constant, T_F is the Fermi temperature ($T_F = E_F/k_B$) and $\beta \approx 1/3$. Extension of the power law dependence all the way to $T = 0$ requires that the system be *precisely* at the critical point. The power law is universal and requires only that the disordered system be in the critical regime. In the insulating regime, transport occurs through variable range hopping (VRH) among localized states (see Section 4.3).

4.5 Electrical conductivity in the critical regime

In the following, we analyze electrical conductivity results obtained from doped polyaniline (PANI) as the prototypical example of the M-I transition in a conjugated polymer. When synthesized with Cl^- as the counter-ion, the emeraldine salt phase of polyaniline is disordered, and in powder form, with transport properties well on the insulating side of the M-I transition. Upon re-dissolving the powder in camphor sulfonic acid (CSA) and recasting a solid film from this "surfactant-like" solution, one obtains homogeneous films with improved structural order.

As shown in Fig. 4.9, such homogeneous films have significantly higher conductivity and a much weaker (more "metallic") temperature dependence. When the data from the PANI-CSA samples are plotted on a log-log plot as in Fig. 4.10, the power law dependence predicted for

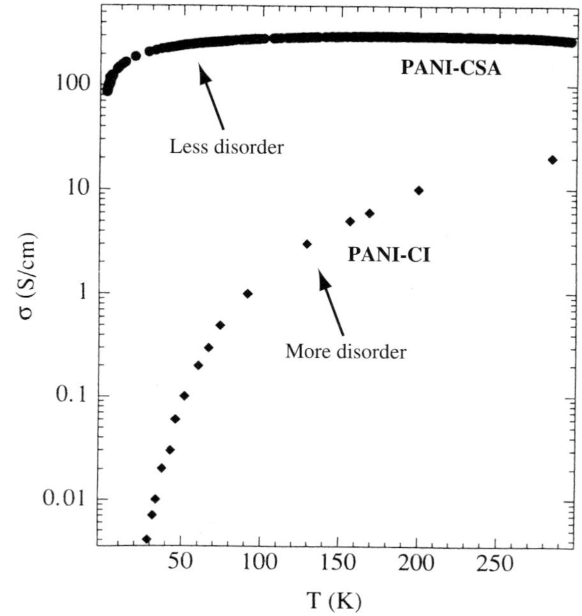

Fig. 4.9 Temperature dependence of the electrical conductivity of the emeraldine salt form of polyaniline with Cl^- as the counter-ion (diamonds) and with camphor sulfonic acid (CSA) as the counter-ion (solid dots). [Reproduced from R. Menon *et al.*, *Phys. Rev. B*, **48**, 17685 (1993) with permission of The American Physical Society.]

the critical regime (see Eqn 4.30) is clearly seen, $\rho(T) = \alpha T^{-\beta} = 0.014 T^{-\beta}$ where $\beta \approx 0.36$. From Eqn 4.30, the coefficient of the power law is given by

$$\alpha^{-1} = (e^2/\hbar)(\pi/2c)(T_F)^{-\beta} \tag{4.33}$$

Substituting $(e^2/h)^{-1} = 25 \times 10^3$ ohms, $c \approx 1$ nm for PANI-CSA, and $T_F \approx 2 \times 10^4 k_B (E_F \approx 2eV)$, one obtains $\alpha \approx 0.006$. From Fig. 4.10, the experimental value for the slope, $\alpha \approx 0.015$, is of the correct order of magnitude.

4.6 Magnetic field induced cross-over from metal to insulator

As shown by Khmel'nitskii and Larkin (see Fig. 4.11), the mobility edge in a disordered conductor can be shifted by an external magnetic field. If the conducting polymer is precisely at the critical point where $E_F = E_C$ and, therefore, $\rho(T) = \alpha T^{-\beta}$ in zero magnetic field, the application of a high magnetic field will shift the mobility edge so that the Fermi energy is in the region of localized states.

As a result of the magnetic field induced shift in E_C, the transport in the magnetic high field regime is expected to cross over from the power law dependence characteristic of the critical

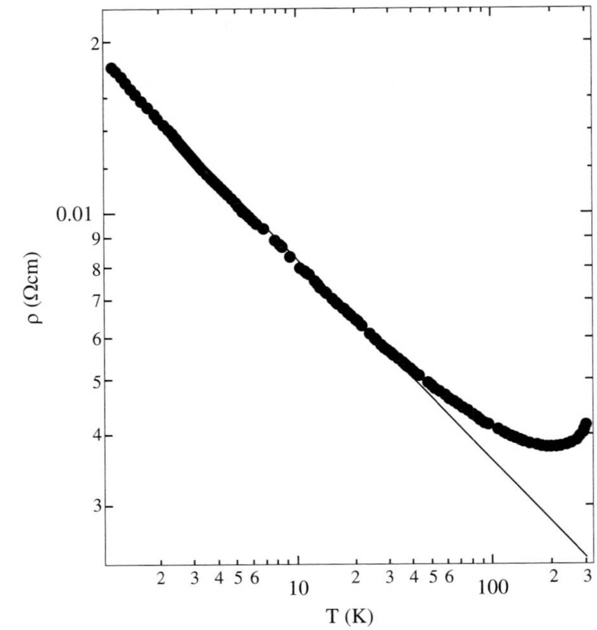

Fig. 4.10 Log-log plot of $\rho(T)$ showing the power law dependence at low temperatures. [Reproduced from R. Menon, *et al.*, *Phys. Rev. B*, **48**, 17685 (1993) with permission of The American Physical Society.]

regime to the variable range hopping characteristic of transport through localized states. As shown in Fig. 4.11, this cross-over has been observed for PANI-CSA.

The zero-field data in Fig. 4.11 are the same as those plotted in Fig. 4.10; the straight line indicates power law behavior with $\rho(T) = \alpha T^{-\beta}$ and $\beta \approx 0.36$. At $H = 8$ T, however, there is a clear deviation of the temperature dependence from that of the power law. To see if the predicted cross-over to VRH transport has occurred, the same data are shown in Fig. 4.12, but the results are presented on a semilog plot of $\rho(T)$ versus $T^{-1/4}$. In Fig. 4.12, the high field data show that $\ln\rho(T)$ is a linear function of $T^{-1/4}$, consistent with VRH transport.

4.7 Zabrodskii plots: Characterization of the metal–insulator transition as a function of magnetic field and pressure

The data can be presented in a form that is particularly useful for identification of the characteristic features of the transport near the M-I transition by defining the quantity W,

$$W = -T[d \ln \rho(T)/dT] \tag{4.34}$$

Log-log plots of $W(T) = (\Delta\ln\sigma/\Delta\ln T)$ versus T are quite sensitive and enable the precise identification of the critical regime. Moreover, the detailed evolution of $\sigma(T)$ in the critical regime at low

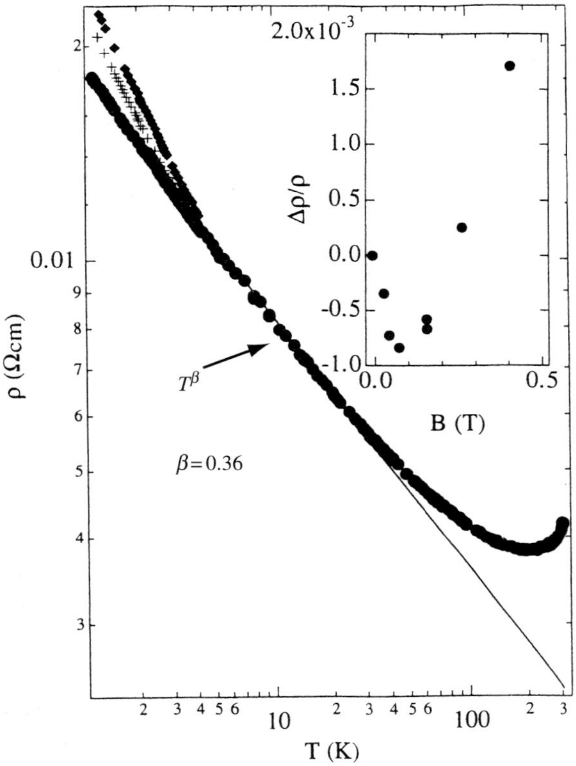

Fig. 4.11 Log-log plot of the resistivity of PANI-CSA at $H = 0$ (solid dots), $H = 4$ T, (pluses) and $H = 8$ T (solid diamonds). The linear dependence at H=0 indicates transport in the critical regime. [Reproduced from R. Menon *et al.*, *Phys. Rev. B*, **48**, 17685 (1993) with permission of The American Physical Society.]

temperatures can be observed in $W(T)$ plots as the conducting polymer is changed from metal to insulator by changing the extent of the disorder or by tuning with pressure or magnetic field.

For variable range hopping through localized states (see Eqn 4.26), W increases as the temperature is lowered;

$$W_{VRH} = (1/4)(T_o/T)^{1/4} \tag{4.35}$$

At the critical point (see Eqn 4.30), W is constant:

$$W_{Crit} = -\beta \tag{4.36}$$

and on the metallic side of M-I transition, W decreases as the temperature is lowered. Different samples with different degrees of disorder can be discriminated using the W function in a well defined way.

Data from the three regimes are shown for PANI-CSA in Fig. 4.13.

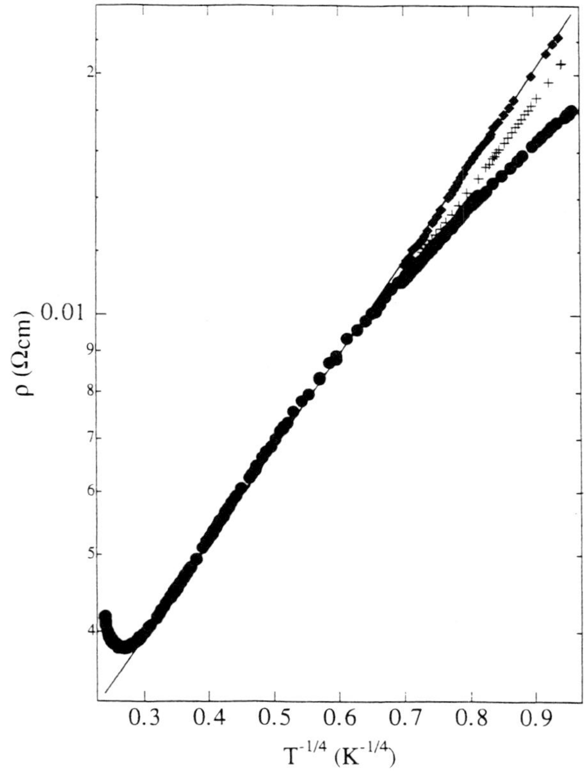

Fig. 4.12 Semilog plot of the resistivity of PANI-CSA versus $T^{-1/4}$ at $H = 0$ (solid dots), $H = 4$ T, (pluses) and $H = 8$ T (solid diamonds). The linear dependence at $H = 8$ T indicates VRH transport. [Reproduced from R. Menon *et al.*, *Phys. Rev. B*, **48**, 17685 (1993) with permission of The American Physical Society.]

The magnetic field induced cross-over from metal to insulator, discussed in connection with the data presented in Figs 4.11 and 4.12, is clearly observable in W plots (see Fig 4.14).

4.8 The metal–insulator transition as observed through infrared measurements

When carried out on metals and semiconductors, reflectance measurements provide information on the electronic structure over a wide spectral range. For conductors (metals or disordered "metals" where there is no energy gap), measurements in the infrared (IR) probe the low energy excitations near the Fermi energy; i.e. intraband free carrier excitations in a metal or localized excitations (such as those involved in variable range hopping transport) in a Fermi glass. Reflectance measurements at higher photon energies probe the interband transitions. The corresponding optical/IR conductivity, $\sigma(\omega)$, provides information on the metal physics, the

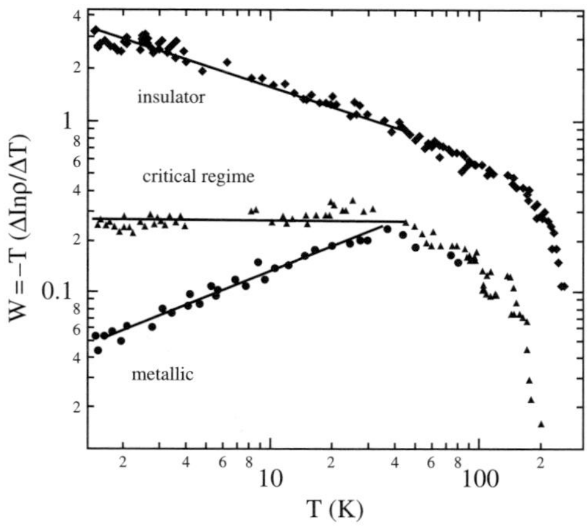

Fig. 4.13 Zabrodski plots, $W(T)$ versus T, from the three regimes are shown for PANI-CSA. [Reproduced from R. Menon *et al.*, *Phys. Rev. B*, **48**, 17685 (1993) with permission of The American Physical Society.]

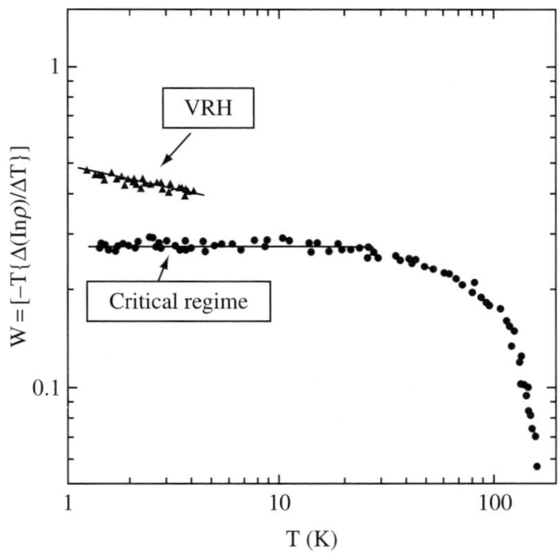

Fig. 4.14 Log-log plots of $W(T)$ versus T for PANI-CSA. The lines are drawn to guide the eye.
(•) Ambient pressure and $H = 0$ and
(▲) Ambient pressure and $H = 8$ T.
The data are the same as those shown in Figs 4.11 and 4.12.

disorder-induced metal–insulator transition, and the joint density of states associated with interband transitions at higher energies. IR reflectance measurements have played an important role in clarifying the metal physics of conducting polymers.

The dielectric function of a material is a complex function of the frequency with real and imaginary parts:

$$\varepsilon(\omega) = \varepsilon_1(\omega) + i\varepsilon_2(\omega) \tag{4.37a}$$

$$= \varepsilon_1(\omega) + i(4\pi/\omega)\sigma(\omega). \tag{4.37b}$$

Since the dielectric functions of metals and semiconductors are directly related to their electronic structure, knowledge of $\varepsilon(\omega)$ over a wide spectral range is essential to understanding the properties of new materials such as conducting polymers. Because $\varepsilon(\omega)$ has real and imaginary parts, one typically needs two independent optical measurements to obtain a complete set of data for analysis. For example, if both absorption (α) and reflectance (R) are measured at each frequency, ω_i, one can obtain both $\varepsilon_1(\omega_i)$ and $\varepsilon_2(\omega_i)$. It is difficult, however, to measure both α and R accurately over a wide spectral range because different optical components (lenses, gratings, etc.) are required for the IR, visible and UV. The alternative approach, and the method of choice, is to carefully measure reflectance over a wide spectral range and to use the Kramers–Kronig (K-K) relations to obtain the K-K conjugate. The ratio of the reflected electric field to the incident electric field is written:[3]

$$E_{ref}/E_{inc} = r(\omega) = \rho(\omega).\exp[i\theta(\omega)] = \frac{n + ik - 1}{n + ik + 1} \tag{4.38}$$

whereas the reflected intensity is measured as:

$$R = E^*_{ref}.E_{ref}/E^*_{inc}.E_{inc} = r^*.r = \rho^2$$

The imaginary parts of $\ln r(\omega) = \ln R^{0.5} + i\,\theta(\omega)$ are related through the K-K relations:

$$\theta(\omega) = -\frac{\omega}{\pi} P \int_0^\infty \frac{\ln R(s)}{s^2 - \omega^2} ds \tag{4.39}$$

where P the principal part of the integral. Thus, from careful measurements of the magnitude of the reflectance over a wide spectral range, one can obtain $\theta(\omega)$ from Eqn 4.39. The real and imaginary parts of the dielectric function can then be obtained from $R(\omega)$ and $\theta(\omega)$:

$$\varepsilon_1(\omega) = n^2 - k^2 \qquad \varepsilon_2(\omega) = 2nk \tag{4.40a}$$

$$\sigma(\omega) = (-i\omega/4\pi)(\varepsilon(\omega) - 1) \tag{4.40b}$$

Reflectance measurements have been carried out for a number of "metallic" polymers, including polyaniline and polypyrrole. By analyzing the resulting $\varepsilon_1(\omega)$ and $\sigma(\omega)$ data, a clear picture of the M-I transition has emerged. The data obtained from polypyrrole are especially clear and complete. When prepared electrochemically, polypyrrolle is doped to the maximum level of approximately one polaron for every four rings with PF_6^- as the counter-ion.

[3] C. Kittel, *Introduction to Solid State Physics*, John Wiley, New York, 1953.

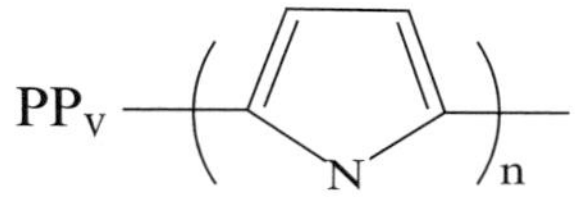

The investigation of the M-I transition through IR-visible reflectance measurements is particularly important. Although disorder is generally recognized to play an important role in the physics of "metallic" polymers, the effective length scale of the disorder and the precise origin of the M-I transition were unresolved issues. In particular, the question of whether disorder is present over a wide range of length scales or whether the properties are dominated by more macroscopic "inhomogeneities" has been a subject of discussion in the conducting polymer literature. When disorder is present over a wide range of length scales, the metallic state and the M-I transition can be described by conventional localization physics (e.g. the Anderson transition). If the disorder is dominated by more macroscopic inhomogeneities, the M-I transition would be better described in terms of percolation between "metallic islands".

During the 1990s, progress in the processing of conducting polymers significantly improved the quality of the materials and reduced the extent of structural disorder with corresponding improvements in the electrical conductivity. Extensive transport studies carried out on PPy-PF6 demonstrated that the improved material is more highly conducting and more homogeneous than that studied earlier. As is typical of conducting polymers, PPy-PF_6 is partially crystalline. The structural coherence length is, however, only $\xi \approx$ 20–50 Å, less than any length used to characterize the electronic properties, i.e. less than the inelastic scattering length ($L_{in} \approx 300$ Å) in the metallic regime, and less than the localization length ($L_{loc} \approx$ 200–300 Å) in the insulating regime near the M-I transition. The corresponding transport data in the critical regime and the cross-over from metal to insulator have been successfully analyzed in terms of conventional disorder-induced localization.

To clarify the nature of the metallic state and to study the M-I transition, high precision reflectance measurements were carried out over a wide spectral range on a series of PPy-PF_6 samples in the insulating, critical, and metallic regimes near the M-I transition. The data demonstrate that metallic PPy-PF_6 is a "disordered metal" and that the M-I transition is driven by disorder; similar results were obtained for polyaniline.

Based on the criteria presented in Fig. 4.13, sample A (in Fig. 4.15) is in the metallic regime, and sample F is in the insulating regime. For samples B through E, the electronic properties gradually evolve from the metallic side of the M-I transition to the insulating side via the critical regime (see Fig. 4.8).

Figure 4.16 shows $R(\omega)$ for samples (A–F) as measured at room temperature. For the most metallic sample (A), $R(\omega)$ exhibits distinct metal-like signatures; a free carrier plasma resonance as indicated by the minimum in R(ω) around 1.5×10^4 cm^{-1} and high $R(\omega)$ in the far-IR ($R \geq 90\%$ for $\omega < 20$ cm^{-1}).

As PPy-PF_6 goes from the metal to the insulator via the critical regime (samples A $\rightarrow$F), R (ω) is gradually suppressed in the IR. In the insulating regime (F), $R(\omega)$ remains well below that of the metallic sample (A) throughout the IR ($R \approx 65\%$ at $\omega = 50$ cm^{-1}); the better the quality of the sample, as defined by smaller ρ_r and higher σ_{dc}(300K), the higher $R(\omega)$ in the IR.

For metals, the far-infrared reflectance can be approximated at low frequencies by the Hagen–Rubens (H-R) relation

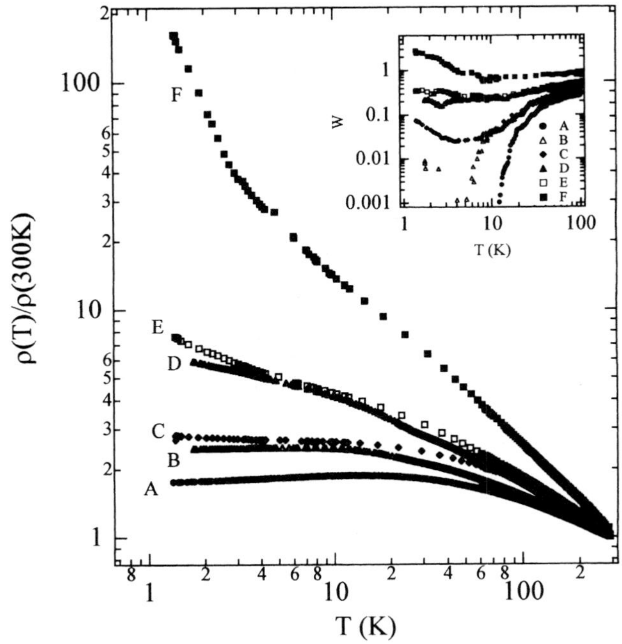

Fig. 4.15 Temperature dependence of the resistivity, $\rho(T)$, for a series of PPy-PF_6 samples with different resistivity ratios. The ρ_r values for samples A through F are summarized in Table 4.1. [Reproduced from K. Lee *et al.*, *Adv. Mater.*., **10**, 456 (1998) with permission of Wiley-VCH.]

$$R_{H-R}(\omega) = 1 - (2\omega/\pi\sigma_{H-R})^{1/2}, \tag{4.41}$$

where $\sigma_{H\text{-}R}$ is the dc conductivity. Thus, applicability of the H-R relation implies classical metallic behavior with $\sigma(\omega) \to \sigma_{dc}$ at low frequencies. For each data set, the complex dielectric function, Eqn 4.40, was obtained by K-K analysis of $R(\omega)$. At the low frequency end, the H-R relation (Eqn 4.41) was used to extrapolate to $\omega \to 0$. As demonstrated in Table 4.1, the $\sigma_{H\text{-}R}$ values obtained from the Hagen–Rubens fits are in remarkably good agreement with the measured values of σ_{dc}(300K). The excellent fits and the agreement between $\sigma_{H\text{-}R}$, σ_{dc}(300K) and $\sigma(\omega \to 0)$ imply a weak ω-dependence in the corresponding optical conductivity, $\sigma(\omega)$, for $\omega < 100$ cm^{-1}.

The frequency dependent conductivity data, $\sigma(\omega)$, are shown in Fig. 4.17. The $\sigma(\omega)$ data are not typical of a Drude metal for which

$$\sigma_{Drude} = (\omega_p^2\tau/4\pi)/[1 + \omega^2\tau^2] \tag{4.42}$$

where ω_p is the π-electron plasma frequency, and τ is the mean scattering time. Even for the most metallic sample (A), $\sigma(\omega)$ decreases with decreasing ω below 2500 cm^{-1} and thus deviates from Drude behavior. On moving toward the insulating regime (from A to F), $\sigma(\omega)$ is suppressed, and the maximum in $\sigma(\omega)$ gradually shifts to higher frequencies (e.g. $\omega_{max} \approx 4500$ cm^{-1} for sample F). For all samples, however, there is a change in slope in $\sigma(\omega)$ at around

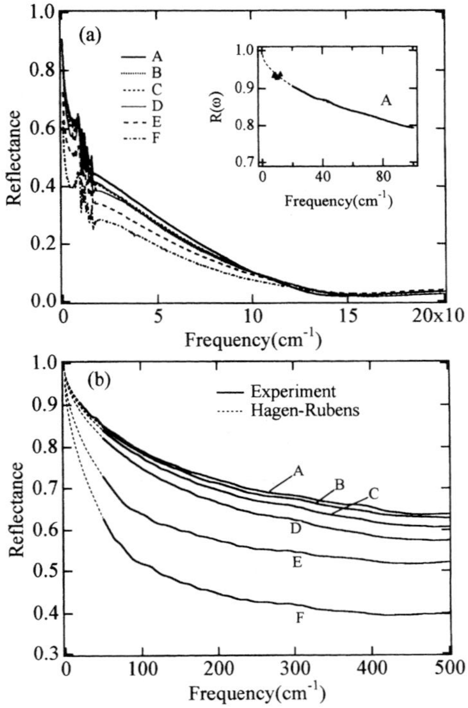

Fig. 4.16 Reflectance spectra, $R(\omega)$, for PPy-PF$_6$ samples (A–F) at 300 K. The inset shows $R(\omega)$ of the metallic sample (A) below $\omega = 100\ \text{cm}^{-1}$: solid curve is $R(\omega)$ obtained with the Michelson interferometer in the far-IR, solid triangles represent $R(\omega)$ data obtained with the submillimeter spectrometer ($8 \leq \omega \leq 13\ \text{cm}^{-1}$), dashed curve is the fit by the Hagen–Rubens formula (Eqn 4.41) with $\sigma_{\text{HR}} = 310$ S/cm. [Reproduced from K. Lee *et al.*, *Adv. Mater.*, **10**, 456 (1998) with permission of Wiley-VCH.]

$\omega \approx 100\ \text{cm}^{-1}$; at lower frequencies, $\sigma(\omega)$ is essentially frequency independent and, in each case, accurately approaches the measured $\sigma_{\text{dc}}(300\text{K})$ value. Such a weak ω-dependence for $\omega \leq 100\ \text{cm}^{-1}$ is expected from the excellent fit of $R(\omega)$ by the H-R approximation in this frequency range.

The $\sigma(\omega)$ data shown in Fig. 4.17 are fully consistent with the "localization-modified Drude model" (LMD) developed for application to conductors near the M-I transition. The suppressed $\sigma(\omega)$ as $\omega \to 0$ arises from localization induced by disorder. Moreover, the gradual evolution of $\sigma(\omega)$ from the insulating regime to the metallic regime implies that the M-I transition proceeds in the context of the Anderson transition. The $\sigma(\omega)$ data from the various regimes were fit with the functional dependence predicted by the LMD model:

$$\sigma_{\text{LMD}}(\omega) = \sigma_{\text{Drude}}\{1 - C[1 - (3\tau\omega)^{1/2}]/(k_F l)^2\} \tag{4.43}$$

where k_F is the Fermi wavenumber, l is the mean free path, and σ_{Drude} is given by Eqn 4.42. In Eqn 4.43, $C = 1$ in the insulating regime, whereas in metallic regime, C is determined by $\sigma 0$). In this model, one additional parameter, $k_F l$, is introduced in the first-order correction term to

Table 4.1. Transport results for the PPy-PF_6 samples in the various regimes. The optical conductivity, $\sigma(\omega \to 0)$, obtained from K-K analysis of $R(\omega)$, and the conductivity obtained from fits to the data using the Hagen-Rubens approximation, σ_{HR}, are included fro comparison.

Sample	σ_{dc}(300 K) [S/cm]	ρ_r	σ_{HR} [S/cm]	$\sigma(\omega \to 0)$ [S/cm]
A	340	1.7	310	330
B	310	2.5	295	310
C	270	2.8	265	280
D	230	6.0	215	235
E	100	7.8	92	90
F	40	160	49	40

σ_{Drude}. Figure 4.17 illustrates the excellent agreement of the fits to the data with the parameters summarized in the inset (except for the phonon features around 400–2000 cm^{-1} which are not included in the LMD model). There are small deviations for $\omega \leq 100$ cm^{-1} (more important in the less metallic samples), below which phonon-assisted hopping makes a measurable contribution to $\sigma(\omega)$ (and to the dc conductivity).

The parameters obtained from the fits are reasonable. The screened plasma frequency, $\Omega_p = \omega_p/(\varepsilon_\infty)^{1/2} \approx 1.5 \times 10^4$cm^{-1}, is in good agreement with the frequency of the minimum in $R(\omega)$, and τ is typical of disordered metals ($\tau \approx 10^{-14}$–10^{-15} s). The quantity k_Fl is of particular interest for it characterizes the extent of disorder and is often considered as an order

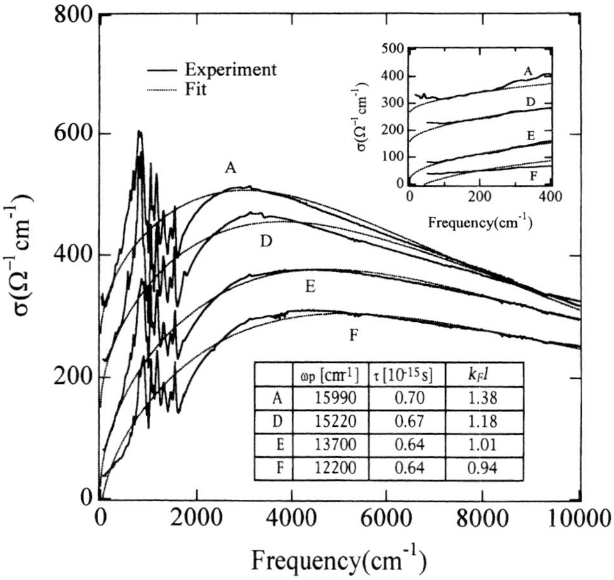

	ωp [cm⁻¹]	τ [10⁻¹⁵s]	k_Fl
A	15990	0.70	1.38
D	15220	0.67	1.18
E	13700	0.64	1.01
F	12200	0.64	0.94

Fig. 4.17 Frequency dependent conductivity, $\sigma(\omega)$, for the various Ppy-PF_6 samples introduced in Figs 4.14, 4.15 and Table 4.1. The solid curves are fits to the LMD model, Eqn 5.42. The inset shows the data and LMD curves in greater detail for $\omega < 400$ cm^{-1}. The table summarizes the parameters obtained from the LMD fits to the data. [Reproduced from K. Lee *et al.*, *Adv. Mater.*, **10**, 456 (1998) with permission of Wiley-VCH.]

parameter in localization theory; see Eqn 4.11. For all four samples represented in Fig. 4.17, $k_Fl \approx 1$, implying that all are close to the M-I transition. As the disorder increases and the system moves from the metallic regime (sample A with $k_Fl = 1.38$) to the critical regime (sample E with $k_Fl = 1.01$), k_Fl approaches the Ioffe–Regel limit, precisely as would be expected. In the insulating regime (sample F) $k_Fl = 0.94 < 1$, consistent with localization of the electronic states at E_F.

As a final step in the evolution to the metallic state, high-quality, metallic PPy-PF_6 films prepared by galvanostatic (electrochemical) polymerization were tensile-drawn to a draw ratio $\ell/\ell_0 \approx 1.5$. The room temperature conductivity increased from $\sigma_{dc} \approx 350$ S/cm (as grown) to $\sigma_{dc//} \approx 800$ S/cm at 300 K. In addition, the partially oriented films showed significant anisotropy, $\sigma_{dc//}/\sigma_{dc\perp} \approx 4.6$ with $\sigma_{dc\perp} \approx 175$ S/cm. Characterization by transport and X-ray diffraction measurements was carried out to ensure sample quality before and after the tensile drawing.

Figure 4.18 shows $R(\omega)$ of partially oriented PPy-PF_6 measured with polarization parallel to and perpendicular to the draw direction. Data from an as-grown sample (nonoriented) are shown for comparison. $R(\omega)$ of the as-grown sample is typical of the highest quality PPy-PF_6 with a well-defined plasma frequency as indicated by the minimum in $R(\omega)$ at approximately 1.5×10^4 cm^{-1} and high $R(\omega)$ in the far-IR. The uniaxial orientation induces anisotropy and causes an improvement in the physical properties of the films; for parallel polarization, $R(\omega)$ is enhanced throughout the IR. For perpendicular polarization, $R(\omega)$ remains below that of the nonoriented sample with a pronounced additional feature near 1.1×10^4 cm^{-1}.

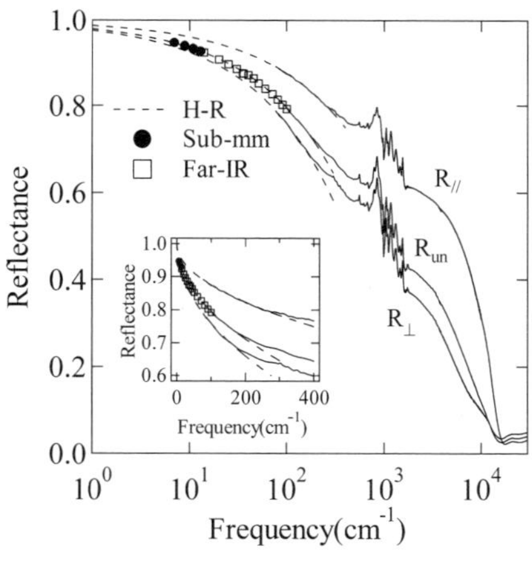

Fig. 4.18 Reflectance spectra, R(ω), of stretch-oriented PPy-PF_6 measured at 300 K with polarization parallel and perpendicular to the stretched direction, together with that of the nonoriented sample. The solid curves represent the measured $R(\omega)$. The dotted lines at the lowest frequencies indicate the extrapolation by the Hagen–Rubens (H-R) approximation. For the nonoriented sample, the H-R approximation is in excellent agreement with low frequency $R(\omega)$ data obtained with the submillimeter spectrometer (•; $8 \leq \omega \leq 13$ cm^{-1}) and the far-IR interferometer (□ $15 \leq \omega \leq 100$ cm^{-1}). The inset shows $R(\omega)$ below 400 cm^{-1} on a linear scale; the Hagen–Rubens curves (dotted lines) are in excellent agreement with the measured R(ω) for $\omega < 250$ cm^{-1}. [Reproduced from K. H. Lee and A. Heeger, *Phys. Rev. B*, **68**, 35201 (2003) with permission of The American Physical Society.]

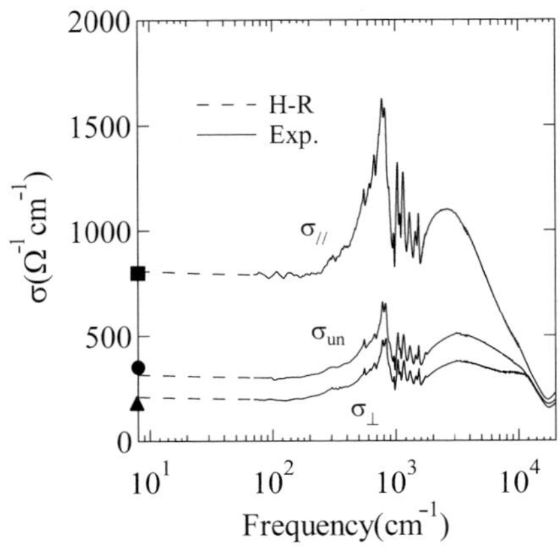

Fig. 4.19 $\sigma(\omega)$ versus ω for oriented samples of Ppy-PF_6 as obtained via K-K transform from the data in Fig. 4.17; the solid curves are fits to the LMD model. [Reproduced from K. Lee and A. Heeger, *Phys. Rev. B*, **68**, 35201 (2003) with permission of The American Physical Society.]

Figure 4.19 displays $\sigma(\omega)$ versus ω as obtained via the K-K transform from the data in Fig. 4.18. Again, the $\sigma(\omega)$ data obtained from the oriented sample are consistent with the dc transport results. For the parallel direction, $\sigma(\omega)$ exhibits significantly enhanced intraband response throughout the IR as compared with that of the as-grown sample. The tensile-drawing

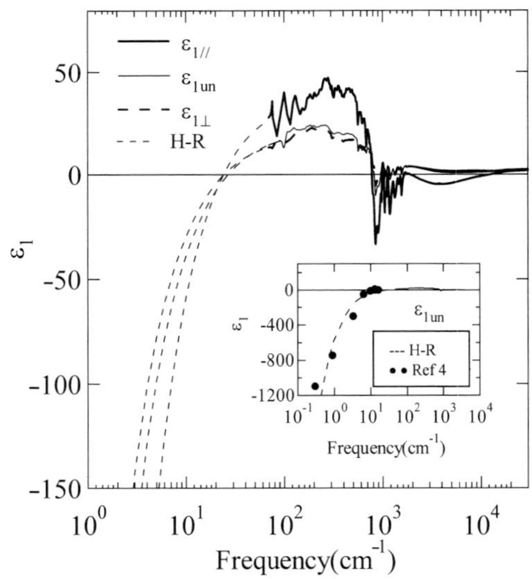

Fig. 4.20 Real part of the dielectric function $\varepsilon_1(\omega)$ of PPy-PF_6 parallel and perpendicular to the stretched direction together with that of the nonoriented sample. The inset compares $\varepsilon_1(\omega)$ of the nonoriented sample and data adapted from Martens *et al.*[4] [Reproduced from K. H. Lee and A. Heeger, *Phys. Rev. B*, **68**, 35201 (2003) with permission of The American Physical Society.]

(chain orientation) does not alter the essential physics of the samples. The decrease in $\sigma(\omega)$ below 2500 cm^{-1} is again consistent with the LMD expression, Eqn 4.43.

As shown in Fig. 4.20, chain orientation has a strong influence on the spectral dependence of $\varepsilon(\omega)$. PPy-PF_6 is a disordered metal close to the M-I transition with $\omega_p\tau \sim 1$, where ω_p is the plasma frequency and τ is the scattering time. On the metallic side of the M-I transition where $\omega_p\tau > 1$, $\varepsilon(\omega)$ is expected to cross zero at the screened plasma frequency, $\Omega_p = \omega_p/(\varepsilon_\infty)^{1/2}$, where ε_∞ is the high-energy contribution to the dielectric constant. This is indeed the case for the parallel direction in the oriented sample; $\varepsilon(\omega)$ crosses zero at 1.3×10^4 cm^{-1} (1.6 eV) and goes to negative values in the near IR. Figure 4.20 shows the real part of the dielectric function $\varepsilon_1(\omega)$ of PPy-PF_6 parallel and perpendicular to the stretched direction together with that of the nonoriented sample. The inset compares $\varepsilon_1(\omega)$ of the nonoriented sample and data adapted from Martens *et al.*[4]

In summary, the data obtained for metallic polymers indicate that they are "disordered metals" near the disorder-induced M-I transition. The evolution form insulator to metal with improved sample quality has been documented and monitored in detail. There is remarkable consistency between the conclusions obtained from transport measurements and those from IR reflectance measurements.

Further reading

R. Menon, C.O. Yoon, D. Moses and A. J. Heeger, in: *Handbook of Conducting Polymers*, edited by T.A. Skotheim, R. L. Elsenbaumer and J. R. Reynolds, Marcel Dekker, New York, 1998, pp. 27–84.

Sir Nevill Mott, M. Pepper, S. Pollitt, R. H. Wallis and C. J. Adkins, *Proc. Royal Soc. London, Series A, Math. Phys. Sci.*, **345**. 169 (1975).

Sir Nevill Mott, *Proc. Royal Soc. London, Series A, Math. Phys. Sci.*, **382**, 1 (1982).

Sir Nevill Mott, *Conduction in Non-Crystalline Materials*, Oxford University Press, Oxford, 1987.

[4] H. C. F. Martens, J. A. Reedijk, H. B. Brom, D. M. De Leeuw and R. Menon, *Phys. Rev.* **B 63**, 73203 (2001).

5

Metallic state of conducting polymers

5.1 Introduction

When carefully processed and doped to sufficiently high carrier densities, conducting polymers demonstrate all the properties of metals:

1. the highest occupied band is partially filled with a well-defined Fermi energy;
2. optical properties are consistent with the physics of metals;
3. transport properties are indicative of "free" electrons.

5.1.1 Highest occupied band is partially filled with a well-defined Fermi energy

The classic signatures of a partially filled band with a well-defined Fermi energy are the temperature independent Pauli spin susceptibility, the electronic contribution to the specific heat that is proportional to the temperature (T) and the corresponding term (also linear in T) in the thermopower.

The temperature-independent Pauli spin susceptibility arises directly from Fermi–Dirac statistics; only the small fraction, k_BT/E_F, of the total number of electrons per unit volume (N) with energies within k_BT of the Fermi energy (E_F) are able to spin-polarize in a magnetic field. The Pauli spin susceptibility of a metal is given by the following expression:

$$\chi_{Pauli} = \mu_B^2 D(\varepsilon_F) = \frac{3N\mu_B^2}{2k_B T_F} \tag{5.1}$$

where μ_B is the Bohr magneton, $D(E_F)$ is the density of electronic states evaluated at E_F (both signs of spin, spin up plus spin down), and T_F is the Fermi temperature, $k_BT_F = E_F$. For a free electron gas, $D(E_F) = 3N/2E_F$.

The linear (in T) contribution to the specific heat arises from the low energy thermal excitations; electrons can be promoted from filled states below E_F to empty states above E_F. For temperatures such that $k_BT << E_F$, the corresponding term in the heat capacity, C_{el}, is given by

$$C_{el}(T) = \gamma T = (\pi^2/3)D(E_F)k_B^2 T = (\pi^2/2)Nk_B T/T_F \tag{5.2}$$

In addition to the electronic contribution, the total specific heat contains a contribution from the lattice vibrations (phonons), $C_{latt}(T) = \beta T^3$ where $\beta \approx (12\pi^4/5)\, N\, k_B(T/\theta)^3$, and θ is the Debye temperature (proportional to the sound velocity). Thus, since $C_{tot}/T = \gamma + \beta T^2$, it is traditional to plot the total specific heat in the form C_{tot}/T versus T^2. In this plot, γ is the intercept at $T = 0$ K.

The thermopower is a measurement of the entropy (S) per carrier. Since the entropy per carrier is related to the electronic contribution to the specific heat, $S_{el}(T) = (1/e) \int [C_{el}(T')/T'] \, dT'$, where the integral is from $T' = 0$ to $T' = T$, the entropy per carrier is given by

$$S_{el}(T) = \gamma T/eN = (\pi^2/3)D(E_F)k_B^2 T/eN = (\pi^2/2)(k_B/e)T/T_F \quad (5.3)$$

Thus, for a metal, the Pauli term in the magnetic susceptibility, the linear term in the specific heat, and the linear term in the thermopower provide independent measurements of the density of states at the Fermi energy and, thereby, demonstrate the existence of the partially filled band. As shown by Eqns 5.1, 5.2 and 5.3, measurements of χ_{Pauli}, $C_{el}(T)$ and $S_{el}(T)$ determine the magnitude of the density of states at E_F.

5.1.2 Optical properties consistent with the physics of metals

Metals have no energy gap between the highest occupied state and the lowest unoccupied state; the density of states in the partially filled metallic band is continuous, but only the states with energies less than E_F are filled. Because there is no energy gap, metals absorb electromagnetic radiation even at long wavelengths in the far infrared. Thus, a signature of metallic behavior is the absence of an energy gap as inferred from far-infrared absorption.

As noted in Chapter 3, a second characteristic feature of the optical properties of a metal is the reflection of light with photon energies below the plasma frequency, ω_p, where

$$\omega p^2 = 4\pi N e^2/m^*. \quad (5.4)$$

and m^* is the effective mass of electrons in the partially filled band (see Eqn 4.4). Thus, metals reflect light at frequencies below the plasma frequency, ω_p.

5.1.3 Transport properties indicative of "free" electrons

Finally, the temperature dependence of the resistivity, $\rho(T)$, is used to characterize the metallic state. Metals are good conductors of electricity even as the temperature approaches zero.

As discussed in detail in Chapter 4, for $k_F l >> 1$, $\rho(T)$ decreases linearly with decreasing temperature as a result of electron–phonon scattering (the number of phonons is proportional to $k_B T$) until the phonon contribution is sufficiently small that scattering by defects and impurities dominates. As $k_F l$ decreases toward the Ioffe–Regel limit, the residual resistivity, $\rho(T = 0) = \rho_o$, increases until the phonon contribution becomes unobservable even at room temperature and above. For $k_F l < 1$, the resistivity goes to infinity as T approaches zero (the metal–insulator transition), while at finite temperatures $\rho(T)$ is activated by variable-range hopping (see Chapter 4, Section 4.3).

5.2 Magnetic susceptibility of doped metallic polymers

The magnetic susceptibility of conducting polymers is often dominated by impurities and disorder. Thus, in typical studies reported in the early literature, the data were a superposition of a Curie term (proportional to $1/T$) and a Pauli term (T-independent). After increasing the order and purity in the samples, a temperature independent susceptibility has been observed in many conducting polymers. Figure 5.1 shows (as an example) the temperature independent

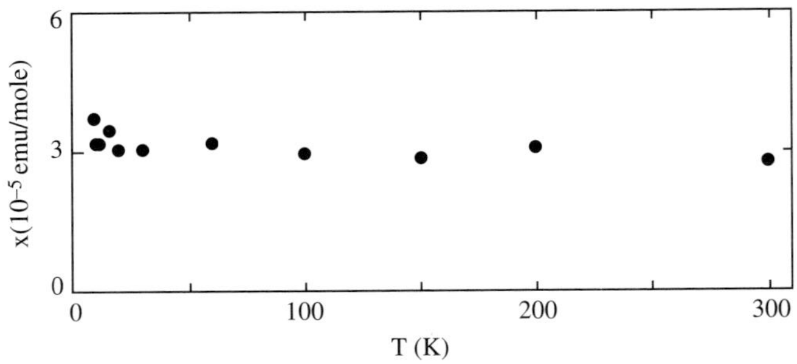

Fig. 5.1 Temperature dependence of the magnetic spin susceptibility (χ) of heavily doped poly(thiophene); 24 mole% AsF_5. [Reproduced from F. Moraes *et al.*, *Synthetic Metals*, **10**, 169 (1985) with permission of Elsevier B.V.]

Pauli spin susceptibility of heavily doped, metallic poly(thiophene). The Pauli spin susceptibility is constant and temperature independent over the entire temperate range (a hint of a T^{-1} tail can be seen below 20 K, corresponding to a small concentration of Curie spins).

From the magnitude of the Pauli term (3 10^{-5} emu per mole of thiophene), one can estimate the density of states at the Fermi energy, $D(E_F) = 0.23$ states per eV per carbon. Assuming the energy dependence is approximately that of a free electron gas (the band is less than half filled since the doping level is approximately one per three thiophene rings), $E_F \approx 6$ eV, consistent with the relatively broad bandwidth expected for the delocalized electrons in the polythiophene π-band (see Chapter 1). The Pauli susceptibility measured in heavily doped metallic polyacetylene yields a density of states equal to 0.12 states per eV per carbon atom. The somewhat narrower bandwidth in polythiophene is reasonable since the unit cell is doubled (there are four carbons in the repeat unit versus two in polyacetylene).

In the presence of a magnetic field, the spin polarization of the electrons (from the Pauli susceptibility), acting through the hyperfine interaction, creates an additional contribution to the magnetic field at the nucleus (i.e. in addition to the external applied magnetic field). As a result, the nuclear magnetic resonance frequency is shifted from that of the same nuclear spin subjected to only the external applied field. This shift, known as the Knight shift, is often used to characterize the metallic state. The Knight shift is defined as follows:

$$K = \frac{\Delta f}{f} = \frac{\Delta H}{H} = \sum_i H^i_{hf} \frac{\chi_{Pauli}}{\mu_B} \tag{5.5}$$

where Δf and ΔH are the measured frequency shift and the measured shift in the magnetic field at resonance, respectively. The shifts are defined in terms of f_o, the resonance frequency in the external field, H_o; $\Delta f = f_r - f_0$ and $\Delta H = H_r - H_0$, where f_r and H_r are the resonance frequency and the field for resonance, respectively. In Eqn 5.5, H^i_{hf} is the hyperfine coupling constant in gauss. Unfortunately, the expected Knight shift in conducting polymers is small (on the order of several ppm) due to indirect hyperfine coupling of the π-electrons to the nucleus via polarization of the s-core. On the other hand, the chemical shift of ^{13}C NMR due to shielding by the p-electrons can be as large as 160 ppm which masks any effects of the metallic Knight

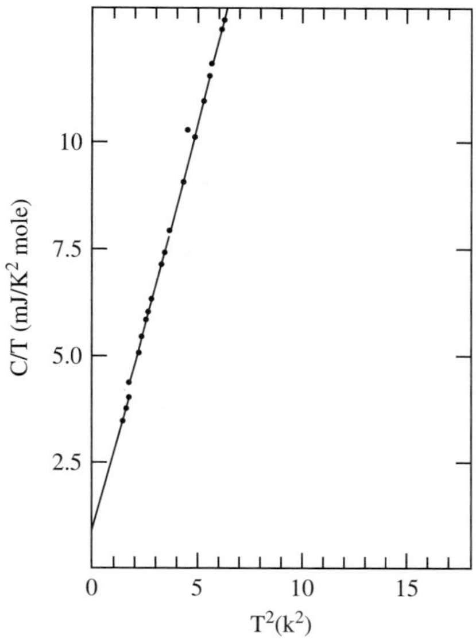

Fig. 5.2 *C/T* versus T^2 for metallic polyacetylene. The intercept at $T = 0$ K yields $\gamma = 0.9$ mJ/mole/K^2. [Reproduced from D. Moses *et al.*, *Solid State Commun.*, **36**, 219 (1980) with permission of Elsevier B. V.]

shift. Thus, the Knight shift is not a useful probe of the metallic state in doped conjugated polymers.

5.3 Specific heat

The only specific heat data available in the literature for metallic polymers were obtained from heavily doped polyacetylene; the C/T versus T^2 plot is shown in Fig. 5.2.

Using Eqn 5.2, $D(E_F) \approx 0.1$ states per eV per C atom, in good agreement with the value obtained from measurements of the Pauli susceptibility in heavily doped polyacetylene.

5.4 Thermopower

The thermopower data obtained from metallic polyacetylene are shown in Fig. 5.3.

From the slope of S versus T (using Eqn 5.3), $D(E_F) \approx 0.2$ states per eV per C atom, approximately twice as large as the corresponding values obtained from the heat capacity and spin susceptibility. The origin of the factor of two discrepency is not known.

As shown in Fig. 5.4, the linear (in T) thermopower is also characteristic of the metallic state of polyaniline.

In polyaniline, however, the effect of disorder has been carefully documented by measurements of the thermopower in a series of polymer blends. As shown in Fig. 5.4, as the

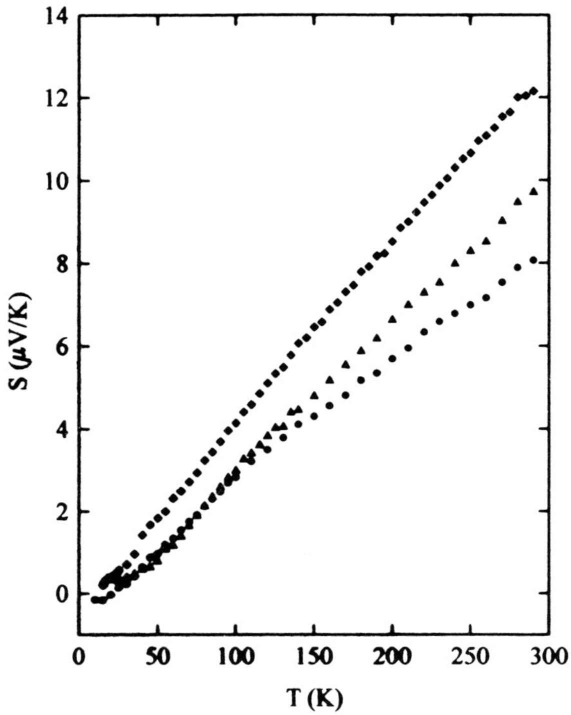

Fig. 5.3 Thermopower of metallic PANI-CSA. [Reproduced from C. O. Yoon *et al.*, *Phys. Rev. B*, **48**, 14080 (1993) with permission of The American Physical Society.]

concentration of polyaniline in the blend decreases, a "hook" appears in the low temperature data. Since the same feature is observed in disordered polyaniline, this feature is a signature of disorder (see Chapter 3). The data obtained from PANI-CSA, 100% (•), increase approximately linearly with temperature. From the slope of the linear term obtained from homogeneous PANI:CSA, $D(E_F) \approx 1.3$ states per eV per two rings, implying a somewhat narrower bandwidth with $E_F \approx 1$ eV.

5.5 Metallic reflection

The two electromagnetic signatures of the metallic state are the following:

(1) the absence of an energy gap as inferred from far-infrared absorption;
(2) metallic reflection for frequencies below the plasma frequency; see Chapter 3, Section 3.4.

Both of the electromagnetic signatures of the metallic state have been demonstrated in metallic polyacetylene; Measurements of the "metallic reflection" of metallic polyacetylene are shown in Fig. 5.5. The anisotropy is indicative of a quasi-one-dimensional metal.

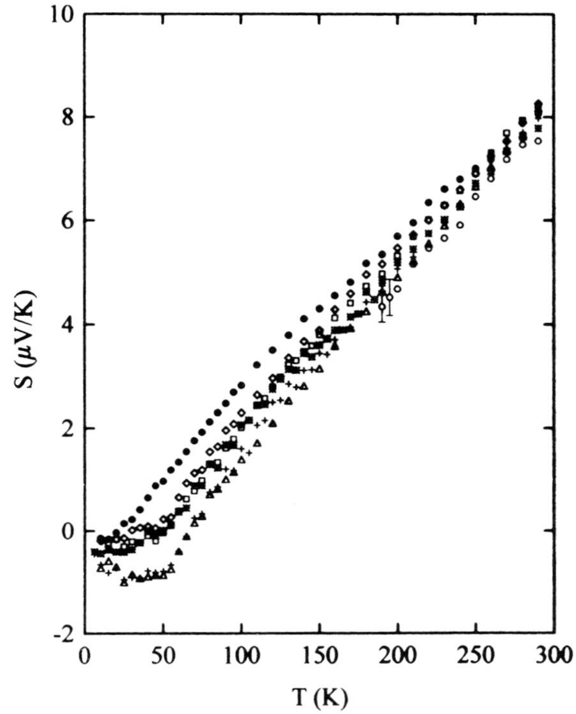

Fig. 5.4 Temperature dependence of the thermopower of metallic PANI-CSA/PMMA blends at various concentration of PANI-CSA in PMMA: 100% (•), 66.6% (◊), 33.3% (□), 9.09% (*), 4.76% (+), 2.43% (▲) and 1.24%(o). [Reproduced from C. O. Yoon *et al.*, *Phys. Rev. B*, **48**, 14080 (1993) with permission of The American Physical Society.]

5.6 Transport measurements

Figure 4.8 shows the general phase diagram that describes the metal–insulator transition. The metallic state is established when the electrical conductivity remains finite as the temperature approaches $T = 0$ K. With this as the criterion, the data in Fig. 5.6, obtained in the mK temperature regime, demonstrate that PANI-CSA is on the metallic side of the metal insulator transition.

With polyaniline samples prepared using self-stabilized dispersion polymerization (SSDP), σ_{dc} (RT) > 1000 S/cm can be obtained. In these materials, $\rho(T)$ decreases monotonically as the temperature is lowered, and the IR spectra are characteristic of the conventional Drude model (rather than the "localization-modified Drude model", see Chapter 4) even at the lowest frequencies in the far-IR.

The SSDP samples exhibit significantly enhanced transport properties compared to those of conventional samples. The temperature dependence of the resistivity is shown for a series of samples with different room temperature resistivities in Fig. 5.7. As ρ(300 K) decreases, $\rho(T)$

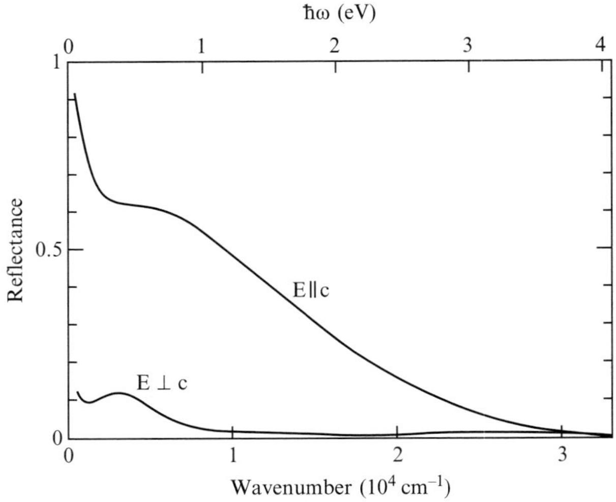

Fig. 5.5 Reflectance of oriented films of metallic polyacetylene: When the light is polarized with electric field parallel to the chain direction, well defined metallic reflectance is observed with the plasma frequency at $\hbar\omega_p \approx 1.2$ eV, in the infrared. See also Fig. 3.14. [Reproduced from G. Leising *et al.*, *Synthetic Metals*, **21**, 267 (1987) with permission of Elsevier B.V.]

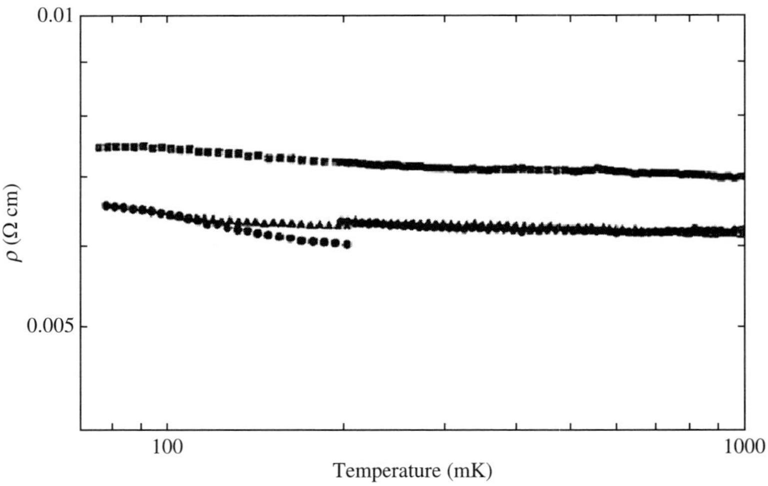

Fig. 5.6 Temperature dependence of the electrical conductivity in PANI:CSA in the millikelvin regime. The conductivity remains finite as T approaches zero. [Reproduced from J. C. Clarka *et al.*, *Synthetic Metals*, **69**, 215 (1995) with permission of Elsevier B.V.]

remains progressively lower (σ_{dc} is higher) over the entire temperature range, the minimum in $\rho(T)$ shifts to lower temperatures, and the low temperature up-turn becomes less pronounced. For the highest conductivity samples, S1 [σ_{dc}(300 K) $\approx$ 1300 S/cm] and S2 [σ_{dc}(300 K) $\approx$ 1100 S/cm], $\rho(T)$ exhibits a positive temperature dependence ($d\rho/dT > 0$) down to 5 K with $\rho_r \approx 0.4$ for S1 (10 K for S2), as shown in the inset of Fig. 5.7.

The metallic nature of SSDP samples was confirmed by measurements of the reflectance, $R(\omega)$, at room temperature as shown in Fig. 5.8(a). The data obtained from S1 and S2 (Fig. 5.7) were almost identical; the data shown in Fig. 5.8(a) were obtained from S2 (see Fig. 5.7). The weak structure in the mid-IR (1000–1500 cm^{-1}) arises from the infrared-active vibration (IRAV) modes characteristic of conducting polymers. Note, however, that the IRAV modes of these highest conductivity samples are particularly weak compared to those reported earlier for "metallic" polyaniline. The IRAV modes in conducting polymers are characteristic of localized states, either self-localized states associated with solitons and polarons or localized states induced by

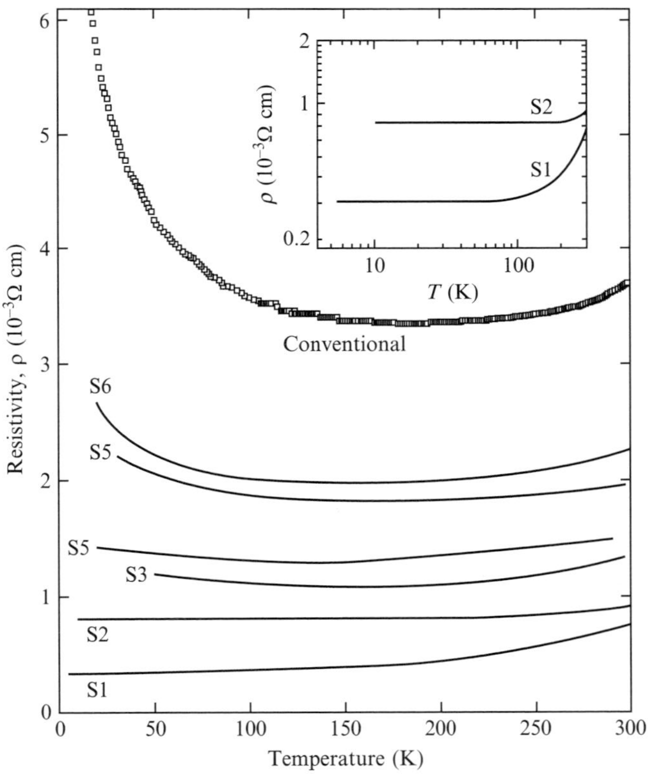

Fig. 5.7 Temperature dependence of the resistivity, $\rho(T)$. Data from a conventional PANI-CSA film are shown for comparison. For the more highly conducting samples (S6 → S1), the resistivity minimum weakens and shifts down to lower temperature, and eventually disappears in the S1 and S2 samples (as shown more clearly in the inset). [Reproduced from K. Lee *et al.*, *Nature*, **441**, 65 (2006) with permission of Nature Publishing Group.]

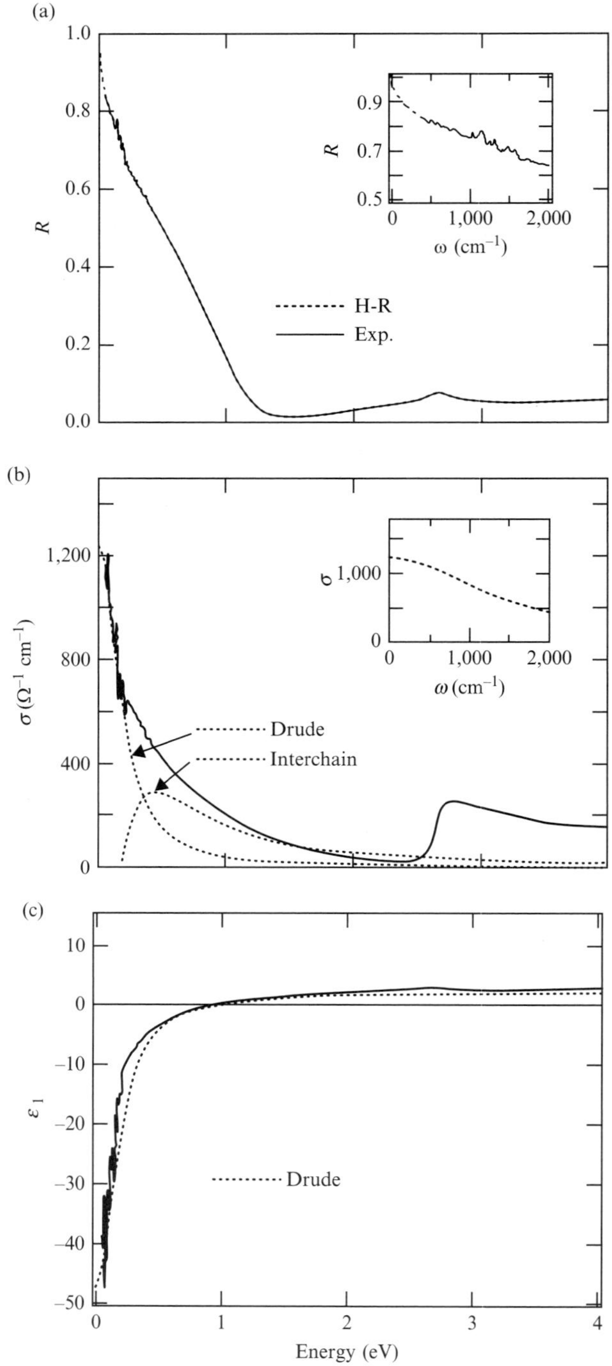

Fig. 5.8 (a) Reflectance spectrum, $R(\omega)$ of the high conductivity SSDP PANI-CSA. The Hagen–Rubens approximation (dotted line) provides an excellent fit to R(ω) in the far-IR. The inset shows $R(\omega)$ below 2000 cm^{-1} with an expanded scale. (b) Frequency dependent conductivity, $\sigma(\omega)$, as obtained from Kramers–Kronig analysis of $R(\omega)$. The dotted line represents the theoretical fit by the Drude model, $\sigma_D(\omega) = (\omega_p^2\tau/4\pi)(1+\omega^2\tau^2)^{-1}$, with $\omega_p = 10500$ cm^{-1} and relaxation time $\tau = 2.2\times10^{-14}$ s. (c) The real part of the dielectric function, $\varepsilon_1(\omega)$. The dotted line also represents the theoretical fit by the Drude model, $\varepsilon_{1D}(\omega) = \varepsilon_\infty - \omega_p^2\tau^2/(1+\omega^2\tau^2)^{-1}$, using the same parameters as in (b). [Reproduced from K. Lee, *Nature*, **441**, 65 (2006) with permission of Nature Publishing Group.]

disorder. When charges occupy such localized states, they move in response to the electric field of the IR radiation and indirectly drive the amplitude modes, causing the Raman-active modes to become IR-active. In the metallic regime, however, where the Fermi velocity is much greater than sound velocity, one does not expect to observe resonant IRAV modes. Thus, the reduced oscillator strength of the IRAV modes is consistent with the $\rho(T)$ data and implies that the metallic PANI-CSA characterized by Figs 5.7 and 5.8 is well beyond the insulator-to-metal transition.

Although the spectral features in $R(\omega)$ are similar to those of conventional PANI-CSA (see Chapter 3, Section 3.8), the higher quality of the samples is evident from the optical conductivity, $\sigma(\omega)$, and real part of the dielectric function, $\varepsilon_1(\omega)$ as obtained through the Kramers–Kronig (K-K) analysis. Figure 5.8(b) shows $\sigma(\omega)$ for S2. In contrast with the results described in Chapter 3, Section 3.8 in which $\sigma(\omega)$ decreases below $\omega < 2500\ \mathrm{cm}^{-1}$, the data in Fig. 5.8(b) follows the typical Drude frequency dependence even at frequencies below 2000 cm^{-1} and approaches the measured $\sigma_{dc}(300\ \mathrm{K})$. The excellent fit to $\sigma(\omega)$ of the classical Drude model, $\sigma_D(\omega) = (\omega_p^2\tau/4\pi)(1+\omega^2\tau^2)^{-1}$, with the plasma frequency $\omega_p = 10500\ \mathrm{cm}^{-1}$ (≈ 1.3 eV) and relaxation time $\tau = 2.2 \times 10^{-14}\mathrm{s}(1/\tau = 1500\ \mathrm{cm}^{-1} = 0.19\ \mathrm{eV})$, is shown in the inset to Fig. 5.8(b). For $\omega > 2000\ \mathrm{cm}^{-1}$, the interchain contribution starts to dominate $\sigma(\omega)$.

The $\varepsilon_1(\omega)$ spectrum is also consistent with the Drude model and true-metallic behavior. The fit to $\varepsilon_1(\omega)$ of the Drude model, $\epsilon_{1D}(\omega) = \epsilon_\infty - \omega_p^2\tau^2/(1+\omega^2\tau^2)^{-1}$, with the same values of ω_p andas in $\sigma_D(\omega)$ and with the high frequency dielectric constant $\varepsilon_\infty \approx 2$, shows again good agreement with the data as shown in Fig. 5.8(c). In the reflectance studies described in Chapter 4, Section 4.8, ε_1 crosses zero and becomes negative at the screened plasma frequency given by $\Omega_p = \omega_p/(\varepsilon_\infty)^{1/2}$, as expected for a typical metal. However, in contrast to simple Drude behavior, $\varepsilon_1(\omega)$ crosses zero again to positive values around $\omega \approx 2500\ \mathrm{cm}^{-1}$ and remains positive down to $\sim 20\ \mathrm{cm}^{-1}$. The $\varepsilon_1(\omega)$ data for PANI-CSA shown in Fig. 5.8(c), however, crosses zero at $\omega = 7800\ \mathrm{cm}^{-1}$ (= 0.97 eV) and remains negative at all frequencies below 7800 cm^{-1}, again consistent with the classic Drude model.

The plasma frequency, $\omega_p \approx 1.3$ eV, is in good agreement with the minimum in $R(\omega)$ around 1.4 eV, and yields $\Omega_p \approx 1$ eV with $\varepsilon_\infty \approx 2$, consistent with the zero-crossing frequency in $\varepsilon_1(\omega)$. Using these parameters, $\sigma_D(\omega{=}0) = \omega_p{}^2\tau/4\pi \approx 1200$ S/cm, in good agreement with the measured $\sigma_{dc}(300\ \mathrm{K}) \approx 1100$ S/cm. Moreover, from the relation $\omega_p = (4\pi ne^2/m^*)^{1/2}$ where n is the charge carrier density and m^* is the effective mass, one obtains $n = 2.5 \times 10^{21}\ \mathrm{cm}^{-3}$ with the reasonable assumption of $m^* \approx 2m_e$ (m_e is the free electron mass) for this class of materials. This result is consistent with that obtained from Hall measurements on this sample; the Hall effect data gave $n_{\mathrm{Hall}} = 2.1 \times 10^{21}\ \mathrm{cm}^{-3}$. In earlier IR studies of conducting polymers (see Chapter 3, Section 3.8), the proper description of the charge dynamics was achieved only after "modifying" the Drude model in the context of weak localization. However, the excellent agreement established in Fig. 5.8 demonstrates that high conductivity PANI-CSA can be successfully described by the simple Drude model without incorporating any contributions from disorder-induced localization theory.

5.7 Striving toward higher conductivities in the metallic state

A continuing goal of the field of conducting polymers is the creation of materials with high electrical conductivity *and* with the excellent mechanical properties of polymers. This goal has been achieved for polyacetylene; more importantly the conditions for realizing this combination of properties are understood and can be applied to other metallic polymers.

Electrical conductivity results from the existence of charge carriers and the ability of those carriers to move. Thus

$$\sigma = Ne\mu \tag{5.6}$$

where is the carrier mobility. Consequently, doped conjugated polymers, "conducting polymers", are good conductors for two reasons:

1. Doping introduces carriers into the π-electron system. Since every repeat unit is a potential redox site, conjugated polymers can be doped to a relatively high density of charge carriers.
2. The π-bonding leads to π-electron delocalization along the polymer chains and, thereby, to the possibility of charge carrier mobility, which is extended into three-dimensional transport by the interchain electron transfer. In principle, broad π-electron bandwidths (often several eV) can lead to relatively high carrier mobilities.

As a result of the same intrachain π-bonding and the relatively strong interchain electron transfer interaction, the mechanical properties (Young's modulus and tensile strength) of conjugated polymer are potentially superior to those of saturated polymers. Thus, metallic polymers offer the promise of truly high performance; high conductivity plus superior mechanical properties. For conducting polymers, the electrical properties and the mechanical properties improve together, in a correlated manner, as the degree of chain extension and chain alignment are improved. Polyacetylene remains the prototypical example.

Figure 5.9 shows the correlation between the electrical conductivity (σ) and the draw ratio (λ) for iodine doped polyacetylene films; the conductivity increases approximately linearly with the draw ratio. The slope of σ versus λ is approximately a factor of 2 larger for the thinner films (evidently the details of polymerization and/or doping result in more homogeneous, higher quality material for the thinner films).

X-ray diffraction studies of these drawn films demonstrate a high degree of structural order that improves with the draw ratio; the structural coherence length perpendicular to the draw direction increases by about a factor of 2 as the films are drawn, from 10 nm at $\lambda = 4$ to 20 nm at $\lambda = 15$. Consistent with the chain orientation and the improved structural order, the anisotropy ($\sigma_{\parallel}/\sigma_{\perp}$) in the electrical conductivity increases with the draw ratio, $\sigma_{\parallel}/\sigma_{\perp} \rightarrow 250$ as $\lambda \rightarrow 15$ (although $\sigma_{\parallel}$ increases dramatically as a function of λ, $\sigma_{\perp}$ remains essentially constant). These data set a lower limit on the intrinsic anisotropy and demonstrate that heavily doped polyacetylene is a highly anisotropic metal with relatively weak interchain coupling.

Does the weak interchain coupling imply poor mechanical properties? The answer is "No". For films with $\lambda = 15$, Young's modulus reaches 50 Gpa, and the tensile strength approaches 1 Gpa; i.e. mechanical properties which are characteristic of high performance materials. More importantly, the data in Fig. 5.9(a) demonstrate a direct correlation between the electrical conductivity and the mechanical properties. The linear relationship implies that the increase in both the conductivity and the modulus (or tensile strength) with draw ratio result from increased uniaxial orientation, improved lateral packing and enhanced interchain interaction.

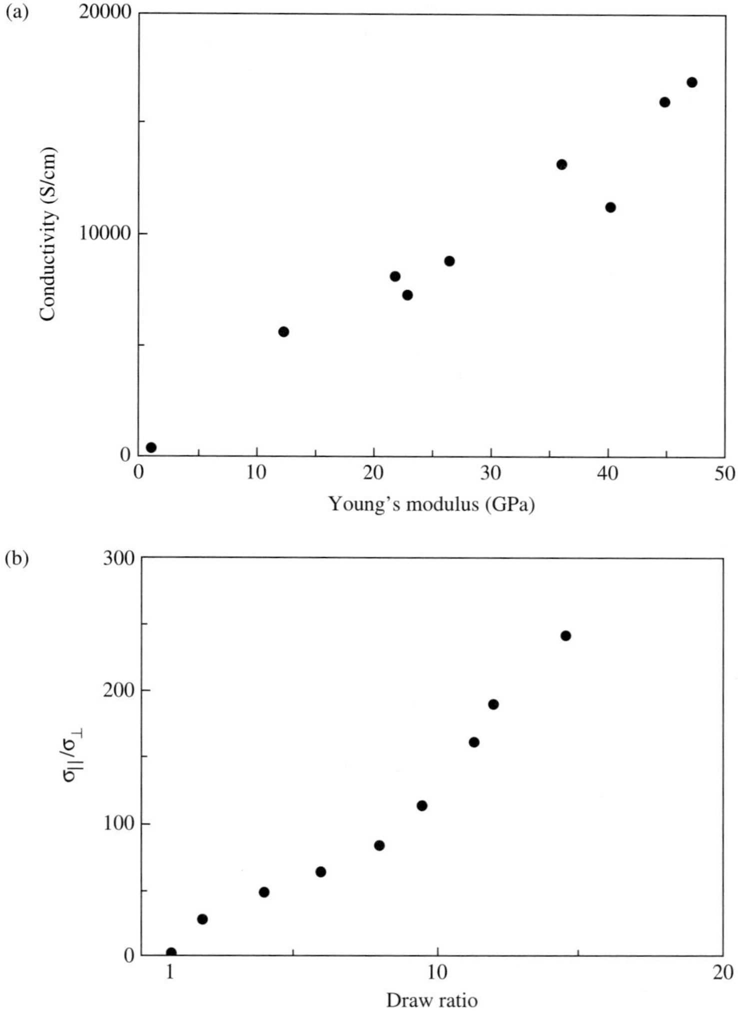

Fig. 5.9 (a) Correlation between the electrical conductivity (π) and the draw ratio (λ) for iodine doped polyacetylene films; the conductivity increases approximately linearly with the draw ratio. (b) The anisotropy ($\sigma_{\|}/\sigma_{\perp}$) shows a dramatic increase with draw ratio. [Reproduced from Y. Cao *et al.*, *Polymer.* **32**. 1210 (1991), with permission of Elsevier B.V.]

The correlation between the electrical conductivity and the mechanical properties has been demonstrated for several different macromolecular structures and can be understood as a general feature of conducting polymers.

Although the electrical conductivity is enhanced by the relatively high mobility associated with intrachain transport, one must have the possibility of interchain charge transfer to avoid the localization inherent to systems with a one-dimensional electronic structure; see Chapter 4.

The electrical conductivity becomes three-dimensional (and thereby truly metallic) only if there is high probability that an electron will have diffused to a neighboring chain prior to traveling between defects on a single chain. For a well-ordered crystalline material in which the chains have precise phase order, the interchain diffusion is a coherent process. In this limit, the condition for extended anisotropic transport is that

$$L/a >> (t_0/t_\perp) \tag{5.7}$$

where L is the coherence length, a is the length of the chain repeat unit, t_0 is the intrachain π-electron transfer integral, and $t_\perp$ is the interchain π-electron transfer integral.

A precisely analogous argument can be constructed for achieving the intrinsic strength of a polymer material; i.e. the strength of the main-chain covalent bonds. If E_o is the energy required to break the covalent main-chain bond and E_{3d} is the weaker interchain bonding energy (from Van der Waals forces and hydrogen bonding), then the requirement is coherence over a length L such that

$$L/a >> E_o/E_{3d} \tag{5.8}$$

In this limit, the large number, L/a, of weak interchain bonds add coherently such that the polymer fails by breaking of the covalent bond rather than by interchain slipping. The direct analogy between Eqns 5.7 and 5.8 is evident. Thus, the achievement of high performance materials requires chain extension, chain alignment and interchain order over many repeat units.

When the interchain structural order is "nematic" (i.e. without precise phase order), the conditions expressed by Eqns 5.7 and 5.8 require that $L/a >> (t_0/t_\perp)^2$ and $L/a >> (E_o/E_{3d})^2$, respectively. Thus, the achievement of high performance materials with nematic order will require chain extension and chain alignment over even longer distances.

For conjugated polymers, E_o results from a combination of σ and π bonds (the latter being proportional to the transfer integral, t_0), and E_{3d} is dominated by the interchain transfer integral, $t_\perp$. Thus, the inequalities 5.7 and 5.8 imply, quite generally, that the conductivity and the mechanical properties will improve in a correlated manner as the degree of chain alignment is increased. This prediction is in excellent agreement with data obtained from studies of the poly(3-alkylthiophenes), the poly(phenylene vinylenes), poly(thienylene vinylene) and polyacetylene.

5.8 Estimating the intrinsic electrical conductivity of metallic polymers

For doped, highly oriented polyacetylene, the electrical conductivity is approximately 10^5 S/cm at room temperature. Even with chain extension and chain alignment consistent with Eqn 5.7, this is a surprising result since the dopant counter-ions diffuse into the structure during the doping process and are therefore not perfectly ordered with long coherence lengths. Why, then, is the high concentration of partially disordered dopant counter-ions so ineffective in scattering the electrons?

The answer is simple, yet profound. Because the charged dopant ions are intercalated between the chains, they are spatially separated from the quasi-1D conduction path (see

Fig. 2.2). As a result, the scattering potential from any disorder in the dopant ions is exponentially small. The physics is analogous to that which occurs in the "modulation doping" of inorganic semiconductors; the carriers move in regions separated from the dopant ions.

Thus, one can anticipate metallic polymers in which the chain extension and interchain order will be sufficiently good that the inequalities 5.7 and 5.8 are valid, and in which the counter-ion disorder is not important. In such high performance materials, the electrical conductivity will be limited by phonon scattering. Again, however, there is a nice surprise; the conductivity is predicted to increase exponentially as the temperature is lowered, rather than linearly as in conventional metals! In an ordered metallic polymer, the back-scattering rate from thermal phonons can be written as

$$\frac{1}{\tau_{\text{ph}}} = \frac{8\alpha^2}{m\omega_0 t_0} \exp\left(-\frac{\hbar\omega_0}{k_B T}\right) \tag{5.9}$$

where m is the mass of the repeat unit (for polyacetylene, $m = m_{\text{C-H}}$), $t_0 \cong 2.5$–3.0 eV (see Chapter 1), α describes the magnitude of the electron–phonon interaction, and ω_o is the phonon frequency at $q = 2k_F$. Since back-scattering results in a change in electron momentum by $\hbar(2k_F)$, the phonons which cause the scattering must take up the momentum. Thus, as shown in Fig. 5.10, the phonons which cause the resistive back-scattering have energy $\hbar\omega_0 = \hbar\omega_{\text{ph}}(q)$ where $q = 2k_F$.

For the half-filled electronic band of metallic polyacetylene, $k_F = \pi/2a$ where a is the carbon–carbon spacing. For polyacetylene, $\hbar\omega_0 = 0.12$ eV and $\alpha = 4.1$ eV/Å. Thus, using Eqns 5.6 and 5.9, with $\mu = e\tau/m^*$, $\tau = \tau_{\text{ph}}$ and $m^* = \pi\hbar/aV_F$ (as calculated with the tight-binding band structure, where V_F is the Fermi velocity), the expression of the temperature dependence of the conductivity is given by

$$\sigma_P(T) = \left(\frac{e^2}{4\pi\hbar a}\right) na^3 \left(\frac{m\omega_0 t_0^2}{\alpha^2\hbar}\right) \exp\left(\frac{\hbar\omega_0}{k_B T}\right) \tag{5.10}$$

The intrinsic electrical conductivity of metallic polyacetylene can be estimated from Eqn 5.10. At room temperature and parallel to the chain direction, $\sigma_{||}$ (300 K) $\approx 2\times10^6$ S/cm; i.e. significantly greater than that of metallic copper. Moreover, Eqn 5.10 indicates that $\sigma_{||}(T)$ will increase exponentially as T is lowered.

In metallic polyacetylene, the highest values obtained for the electrical conductivity are approximately 10^5 S/cm with no indication of the predicted exponential increase as the

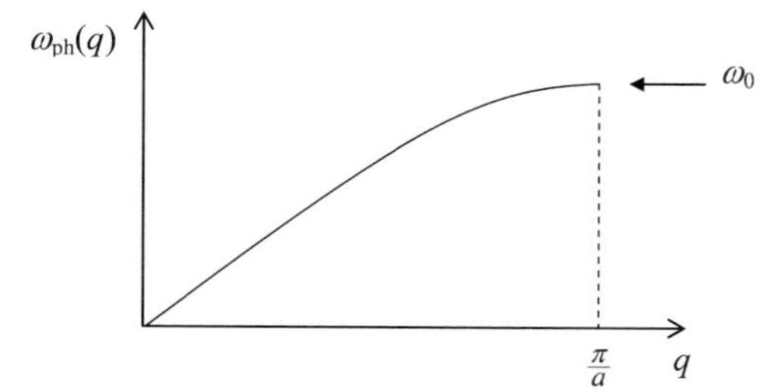

Fig. 5.10 The phonon spectrum of the polyacetylene chain.

temperature is lowered. In other polymers, the highest values obtained are an order of magnitude (or even more) lower. Thus, the conductivity in existing materials is still severely limited by defects and disorder. These limitations are not intrinsic; the materials can be improved through a combination of better chemistry and improved material processing.

Further reading

N. E. Ashcroft and N. D. Mermin, *Solid State Physics*, Phildadelphia, 1976.

G. Grüner, *Density Waves in Solids*, Addison–Wesley, Reading, 1994.

H. Kuzmany, M. Mehring and S. Roth (eds.), *Electronic Properties of Polymers*, Springer Series in Solid State Sciences, Vols. **66, 76, 91,** Springer-Verlag, Berlin, 1985, 1987, 1989.

6

Nonlinear excitations in conjugated polymers: Solitons, polarons and bipolarons

6.1 Polyacetylene: Semiconductor with degenerate ground state

When synthesized at low temperatures, polyacetylene has the *cis*-structure shown in Fig. 6.1(a). The *cis*-polyacetylene structure is unstable and readily converts at elevated temperatures to thermodynamically stable *trans*-polyacetylene with structure shown in Fig. 6.1(b).

Since π-bonds can form between neighboring carbons in two energetically equivalent ways, there is a novel and interesting two-fold degeneracy in the structure of in *trans*-$(CH)_x$ as shown in Fig. 6.2.

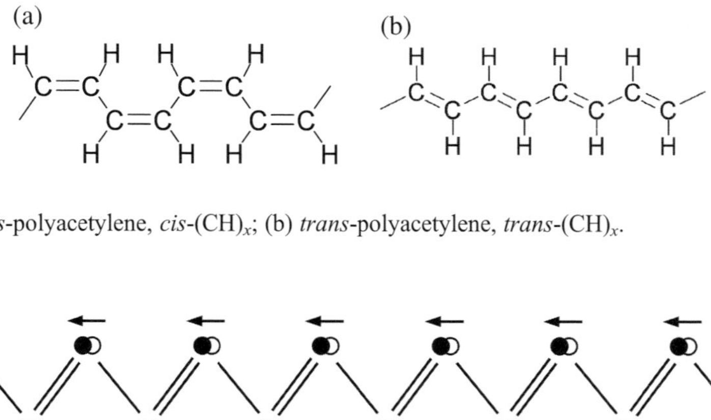

Fig. 6.1 (a) *cis*-polyacetylene, *cis*-$(CH)_x$; (b) *trans*-polyacetylene, *trans*-$(CH)_x$.

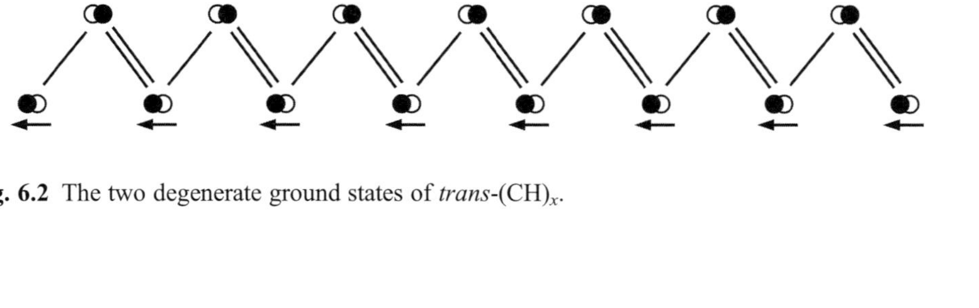

Fig. 6.2 The two degenerate ground states of *trans*-$(CH)_x$.

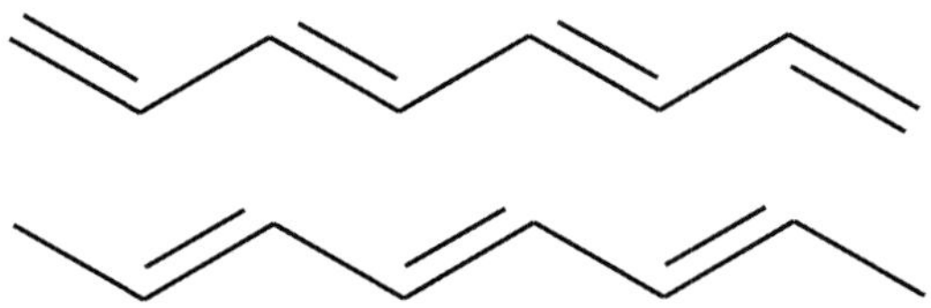

Fig. 6.3 The two degenerate A and B phases of *trans*-polyactylene.

The two structural phases sketched in Fig. 6.2 are equivalent; a sample of the upper form cannot in any way be distinguished from a sample in the lower form. Because these two phases have the same ground state energy, polyacetylene has a "two-fold degenerate ground state". If one starts with the hypothetical structure of identical bond lengths in *trans*-polyacetylene (open circles in Fig. 6.2) and moves the carbon atoms towards the left or right, respectively, by a distortion with magnitude u (as sketched in Fig. 6.2), the total energy will decrease.

As noted in Chapter 1, polyacetylene with uniform bond lengths would be a metal and, as shown by Peierls, such a one-dimensional metal is unstable. The two structures sketched in Fig. 6.2 (and repeated schematically in Fig. 6.3) are the Peierls distorted ground states of semiconducting *trans*-$(CH)_x$.

The energy of a polyacetylene macromolecular chain as a function of the distortion is shown in Fig. 6.4. The double minimum in the ground state energy arises from the two-fold degeneracy sketched in Figs 6.2 and 6.3.

The distortion with minima at $\pm\, u_o$ corresponds to the formation of a π-bond between neighboring carbon atoms. The associated energy of the π-bond, E_π, is of the order of 1–2 eV per pair of carbon atoms (Chapter 1). However, the energy barrier between the two ground state minima in Fig. 6.4 is proportional to the *number* of π-bonds in the chain:

$$E_{barrier} = [E(u = 0) - E(u = \pm u_o)] = (1/2)NE_\pi. \tag{6.1}$$

As a result, fluctuations between the two phases is suppressed; the system is either in the A-phase or the B-phase. The Pauling resonance that is so familiar in the structure of small

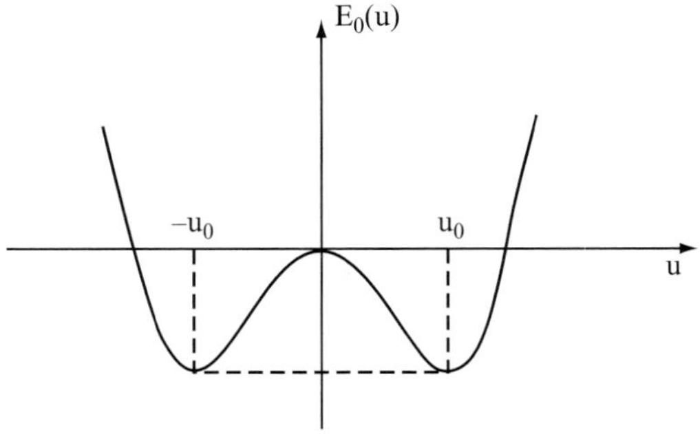

Fig 6.4 The total energy (electronic plus lattice distortion) as a function of the distortion, u. Note the double minimum associated with the spontaneous symmetry breaking and the two-fold degenerate ground state.

aromatic molecules does not occur in macromolecular polyacetylene chains. This suppression of the Pauling resonance is an example of the general concept of broken symmetry. Although the A and B phases are identical, a *trans*-polyacteylene chain is in one or the other of the two phases. In the analogous situation of an antiferromagnet, Anderson[1] showed in 1952 that because the energy barrier is proportional to N, the time required for the system to flip between A and B approaches the age of the universe as N becomes large.

The broken symmetry has been verified for polyacetylene. There is bond alternation, alternating single- and double-bonds, with corresponding changes in the bond lengths; the double bonds are slightly shorter and the single bonds are slightly longer. The exact lengths have been measured by X-ray diffraction and by analysis of nuclear magnetic resonance data; the shorter bond length is 1.35 Å and the longer bond length is 1.45 Å. The proper chemical formula for the polymer is not $(CH)_n$, but rather $(CH{=}CH)_n$; the Peierls distortion has broken the symmetry.

6.2 Solitons in *trans*-polyacetylene

If one constructs a polyacetylene chain in a "Gedanken" experiment where the left end of the polymer chain is in the A-phase and the right end of the polymer chain is in the B-phase, the resulting structure will contain a topological defect – a "kink" – a boundary between A and B phases. The topological kink shown in Fig. 6.5 has the mathematical properties of a soliton; the soliton in polyacetylene is an example of a nonlinear elementary excitation in a system with a two-fold degenerate ground state. The soliton in *trans*-polyacetylene is particularly interesting because the structural kink is coupled to the electronic structure with remarkable consequences.

As shown in Fig. 6.5, the soliton which forms the phase boundary between the A phase and the B phase is a neutral radical. The existence of the unpaired electron at the boundary between the A and B phases can be inferred by simply counting the number of electrons on each carbon atom. Far to the left and far to the right of the kink, each carbon atom has three valence electrons in covalent bonds (to one hydrogen and two carbon atoms) with the fourth valence electron in the π-bond. Right at the site of the kink, however, where the bond alternation reverses, there is no π-bond. As a result, the fourth electron is unpaired and exists in a nonbonding state. Since the π-band is the bonding band and the π*-band is the antibonding

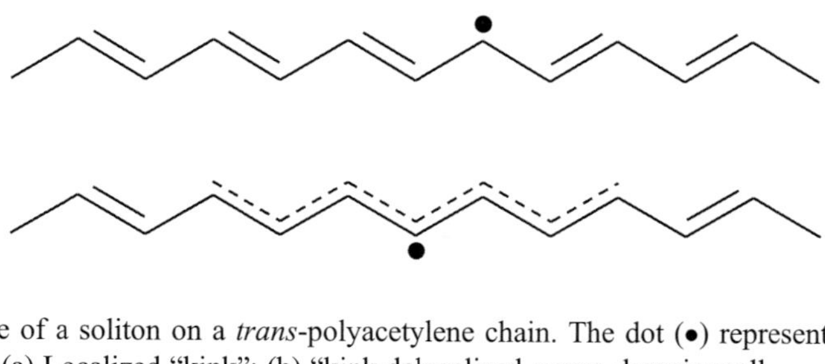

Fig. 6.5 Structure of a soliton on a *trans*-polyacetylene chain. The dot (•) represents an electron in a nonbonded state. (a) Localized "kink"; (b) "kink delocalized over a domain wall.

[1] P.W. Anderson, *Phys. Rev.*, **86**, 694 (1952).

band, the unpaired electron associated with the kink must be in a nonbonding state at the center of the energy gap.

The soliton cannot be localized onto a single carbon atom (upper sketch of Fig. 6.5); it is extended spatially into a domain wall since, as is well-known from the Heisenberg uncertainty principle, localization of this unpaired electron onto a single carbon atom would increase the energy. Therefore, the soliton is delocalized over several carbon atoms. As shown by SSH and verified by ENDOR experiments, the full width of the soliton (both the wavefunction of the localized electron and the distance over which the bond-alternation switches) is approximately 14 carbon atoms.

The magnitude of the distortion ($\o_n$) as a function of the site number (n) from the center of the soliton is

$$\phi_n = -\mathrm{u_o tanh(na/\xi)} \tag{6.2}$$

where a is the repeat length in the uniform polyacetylene chain, and u_o is the value of the distortion that minimizes the energy (see Fig. 6.4). The width of the domain wall is determined by ξ; for the parameters appropriate to *trans*-polyacetylene, $\xi \approx 7$. The energy required to create a soliton–antisoliton pair was calculated in the SSH model:

$$E_S = (2/\pi)E_g \tag{6.3}$$

where E_g is the energy gap of *trans*-polyacetlyene.

The existence of the soliton distortion along the chain changes the potential energy experienced by the electrons in the vicinity of the kink. If one solves the Schrdinger equation in the presence of this modified potential, one finds that a new eigenstate appears with energy at the center of the energy gap.

Thus, the soliton in polyacetylene is a self-localized electronic state. There is a structural distortion (see Fig. 6.5). As a result of this structural distortion, there is a change in the local electronic potential. This change in the local electronic potential leads to the formation of an electronic state in the energy gap, the "midgap" state. The midgap state is a true eigenstate of the system, stable when the state is occupied by zero, one or two electrons.

As shown in the band diagram of Fig. 6.6, this state is singly occupied when the soliton (kink) is neutral, as described above. When singly occupied, the soliton has zero charge, but it has an unpaired spin. Upon oxidation of the neutral soliton, a positive soliton is created with no electrons

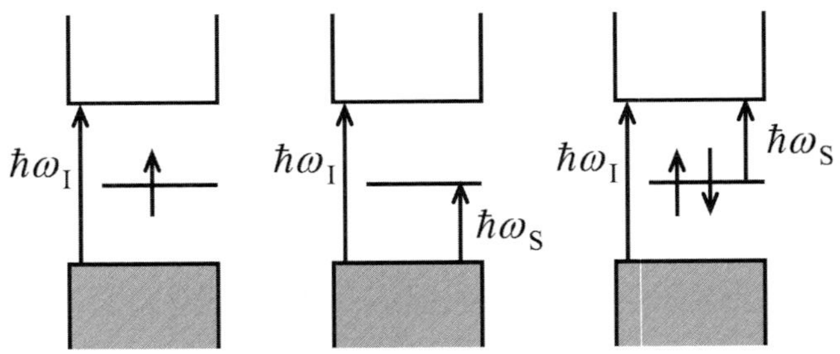

Fig. 6.6 Band diagram showing the midgap state associated with the soliton, and the interband ($\hbar w_I$) and midgap ($\hbar w_s$) transitions: left, neutral soliton; center, positively charged soliton; right, negatively charged soliton.

in the midgap electronic state. The positive soliton has charge $+e$ and zero spin. Symmetrically, upon reduction of the neutral soliton, an electron is added to form a doubly occupied soliton energy level at midgap. This negative soliton has a net charge of $-e$ and zero spin.

The spin-charge relations of the soliton can be summarized as follows:

Charge	Spin
0	½
+1	0
−1	0

Thus, although the electrons which occupy the midgap state are fermions, the soliton has spin-charge relations that are reversed from those of fermions. As discussed in detail in the *Reviews of Modern Physics* article on this subject,[2] the reversed spin-charge relations arise fundamentally from splitting the electron charge: to form the midgap state, half of a state for each sign of spin is taken from the π-band and half of a state for each sign of spin is taken from the π*-band.

Because the transitions involving the midgap state shown in Fig. 6.6 determine the optical absorption of *trans*-$(CH)_x$ at photon energies below the energy gap, one would expect that the quantum efficiency for charged soliton pair photogeneration ($\o_{ch}$) to be unity for $2E_g/\pi < h\nu < E_g$. and to remain near unity well above E_g. Su and Schrieffer[3] described the process of charged soliton pair formation subsequent to photoexcitation via the π-π* interband transition in *trans*-$(CH)_x$ with photon energies greater than or equal to E_g. They showed that within ∼ 100 fs, the lattice around a photoexcited electron–hole pair relaxes to form a well separated $S^+ - S^-$ pair. Thus, $\o_{ch} \approx 1$ for photon energies larger than E_g.

6.3 Nondegenerate ground state polymers

The degenerate ground state of *trans*-polyacetylene is not typical of semiconducting polymers. In poly-*para*-phenylene (PPP), for example, there are also two different phases which are analogous to the A phase and the B phase of *trans*-polyacetylene. In PPP, these phases are denoted as the "benzenoid" and "quinoid" structures (Fig. 6.7). In a linear chain, the benzenoid

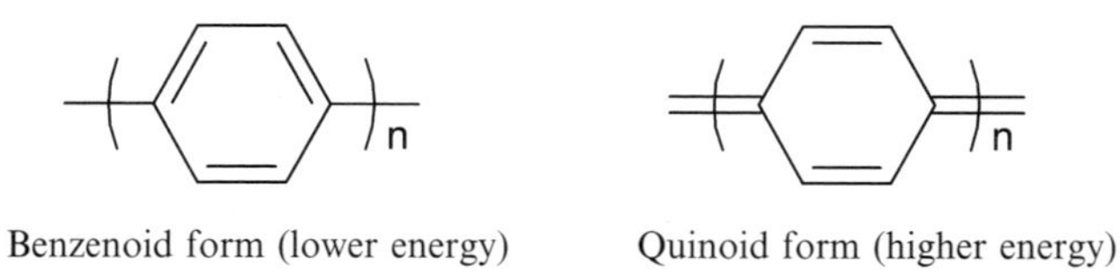

Fig. 6.7 Benzenoid and quinoid structures of poly-*para*-phenylene (PPP).

[2] A. J. Heeger *et al.*, Rev. Mod. Phys. 60, 781 (1988).
[3] W.P Su and J. R. Schrieffer, Proc. Natl. Acad. Sci., **77**, 5626 (1980).

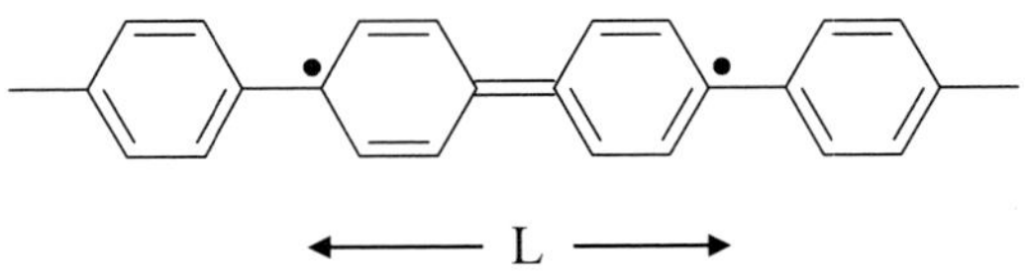

Fig. 6.8 Two neutral kinks separated by a length L in a PPP chain. This structure can be thought of as a "neutral bipolaron".

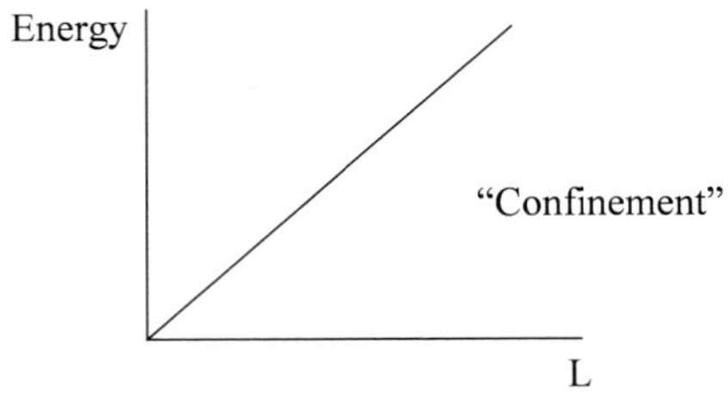

Fig. 6.9 In semiconducting polymers with nondegenerate ground state, the linear increase in the energy of two solitons in a polymer chain with nondegenerate ground state; the linear increase with L leads to confinement of the soliton–antsoliton pair.

structure with fully aromatic rings has a lower ground state energy than the double bonded quinoid ring. As a result of the inequivalence of the two different ground state energy configurations, PPP is an example of a semiconducting polymer with a "nondegenerate ground state". In fact, all conjugated polymers with aromatic groups in the repeat unit belong to the class of polymers in which the ground state degeneracy has been lifted.

Consider the insertion of a pair of neutral solitons (a soliton and an antisoliton) into such a nondegenerate ground state polymer, with benzenoid phase on the left, quinoid phase in the center and benzenoid phase on the right, as sketched in Fig. 6.8.

As the separation, L, between the two neutral solitons increases, the higher energy quinoid structure between the two kinks will cause the total energy to increase linearly with L as shown in Fig. 6.9. Consequently, it will be impossible to separate the two kinks; the soliton and the antisoliton are "confined".

For large L, one can describe the two separated objects as two neutral solitons with two midgap states as sketched in Fig. 6.10. As shown in Fig. 6.8, however, the energy of the separated pair will grow linearly with L.

Because of the confinement energy, the two separated solitons are attracted to one another. As they approach one another such that $L \sim \xi$, the two electronic wavefunctions begin to overlap, and the states evolve into a bonding/antibonding pair. The resulting energy diagram (Fig. 6.11) has a pair of states split symmetrically with respect to the center of the gap.

The "neutral bipolaron" (Fig. 6.11a) is unstable because the doubly occupied lower energy state approaches the top of the π-band as L approaches zero. The confinement energy (Fig. 6.10), causes the neutral bipolaron to disappear into the undisturbed poly(benzenoid) structure and the gap states to vanish.

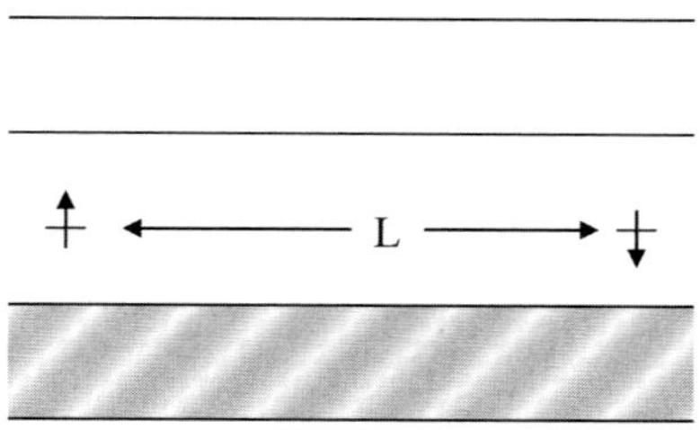

Fig. 6.10 Two separate midgap states associated with a confined soliton pair separated by $L >> \xi$ where ξ is the width of each of the kinks.

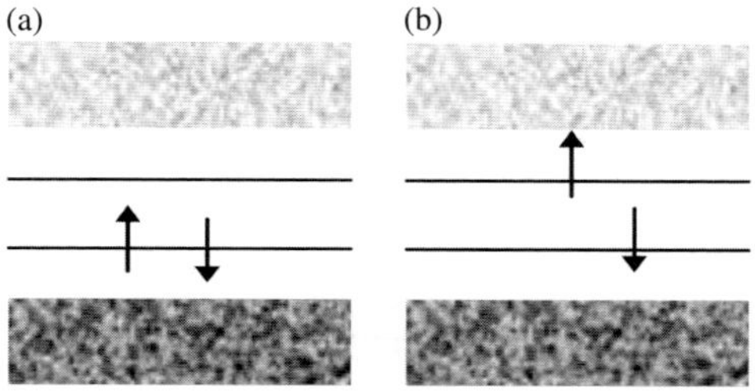

Fig. 6.11 Bonding/antbonding pair of gap states associated with confined soliton pairs separated by $L >> \xi$ where ξ is the width of each of the kinks. (a) Lower gap state is doubly occupied; (b) both gap states occupied with electrons of opposite spin.

If, however, the two states were occupied as in Fig. 6.11(b) with one electron in each of the two gap states (one in the upper state with up-spin and one in the lower state with down-spin), the resulting structure would be a stable excited state of the conjugated polymer; analogous to a neutral exciton, except that the electron and the hole are bound by the overlap of their lattice distortions rather than by the Coulomb attraction.

6.4 Polarons and bipolarons in conjugated polymers with nondegenerate ground state

Consider two solitons on a nondegenerate ground state polymer, where one of the solitons is charged and the other is neutral. Because of the confinement energy, the two will form a stable "polaron". The polaron is a confined soliton pair with charge ± 1 and with spin ½. The electronic structures and gap state occupancies of the positive and negative polaron are shown in Fig. 6.12.

The spin-charge relations of the polarons are summarized as follows:

Charge	*Spin*
0	0
+1	½
−1	½

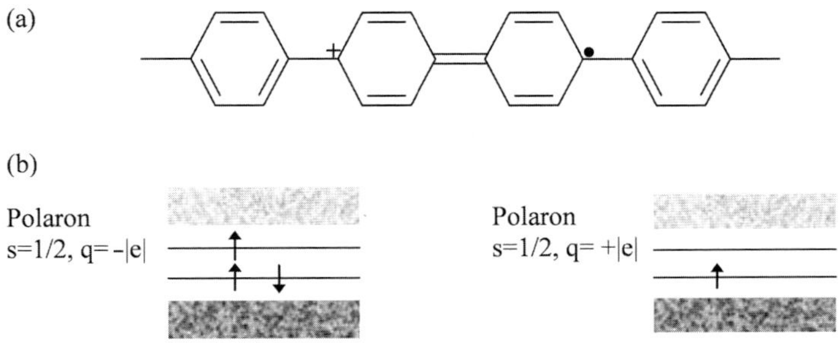

Fig. 6.12 (a) The structure of a positively charged polaron shown as a pair of confined positive and neutral solitons on poly(*para*-phenylene). A negative polaron would be identical but shown as a confined pair of negative and neutral solitons. (b) The electronic structures and gap state occupancies of the positive and negative polaron.

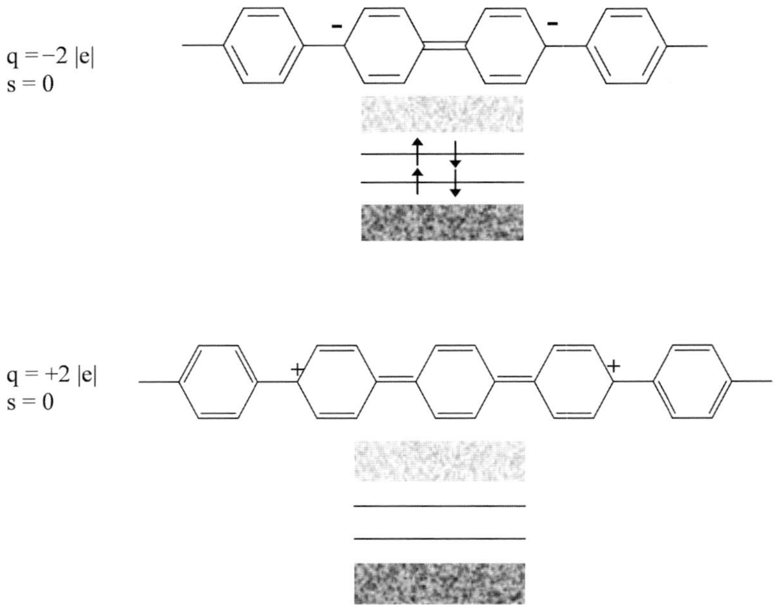

Fig. 6.13 Bipolarons with charge ± 2 and with spin = 0.

A confined soliton pair with both kinks having the same sign of charge is called a "bipolaron"; a stable excitation with charge ±2 and with spin 0, see Fig. 6.13. Doubly charged bipolarons can be formed in semiconducting polymers with a nondegenerate ground state by bringing two polarons together. In the bipolaron, the lattice distortions associated with the two polarons overlap to form a region in which the changes in bond lengths are shared by the two polarons. Thus, in analogy with two bowling balls on a mattress, the energy is lowered by sharing the distortions. This overlapping of the distortions can provide enough energy to stabilize the bipolaron. Such boson-like charge carriers are quite interesting since, in principle, Bose–Einstein condensation would result in a superconducting ground state (analogous to real space Cooper pairs).

Whether or not the fusing of two polarons to form a bipolaron is energetically favorable is not a priori clear. Because of the competition between the Coulomb energy (repulsive) and lattice energy (attractive), semiconducting polymers with different structural repeat units could have either the bipolaron or two separated polarons as lowest energy excitations.

The total charge is $|Q| = |e|(N_p + 2N_b)$ where N_p and N_b are the number of polarons and the number of bipolarons, respectively. Thus, the number of particles is doubled for polarons with respect to bipolarons. Therefore, the total entropy of the system is significantly higher if at least some of the charge is stored in polarons. If the energy difference between a bipolaron (E_B) and two polarons ($2E_P$) is small, then the polaron and bipolaron densities will be determined by the difference in free energy; i.e. by the entropy.

This balance and the role of the entropy are summarized in Fig. 6.14. Because polarons have spin = ½ and bipolarons have spin = 0, the magnetic susceptibility is determined by the density of polarons. The best fit to the magnetic susceptibility data yields $(2E_P - E_B) \approx 0.16$ eV. Although bipolarons have lower energy, polarons are generated at low doping concentrations because of the higher entropy associated with the increased number of degrees of freedom.

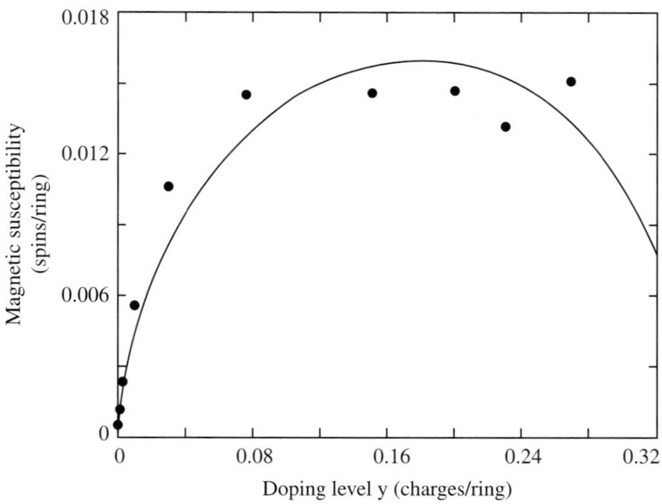

Fig. 6.14 Magnetic susceptibility versus dopant level at a fixed concentration $c = c_0$ of semiconducting polymer in solution. The magnetic susceptibility is proportional to the number of polarons. The solid line is the best fit given by the model. [Reproduced from M. J. Nowak *et al.*, *Macromolecule*, **22**, 2917 (1989) with permission from the American Physical Society.]

Further reading

J. M. Andre, J. Delhalle and J. L. Bredas, *Quantum Chemistry Aided Design of Organic Polymers*, World Scientific, Singapore, 1991.
A. J. Heeger, S. Kivelson, J. R. Schrieffer and W. P. Su, *Rev. Mod. Phys.*, **60**, 781 (1988).
M. J. Nowak *et al.*, *Macromolecule*, **22**, 2917 (1989).
T. A. Skotheim (ed.), *Handbook of Conducting Polymers*, Marcel Dekker, New York, 1986.

7

Solitons, polarons and bipolarons: Experiments

7.1 Doping induced optical absorption in polyacetylene

The existence of midgap states, as shown in Figs 6.6 and 6.10, implies that new absorption bands should be generated upon doping or upon photoexcitation. Since the total oscillator strength of the π-π* interband transition must be conserved, the oscillator strength in the midgap transitions will be correspondingly removed from the interband transition. The changes in absorption can be monitored *in situ* during electrochemical doping; the apparatus is sketched in Fig. 7.1. After making a step-change increase in the applied voltage, sufficient time must be allowed for the system to come to electrochemical equilibrium at a new doping level before the next spectrum is taken.

As shown in Fig. 7.2, the doping-induced absorption observed in *trans*-polyacetylene is consistent with the formation of a nonbonding state near the center of the gap, as sketched in the band diagram shown on the right. Initially, the increase in the strength of the midgap absorption is associated with a decrease in the π-π* absorption. At high doping levels, > 6%, the original π-π* interband absorption characteristic of the semiconducting state is no longer observed. At high doping levels, one sees only the broad absorption of a disordered metal, as described in detail in Chapter 4.

7.2 ESR measurements of the spin susceptibility of doped polyacetylene

The reversed spin-charge relationship of solitons in *trans*-polyacetylene is the principal distinguishing feature that is capable of verifying the proposal of charge storage in solitons. Because of the importance of the spin-to-charge ratio, doping-induced and photoinduced electron spin resonance (ESR) experiments played an important role in sorting out the physics of the nonlinear excitations in semiconducting polymers.

In situ electron spin resonance (ESR) measurements during electrochemical doping of *trans*-polyacetylene prove that the doping proceeds initially via spinless charges, i.e. consistent with doping via charged solitons. The apparatus used for *in situ* ESR measurements during electrochemical doping is sketched in Fig. 7.3. By placing only that portion of the long ESR tube that contains the polymer on the ITO glass electrode into the microwave cavity, it is possible to maintain a sufficiently high Q to enable relatively high sensitivity.

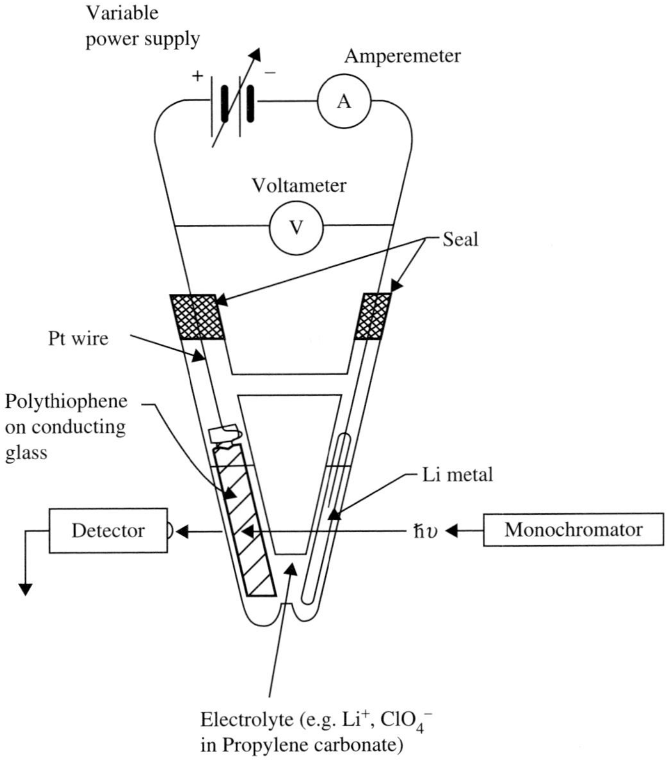

Fig. 7.1 Diagram of apparatus used for *in situ* visible-IR-absorption measurements during electrochemical doping.

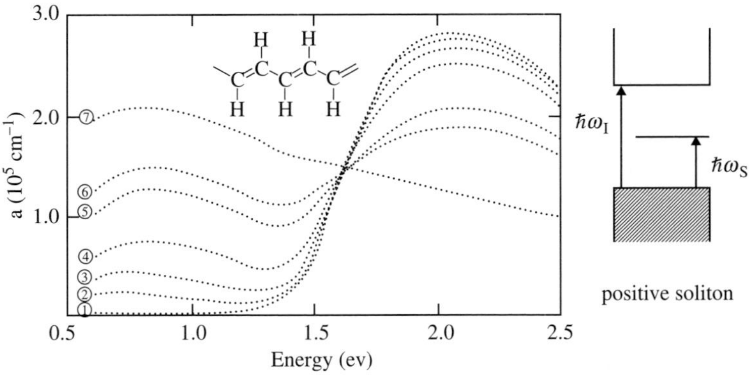

Fig. 7.2 *In situ* absorption data for *trans*-$(CH)_x$ during electrochemical doping with $(ClO_4)^-$. The applied cell voltages (relative to Li) and corresponding concentrations are as follows: curve 1, 2.2 V (y=0); curve 2, 3.28 V ($y \sim 0.003$); curve 3, 3.37 V ($y \sim 0.0065$); curve 4, 3.46 V ($y \sim 0.012$); curve 5, 3.57 V ($y \sim 0.027$); curve 6, 3.64 V ($y \sim 0.047$); curve 7, 3.73 V ($y \sim 0.078$). [Reproduced from A. Feldblum *et al.*, *Phys. Rev. B*, **26**, 815 (1982) with permission from American Physical Society.]

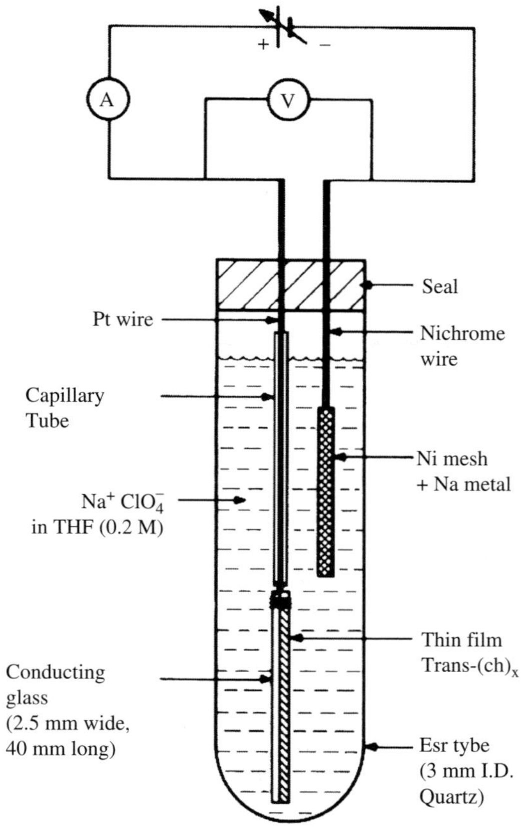

Fig. 7.3 Sketch of apparatus used for the *in situ* measurement of electron spin resonance during electrochemical doping.

The data obtained from *trans*-polyacteylene during n-type doping (left) and during p-type doping (right) are shown in Fig. 7.4. Each successively higher value of the applied voltage corresponds to a higher doping concentration. For doping concentrations less than the critical concentration, y_c, where y_c, $\approx$ 6% per carbon atom, the susceptibility remains small. For example, at $y = 0.8\%$, the number of spins per charge, $N_S/N_{\rm ch}$, was found to be $N_S/N_{\rm ch} = 3 \times 10^{-4}$. For $y = 4.0\%$, $N_S/N_{\rm ch} = 1.5 \times 10^{-4}$ spins/charge. Thus, for $y < y_c$, the charge carriers in *trans*-polyacetylene generated by doping are spinless. These data established the reversed spin-charge relation predicted for solitons (see Chapter 6) for both for n-type doping and p-type doping. The data shown in Fig. 7.4 are fully consistent with the expected charge conjugation symmetry implied by the basic electronic structure of polyacetylene.

There is clear evidence of a transition in the doping induced spin susceptibility for doping concentrations $y > y_c$, $\approx$ 6% per carbon atom. In this regime, the spin susceptibility is finite and relatively insensitive to the dopant level. Measurements of the temperature dependence in the high concentration regime indicated that the spin susceptibility is temperature independent

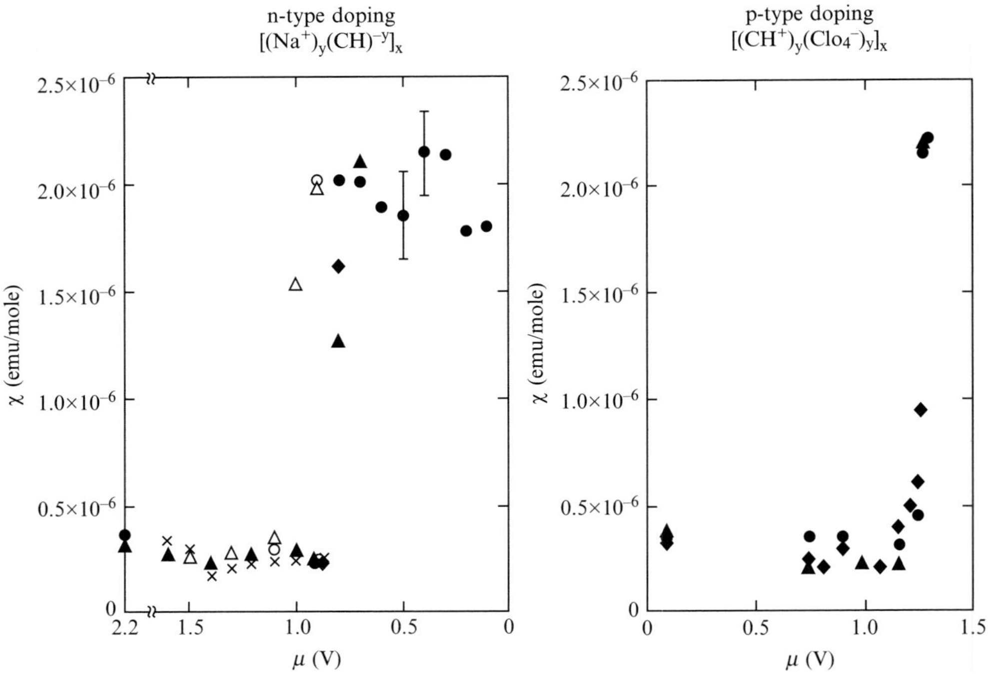

Fig. 7.4 The data obtained from polyacteylene during n-type doping (left) and during p-type doping (right). The strength of the ESR signal is plotted as a function of the voltage applied between the polyacetylene film and the counter-electrode. [Reproduced from J. Chen *et al.*, *Synthetic Metals*, **24**, 311 (1988) with permission from Elsevier Science B.V.]

as expected for the Pauli spin susceptibility of a metal, $\chi_{Pauli} = N\frac{\mu_B^2}{k_B T_F}$, where $T_F =$ Fermi temperature, and $k_B T_F = E_F$. Thus, for $y > y_c$, the spin susceptibility data imply that doped polyacetylene becomes a metal.

7.3 Infrared active vibrational (IRAV) modes

The generation of solitons either by doping or by photoexcitation is accompanied by a new series of characteristic infrared active vibrational (IRAV) mode absorption bands. These IRAV bands are different from the "normal" IR absorption bands of the polymer backbone and are observed at wavelengths that correspond to the totally symmetric Ag modes observed in Raman spectroscopy. For example the "breathing" mode of phenylene is fully symmetric and therefore not IR active (see Fig. 7.5).

Solitons, polarons and bipolarons involve local structural distortions within the polymer chain; i.e. local changes in bond lengths. This distortion couples to the self-localized charge and thereby has a strong effect on the vibrational spectrum in the IR. The electronic charges

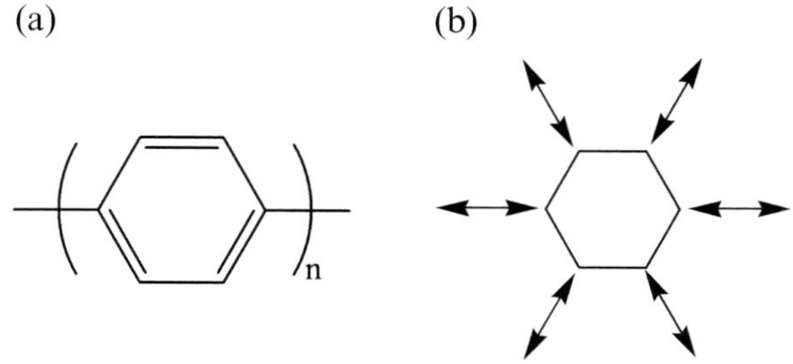

Fig. 7.5 (a) Structure poly(*para*-phenylene); and (b) the symmetric breathing mode.

stored in solitons, polarons or biplarons break the local symmetry of the conjugated backbone, and thereby cause the Raman modes to become IR active.

The high oscillator strength of the IRAV modes results from indirect excitation of the amplitude (Raman) modes via motion of the self-localized charge. The charge stored in the soliton (or polaron) moves in response to the electric field associated with the incident IR beam. When the amplitude mode frequency is in resonance with the incident IR frequency, the moving charge will drive the amplitude mode. Consequently, the amplitude mode becomes IR active with an oscillator strength characteristic of an electronic transition. As a result, the IRAV modes are unusually strong, and they are polarized along the chain direction, for that is the only direction in which the electronic charges can move. Thus, a characteristic feature of doped semiconducting polymers is the appearance of IRAV modes with remarkably large oscillator strengths at IR wavelengths that correspond to totally symmetric amplitude modes that are normally only Raman active.

The IRAV modes are spectral "fingerprints" of solitons (and polarons) and can be generated either by doping or by photoexcitation via the interband transition (across the energy gap). The observation of IRAV modes indicates the following:

(1) the existence of a local structural distortion;
(2) the self-localization of the charges;
(3) the strong coupling of the self-localized charges to the local chain distortion.

Electron–hole symmetry implies that the same spectrum should be observed for positive or negative charges on the polymer chains, in agreement with experiments. The intensity of IRAV bands is proportional to the density of doping-induced or photoinduced charges on the chain.

The IRAV modes observed in lightly doped polyacetylene, both *trans*-$(CH)_x$ and *trans*-$(CD)_x$ are shown in Fig. 7.6. As expected, the frequencies of the IRAV modes are isotope sensitive with corresponding shifts on going from $(CH)_x$ to $(CD)_x$.

The IRAV modes result from charge oscillations along the polymer chain and are therefore denoted as "phase modes". Note, however, that the motion of the electronic charges along the chain is inhibited by disorder. The effect of disorder can be taken into account by defining a "pinning" potential which resists the translational motion of the charges. The IRAV mode with lowest frequency is called the "pinned mode", and its frequency can be related directly to the

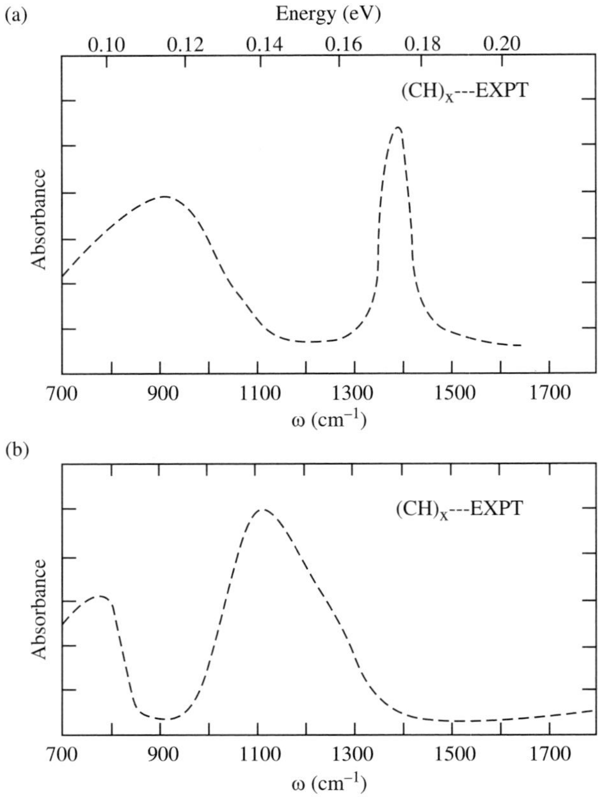

Fig. 7.6 Doping induced IRAV modes for *trans*-polyacetylene (a) *trans*-$(CH)_x$; (b) *trans*-$(CD)_x$. [Reproduced from S. Etemad *et al.*, *Phys. Rev. B*, **23**, 5137 (1981) with permission from the American Physical Society.]

strength of the pinning potential. In particular, in a perfect translationally invariant polymer chain with zero pinning potential, the pinned mode frequency would shift down to zero.

Horovitz[1] and colleagues formulated a theoretical model for the IRAV phase modes. Assuming that the electrons follow the ion displacements adiabatically, they showed that the pinning potential renormalizes the "bare" frequencies, ω_{n0} of the chain, according to

$$D_0(\omega) = \sum_n c_n \frac{\omega_{n0}^2}{\omega^2 - \omega_{n0}^2} = -\frac{1}{1-\alpha} \tag{7.1}$$

where $\alpha(< 1)$ is the "pinning" parameter proportional to the second derivative of the pinning potential, and the c_n are related to the mode geometry ($\Sigma c_n = 1$). The dynamic conductivity, which determines the strength of the IRAV optical absorption, is given by

[1] B. Horovitz, *Solid State Commun.*, **41**, 729 (1982).

$$\sigma(\omega) = i\omega \frac{e^2 \rho_c}{M_c \Omega_0^2} \frac{-D_0(\omega)}{1 + (1-\alpha)D_0(\omega)} \tag{7.2}$$

where ρ_c is the average defect charge density, M_c is the kinetic mass of the charged defect, and

$$\Omega_0^{-2} = \sum_n c_n \omega_{n0}^{-2} \tag{7.3}$$

These formulas are the key to understanding many of the properties of the IRAV modes.

(a) There is a one-to-one correspondence with the Raman modes, but the frequencies are shifted because of the response to the self-localized charges. The IR frequencies are given by the solutions to $D_0(\omega) = -1/(1-\alpha)$. Hence, the number of the IRAV modes is equal to the number of bare modes, and the frequency shifts away from the bare modes are determined by α. In the limit of very strong pinning ($\alpha \to \infty$), the IRAV modes are precisely at the frequencies of the bare Raman modes.
(b) The large oscillator strength of the IRAV results from the very small kinetic mass of the soliton or polaron; the mass is of the order of the electron mass, rather than the atomic mass.
(c) The doping-induced IRAV modes are strongly pinned by the oppositely charged ions inserted between polymer chains (see Chapter 2). The corresponding photogenerated IRAV modes are observed at lower frequencies.
(d) All IRAV frequencies obey the "product rule" relation $\Pi(\omega_n/\omega_{n0})^2 = \alpha$, which can be derived directly from the above equations.
(e) The relative intensities of the various modes are correctly given by

$$I_n \propto \left(\frac{\partial D_0}{\partial \omega_n}\right)^{-1} \tag{7.4}$$

Analysis of the IRAV modes showed that the experimentally observed IRAV frequencies and relative intensities could be quantitatively understood and that quantitative estimates can be obtained for the pinning parameter for either doping induced or photogenerated solitons (see Section 7.4 for details).

7.4 Photoinduced absorption in *trans*-polyacetylene

Charged solitons can be created by photoexcitation; i.e. by absorption of photons and the initial creation of electron–hole pairs via the π-π^* interband transition. These electrons and holes will relax to the band edges and subsequently form a charged soliton pair (one positively charged and the other negatively charged), as shown schematically in Fig. 7.7.

As described in Chapter 6, photoexcitation of an electron–hole pair will result in a soliton–antisoliton pair with states near midgap within a time period of the order of the period of the vibrational mode; i.e. on the order of 10^{-13} s. The oscillator strength shifts accordingly.

The photoinduced absorption spectrum of *trans*-polyacetylene $(CH)_x$ and $(CD)_x$ are shown in Fig. 7.8 (the bleaching of the π-π^* interband transition at higher photon energies is not shown).

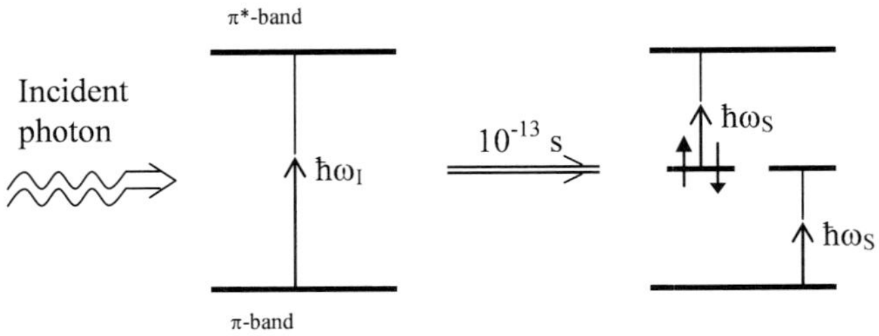

Fig. 7.7 Photogeneration of charged soliton ($S^+ - S^-$) pairs.

All the expected features are observed:

- IRAV modes at photon energies less than 0.15 eV;
- the "midgap" transition;
- sensitivity to the atomic mass as indicated by the different IRAV modes.

For photoexcitation, the lowest frequency mode, the "pinned mode", is particularly interesting because in the absence of counter-ions, the pinning potential is significantly weaker that in the doped polymer. As a result, the frequency of the lowest frequency mode is expected to shift toward lower frequencies (longer wavelengths). This softening is observed in Fig. 7.8. For doped polyacetylene, the lowest frequency IRAV mode is at approximately 900 cm^{-1} ($\alpha = 0.23$) whereas following photoexcitation, the lowest frequency mode is at approximately 450 cm^{-1}($\alpha = 0.06$), lower in frequency by a factor of 2. For doping-induced charged solitons, the pinning arises from a combination of the Coulomb attraction of the counter-ion and the structural disorder. for photoinduced solitons, the weaker pinning arises only from structural disorder (absence of translational symmetry).

The photoinduced "midgap" transition in the spectrum obtained from photoinduced charged solitons (Fig. 7.8) is not in fact at *midgap*. The data show that $\hbar\omega_s \approx 0.45$ eV $< E_g/2$. This shift originates from the Coulomb energy associated with the doubly occupied negative soliton (by symmetry, the same argument holds for the positive soliton). One can estimate the shift from the Coulomb energy by realizing that, as shown in Fig. 7.9, the midgap transition converts a charged soliton to a neutral soliton.

Defining U_S as equal to the Coulomb interaction between two electrons (or two holes) in the midgap energy level of the soliton,

$$\hbar\omega_S^{\text{Photo}} = \frac{1}{2}E_g - U_S. \tag{7.5}$$

For the doping-induced solitons there is, however, an additional contribution to the change in energy associated with the midgap transition; i.e. the binding energy of the charged soliton to the dopant counter-ion, U_{Ci}. These two Coulomb energies nearly cancel one other

$$\hbar\omega_S^{\text{doping}} = E_g/2 - U_S + U_{\text{Ci}} \tag{7.6}$$

Thus, one should expect to observe that:

- For doping-induced solitons, the "midgap" transition will occur at approximately midgap because the two Coulomb contributions approximately cancel.

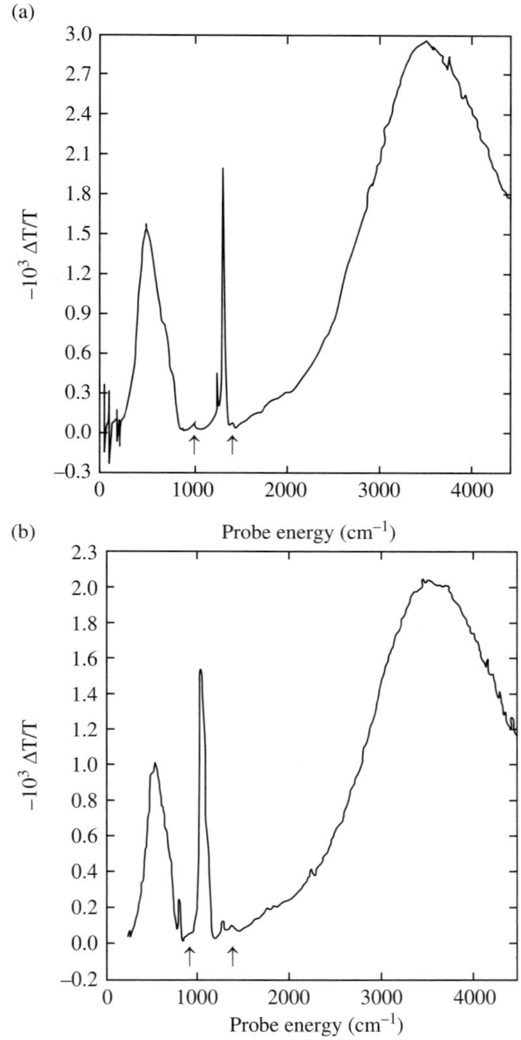

Fig. 7.8 Comparison of the infrared photoinduced absorption spectra for *trans*-$(CH)_x$ and *trans*-$(CD)_x$. (a) Photoinduced absorption of *trans*-$(CH)_x$ between 40 and 4500 cm^{-1}. (b) Photoinduced absorption of *trans*-$(CD)_x$ between 200 and 4500 cm^{-1}. [Reproduced from H.E. Schaffer *et al.*, *Phys. Rev. B*, **36**, 7537 (1987) with permission from the American Physical Society.]

- For photoinduced solitons, the "midgap" transition will be shifted to lower energy by an amount approximately equal to U_s.

Comparison of the photoinduced absorption and the doping-induced absorption provides an estimate of $U_s \approx 0.3$ eV. Note that since $U_s << E_g$, one can include the Coulomb interaction as a perturbation.

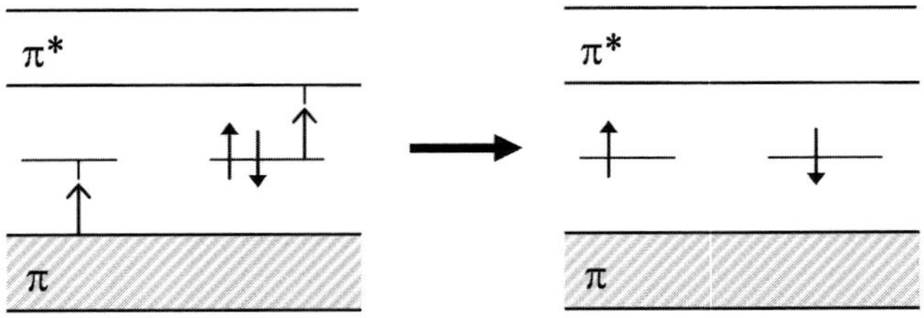

Fig. 7.9 Absorption by the midgap transitions convert charged solitons to neutral solitons.

7.5 Photoinduced ESR in *trans*-polyacetylene: Photogeneration of spinless solitons

Light induced ESR (LESR) experiments demonstrated that the photogenerated positive and negative charged excitations are spinless, in agreement with the reversed spin-charge relations of solitons described in Chapter 6. The methodology is to determine the number of photogenerated spins ($N_s{}^{ph}$) from LESR measurements during photoexcitation and to determine the number of photogenerated charges ($N_{ch}{}^{ph}$) from the strength of the photogenerated IRAV modes and the strength of the photogenerated "midgap" absorption. These photoinduced infrared features are calibrated by direct comparison with the corresponding doping-induced spectral features where the dopant concentration is known (see, for example, Fig. 7.2). The results set an upper limit on the ratio $N_s^{ph}/N_{ch}^{ph} < 3 \times 10^{-3}$, indicating that in *trans*-$(CH)_x$ the photogenerated charge carriers are spinless.[2]

There are two relevant time scales for the photoexcitation and decay processes in *trans*-$(CH)_x$: the subpicosecond time regime during photogeneration of intrachain charged photoexcitations (see the following section for a discussion of the ultrafast time regime) and the long time behavior ($t > 10^{-9}$ s) during which the residual excitations are trapped and slowly decay. At times longer than 10^{-9}, only a few per cent of the initial photoexcitations remain. It is these trapped long-lived excitations that are relevant to the LESR experiments and steady-state photoinduced absorption measurements.

In addition to intrachain excitation, photoexcitation of electron–hole pairs on neighboring chains will occur with finite probability because of the finite interchain transfer interaction. Such *interchain* pairs will form polarons or weakly bound excitons. Although the magnetic properties associated with interchain polarons are not observed,[3] interchain photogenerated polarons might play a transient role in the formation of the charged solitons that are observed at long times. As these polarons diffuse along and hop between the *trans*-$(CH)_x$ chains, they can form solitons in two ways: (a) two polarons with the same charge on a single chain will lead to a pair of charged solitons; (b) a polaron on a chain with a pre-existing defect (neutral soliton) will convert the neutran soliton into a charged soliton. The first process appears to be dominant; it involves no net change in the number of spins and is therefore consistent with

[2] J. D. Flood *et al.*, *Phys. Rev. B*, **28,** 2556 (1983).
[3] F. Moreas *et al.*, *Synthetic Metals*, **13**, 113 (1986).

the LESR data. Note that a soliton–antisoliton pair with like charge is metastable; the pair can only be annihilated by an oppositely charged pair on the same chain. This might account for the long lifetime of the residual charged solitons observed during photoinduced absorption.

7.6 Time resolved photoinduced absorption in *trans*-$(CH)_x$; solitons in < 250 fs

Su and Schrieffer[4] described the process of charged soliton pair formation subsequent to photoexcitation in *trans*-$(CH)_x$. As shown in Chapter 6, the minimum energy required for creation of a soliton pair is $2E_g/\pi$ within the SSH model (see Eqn 6.3). The Su–Schrieffer calculations predicted that within $\sim$ 100 fs, the lattice around a photoexcited electron–hole pair would relax to form a separated $S^+ - S^-$ pair. The process is shown in Fig. 7.7.

Ultrafast photoinduced absorption by IRAV modes, an all-optical technique with subpicosecond time resolution carried out in zero applied electric field, was used to investigate soliton photogeneration in oriented *trans*-polyacetylene.[5] Ultrafast photoinduced IRAV absorption probes the photogeneration of charged solitons. The ultrafast data demonstrated that charged soliton pairs are photogenerated within $\sim$ 250 fs.

The excitation spectrum for ultrafast charged soliton photogeneration is shown in Fig. 7.10, together with the absorption spectrum of *trans*-polyacetylene. The results are plotted as twice the soliton pair photogeneration quantum efficiency ($\o_{ch}$) times the cross-section for the IRAV mode at 1370 cm^{-1} ($2\sigma\o_{ch}$).

Assuming $E_g = 1.4$ eV (consistent with the data in Figs 7.10 and 7.2), the SSH threshold is $E_S = 2E_g/\pi = 0.9$ eV, in excellent agreement with the observed onset ($\sim$ 0.9–1.0 eV); remarkable considering the simplicity of the model.

The absolute value of $\o_{ch}$ was estimated from the data in Fig. 7.10 using the cross-section for the IRAV mode at 1370 cm^{-1} ($\sigma \approx 8.3\times10^{-17}$ cm^2) estimated from doping-induced IRAV absorption (α) data using $\alpha = ½ N\sigma$, where N is the doping density and the factor ½ arises because the measurement was done under unpolarized light whereas the IRAV absorption is polarized along the polymer chain. Considering the measurement uncertainty, $\o_{ch}$ ranges from 50 to 100% for photon energies above 1.0 eV, implying that the photogeneration of charged soliton pairs is the dominant process.

7.7 Doping induced absorption in conjugated polymers with nondegenerate ground state

When doping a semiconducting polymer with nondegenerate ground state such as polythiophene, two distinct absorption bands are observed at photon energies within the energy gap. These doping induced absorption features are characteristic of polarons and bipolarons (Chapter 6).

[4] W.-P. Su *et al.*, *Proc. Natl. Acad. Sci. USA*, **77**, 5626 (1980).
[5] P.B. Miranda *et al.*, *Phys. Rev. B*, **66**, 125202 (2002); P. B. Miranda *et al.*, *Phys. Rev. B*, **64**, 081201 (2001).

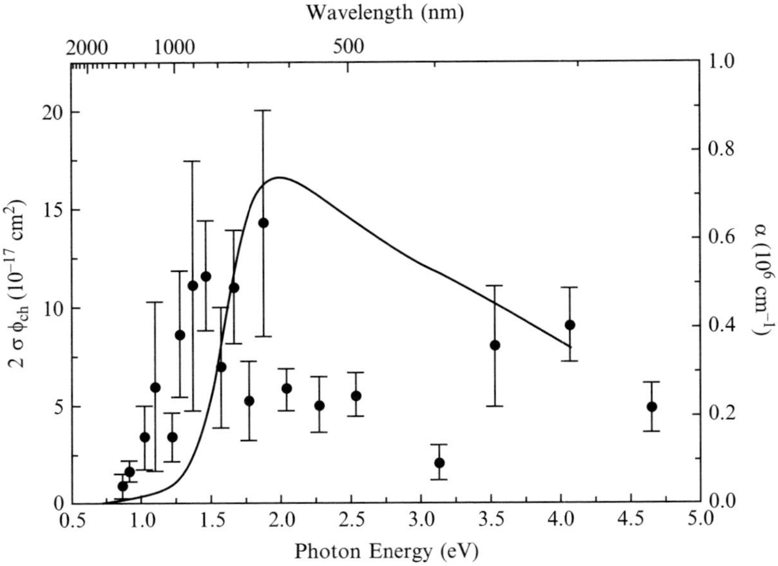

Fig. 7.10 Excitation spectrum for ultrafast charged soliton photogeneration in oriented *trans*-polyacetylene (filled circles). The IR probe was kept at 1360 cm^{-1}. The solid line is the absorption spectrum. [Reproduced from P.B. Miranda *et al.*, *Phys. Rev. B*, **66**, 125202 (2002) with permission from the American Physical Society.]

The changes in absorption can be monitored *in situ* during electrochemical doping using the apparatus sketched in Fig. 7.1. As in polyacetylene, the initial increases in the strengths of the gap-state absorption bands are associated with a corresponding decrease in the π-π^* absorption, as expected from the required conservation of the total oscillator strength. Again, at high doping levels, one sees only the broad absorption of a disordered metal.

Detailed difference spectra, shown in Fig. 7.11 on the right, exhibit two subgap absorption bands. The data indicate a broad joint density of states; the absorption transition involves either the π- or π^*-band as the initial or the final state and a localized gap state as the corresponding final or initial state. The narrow absorption band associated with the transition between the two gap states of the polaron (Chapter 6, Fig. 6.12) is not observed and is assumed here to be a forbidden transition.

The data in Fig. 7.11 are consistent with either polarons (Chapter 6, Fig. 6.12) or bipolarons (Chapter 6, Fig. 6.13). To distinguish between the two, one must have information on the spin; polarons have spin = ½ and bipolarons have spin = 0. In general, we will use the term "polaron" to designate either the polaron or bipolaron except when there is specific information on the corresponding magnetic susceptibility. Thus, in nondegenerate ground state polymers such as polythiophene, charge is stored in polarons upon electrochemical or chemical doping.

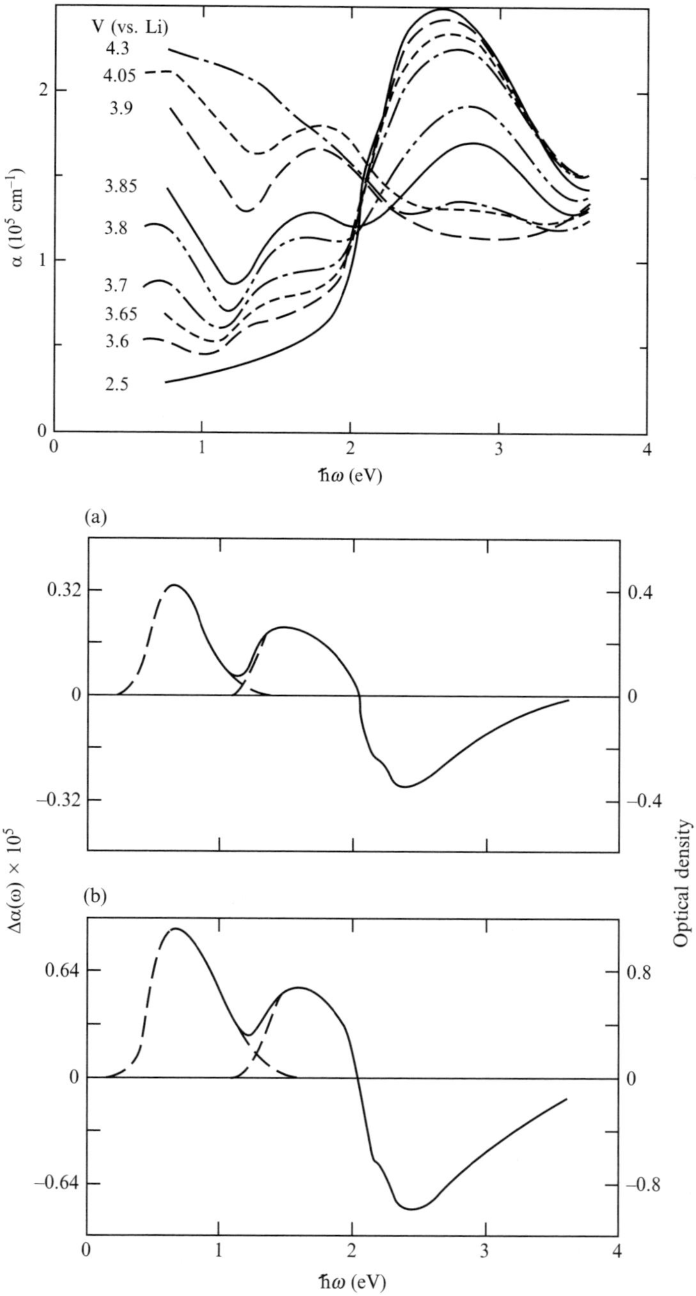

Fig. 7.11 Top: *in situ* absorption curves for polythiophene during electrochemical doping with (ClO4)–. the applied voltages (versus Li) are shown on the left. The corresponding concentrations are as follows (in mol% per thiophene ring): 3.60 V ($y = 0.028$), 3.65 V ($y = 0.04$), 3.70 V ($y = 0.054$), 3.80 V ($y = 0.096$), 3.85 V ($y = 0.12$), 3.90 V ($y = 0.14$), 4.05 V ($y = 0.20$). Bottom (a) and (b): difference spectra obtained with (a) $y = 0.04$; (b) $y = 0.12$. [Reproduced from T.C. Chung *et al.*, *Phys. Rev. B*, **30**, 702 (1984) with permission from American Physical Society.]

7.8 ESR measurements of the spin susceptibility in conjugated polymers with nondegenerate ground state

ESR measurements of the magnetic (spin) susceptibility, χ, of polythiophene as a function of dopant (ClO_4^-) concentration (Fig. 7.12) indicate that the doping proceeds without an increase of paramagnetic susceptibility. At molar concentration of 14%, $N_s/N_{ch}<3 \times 10^{-3}$ where N_s is number of localized spins as obtained from the Curie contribution to χ and N_{ch} is the number of charges (as obtained from dopant concentration). No indication of spins with a Curie law susceptibility was observed even at dilute concentration. The spinless character of charge storage configuration indicates that in polythiophene doping occurs via bipolarons. Therefore, for polythiophene, $2E_{pol} > E_{bipol}$ (Chapter 6, Section 6.4).

There are, however, reports on other polymer systems, where the spin concentration increases in proportion to the concentration of charges (dopant ions). In such cases $2E_{pol} \approx E_{bipol}$. Although the origin of the different results is not understood in detail, the polaron-to-bipolaron ratio would be very sensitive to the quality of the polymer (for example, Mw, PDI) when $2E_{pol} \approx E_{bipol}$; see, for example, Fig. 6.14.

If there is indeed an evolution for polarons to bipolarons as a function of the doping concentration as described in Chapter 6, this evolution should be seen in the ESR measurements. This evolution has been observed for P3HT in solution in CH_2Cl_2 and doped with $NO^+PF_6^-$ (Fig 7.13).

Similar results were reported for the nondegenerate ground state polymer, polypyrrolle. The initial doping proceeds with a spin per charge ratio of near unity indicating charge storage in polarons (rather than bipolarons). Measurements at higher doping levels showed a reduction in the spin susceptibility consistent with the evolution from polarons to bipolarons upon increasing the doping level. The results are strikingly similar to the theoretical curve in Fig. 6.14.

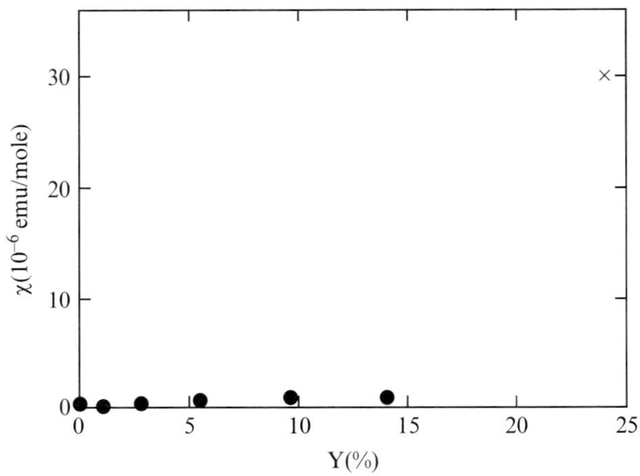

Fig. 7.12 Magnetic (spin) susceptibility χ of polythiophene as a function of dopant (ClO_4^-) concentration. [Reproduced from J. Chen *et al.*, *Solid State Commun.*, **58**, 251 (1986) with permission from Pergamon Press.]

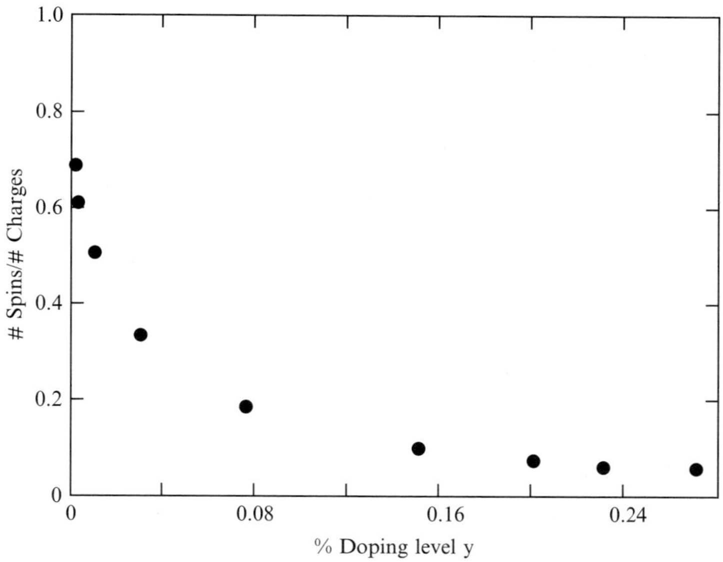

Fig. 7.13 Spins per charge observed for P3HT in solution in CH_2Cl_2 and doped with $NO^+PF_6^-$ versus dopant level at a fixed polymer concentration c=c_0 in dichloromethane. The number of spins in the figure is the number of additional spins upon doping; i.e. the background (pristine) signal has been subtracted away. [Reproduced from M.J. Nowak *et al.*, *Macromolecules*, **22**, 2917 (1989).]

7.9 IRAV modes observed in semiconducting polymers with nondegenerate ground state

IRAV modes have been observed in many conjugated polymers, generated either by chemical doping or by photoexcitation. Typically, the IRAV modes have been used as "fingerprints" and, following photoexcitation, for determining whether the excitation generates charge carriers (polaron or bipolaron) or neutral excitons. However, there have been few attempts to quantify the mechanism of pinning.

For polythiophene and the substituted polythiophenes, the IRAV data were analyzed using the phase mode model. The pinning parameter obtained for doping-induced and photoinduced polarons/bipolarons was approximately 0.3 (for polythiophene, the IRAV appear at nearly the same frequencies).

7.10 Photoinduced absorption in polythiophene (nondegenerate ground state)

The photoinduced absorption spectrum of polythiophene is shown in Fig. 7.14. Again, all the features expected for charged polarons or bipolarons are observed:

- IRAV modes at photon energies less than 0.2 eV;
- two midgap transitions at photon energies below 2 eV; i.e. at energies less than the energy gap;

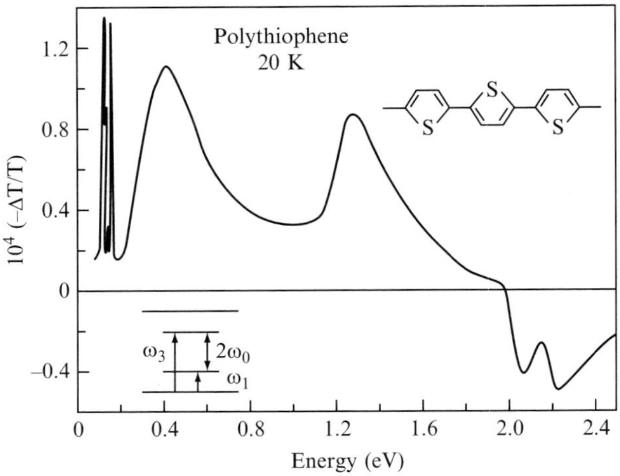

Fig. 7.14 Photoinduced absorption of polythiophene at 20 K. The inset shows the band diagram for a charged (+2) bipolaron. [Reproduced from Z. Vardeny *et al.*, *Phys. Rev. Lett.*, **56**, 671 (1986) with permission from the American Physical Society.]

- bleaching in the region of the interband transition consistent with the conservation of oscillator strength.

Assuming the photogeneration of bipolarons, consistent with the data obtained from doping studies (Fig. 7.12), the results in Fig. 7.14 can be understood with the simple band scheme shown in the inset. One would expect to observe two transitions involving the gap states at $\hbar\omega_1$ and $\hbar\omega_2$ where $\hbar\omega_1 + \hbar\omega_2 = E_g$. As discussed on Section 7.3 for the midgap transition in *trans*-polyacetylene, however, with the Coulomb interaction energy included, the gap state transitions are shifted such that:

$$\hbar\omega_1 + \hbar\omega_2 = E_g - 2U_B \qquad (7.7)$$

where U_B represents the Coulomb energy associated with $\hbar\omega_1 + \hbar\omega_2 = 1.7eV$ whereas the energy gap, as obtained from the interband absorption, is $E_g \cong 2.1$–2.2 eV. Thus, $U_B \approx 0.25$ eV, similar in magnitude to that inferred for polyacetylene. As is also true for polyacetylene, $U_B / E_g \approx 10^{-1}$ implying, once again, that one can treat the Coulomb interaction with perturbation theory. Although the Coulomb interaction is important and must be taken into account to understand the gap-state spectroscopy, it does not dominate the photophysics of conjugated polymers.

7.11 Time resolved photoinduced absorption: Nondegenerate ground state

Ultrafast photoinduced absorption by IRAV modes was used to study the photogeneration of polarons on semiconducting polymer chains in dilute solutions and in solid films of a soluble derivative of poly(*para*-phenylene vinylene).[6] In dilute polymer solutions, polaron pairs are photogenerated within less than 250 fs with quantum efficiency of about one-third of that for polymer films. The excitation spectra of ϕ_{ch} for both solutions and films showed that ϕ_{ch} is weakly dependent on photon energy between 2.2 eV (the onset of absorption) and 4.7 eV. The recombination dynamics of polarons is very fast and highly dependent on the excitation density for polymer films, but it is significantly slower and less sensitive to pump intensity for the semiconducting polymer in dilute solution. Positive and negative polarons on a single chain in solution are typically separated by hundreds of monomer repeat units; the 1D diffusion along the chain is inhibited by the intervening excitons.

Further reading

A. J. Heeger, S. Kivelson, J. R. Schrieffer and W. P. Su, *Rev. Mod. Phys.*, **60**, 781 (1988).
T. A. Skotheim (ed.), *Handbook of Conducting Polymers*, Marcel Dekker, New York, 1986.
T. A. Skotheim and J. R. Reynolds (eds.), *Handbook of Conducting Polymers*, 3rd edition, CRC Press, Taylor and Francis, Boca Raton, 2007.

For the theory of IRAVmodes:
A. Girlando *et al.*, *J. Chem. Phys.*, **98**, 7459 (1993).
A. Girlando *et al.*, *Chem. Phys.*, **184**, 139 (1994).
Z. G. Soos *et al.*, *J. Chem. Phys.*, **100**, 7144 (1994).

[6] P. B. Miranda *et al.*, *Phys. Rev. B*, **64**, 081201 (2001).

8

Conjugated polymers as semiconductors

8.1 Solution cast thin film diodes

The single layer, thin film diode first described in1987 provides a simple example of a semiconductor device: a metal–semiconductor–metal (MSM) structure comprising a single layer of semiconducting polymer cast from solution and sandwiched between metal contacts with different work functions operates as a diode with relatively high rectification ratio. The built-in potential of the diode originates from the difference in the work functions of the two metallic contacts; i.e. indium/tin-oxide (ITO) as the larger work function material for one electrode and either In or Al as the smaller work function material for the other electrode. Poly (3-hexylthiophene), a soluble alkyl derivative of poly(thiophene), was used for the semiconductor layer (thicknesses of 100–200 nm). The current–voltage characteristics obtained from this simple MSM structure exhibited rectification of 100:1 to 1000:1; the forward current increases exponentially over several decades, whereas a relatively small leakage current flows under reversed bias (see Fig. 8.1).

Even in these early experiments, capacitance versus voltage data indicated uniform films with a remarkably small concentration of electronically active impurities. For such an asymmetric MSM structure, the capacitance per unit area versus voltage is given by the Schottky equation:

$$\frac{1}{C^2} = 2\frac{V_{\text{int}} - V}{e\varepsilon\varepsilon_0 N_a} \tag{8.1}$$

where$V_{\text{int}} = V_{\text{bi}} - k_B T/e$ is the intercept voltage, V_{bi} is the built-in voltage, ε is the dielectric constant (ε_o is the permittivity of the vacuum), and N_a is the number of impurities (donors) per unit volume.

The capacitance versus voltage data obtained from these early MSM diodes are shown in Fig. 8.2.

The impurity concentration can be obtained from the data in Fig. 8.2 using Eqn 8.1:

$$N_a(x) = -\frac{2}{e\varepsilon\varepsilon_o} \Big/ \frac{d(C^{-2})}{dV} \tag{8.2}$$

The linear dependence of C^{-2} in reverse bias implies a nearly uniform impurity concentration within the semiconducting polymer. With $\varepsilon \approx 3$, the density of electronically active impurities inferred from the data is surprisingly small, of order 10^{17} cm^{-3}. Further studies showed that in polythiophenes this extrinsic impurity concentration is sensitive to exposure of the polymer to

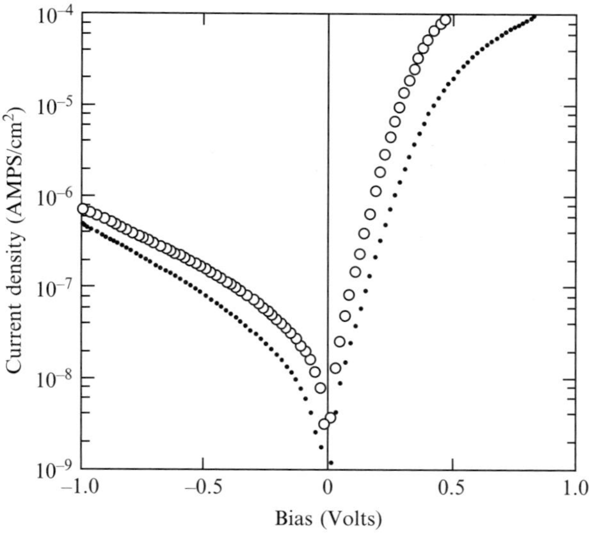

Fig. 8.1 Current–voltage characteristics in In/P3HT (•) and In/P3MT (°) diodes. [Reproduced from H. Tomazawa *et al.*, *Synthetic Metals*, **22**, 63 (1987) with permission from Elsevier B.V.]

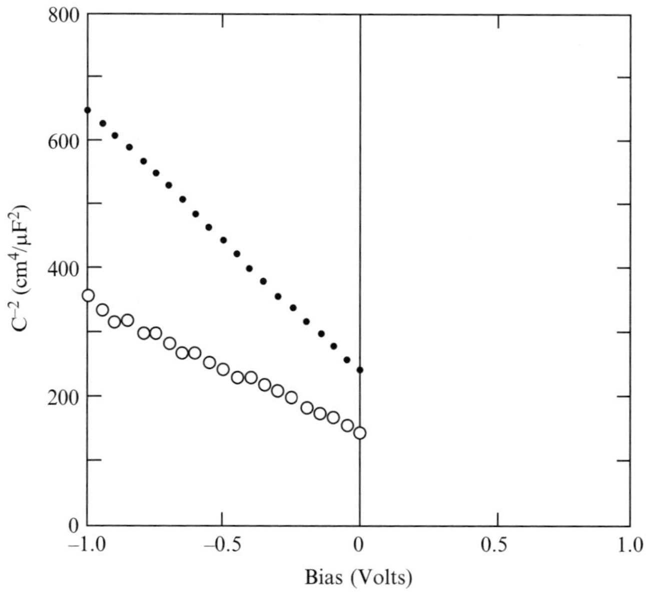

Fig. 8.2 Capacitance–voltage characteristics of In/P3HT (•) and In/P3MT (°) diodes; the data are plotted as C^{-2} versus V. [Reproduced from H. Tomazawa *et al.*, *Synthetic Metals*, **22**, 63 (1987) with permission from Elsevier B.V.]

the oxygen. The zero bias depletion width, $W = [2\varepsilon\varepsilon_0 V_{\mathrm{int}}/eN_i]^{1/2}$, implied by the data in Fig. 8.2, is $W \approx 100$ nm, i.e. comparable to the thickness of the semiconducting polymer.

When the zero bias depletion width is significantly less than the film thickness (d), the MSM is described as a Schottky diode. In this limit, the energy bands in the semiconductor bend to self-consistently enable the depletion of electronic charge from the impurities; the voltage dependent band bending is responsible for the voltage dependence of the capacitance.

Subsequent studies of MSM light-emitting diodes made with higher purity luminescent polymers showed significantly higher rectification ratios and lower concentrations of electronically active impurities (~10^{15} cm^{-3}). For such high quality materials, C becomes independent of V implying a zero bias depletion width larger than the thickness of the semiconducting polymer film (Fig. 8.3). In this higher purity limit, there is no band bending; there is a uniform electric field within the semiconducting polymer layer as shown in the inset to Fig. 8.3 and in Fig. 9.6 of Chapter 9 on polymer LEDs.

The band structure of the MSM diode is shown in Fig. 8.4 in the two limits, $W >> d$ and $W << d$.

The MSM diodes are significant for two reasons. First, they demonstrate that potentially useful electronic components can be fabricated by processing semiconducting polymers from solution. Because solutions of conjugated polymers can be formulated as "inks" with electronic functionality, "printed plastic electronics" has become a focal point of interest in the science and technology of semiconducting polymers. Second, although semiconducting polymers cannot be considered materials with ultra-high chemical purity, the data obtained from MSM diodes demonstrated that the density of *electronically active* impurities can be small, in fact comparable even to the impurity levels in inorganic semiconductors.

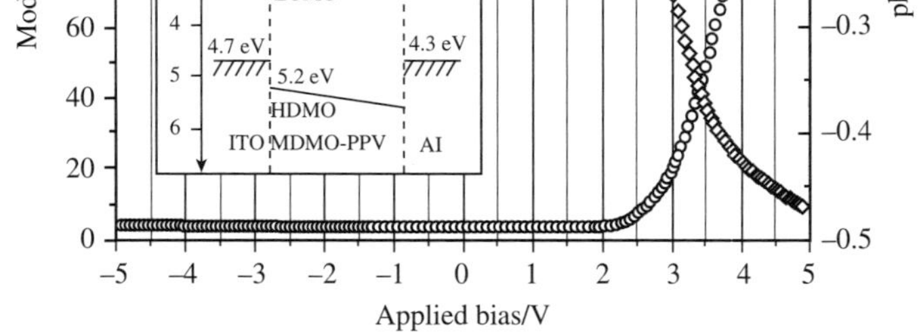

Fig. 8.3 Absolute magnitude and phase of the impedance of an ITO/PEDOT-PSS/MDMO-PPV/LiF/Al diode as a function of applied voltage. In reverse bias, the impedance (and thus the capacitance) is independent of voltage. In forward bias for voltages larger than the built-in voltage, carriers are injected and the impedance is dominated by the resistance of the semiconducting polymer.

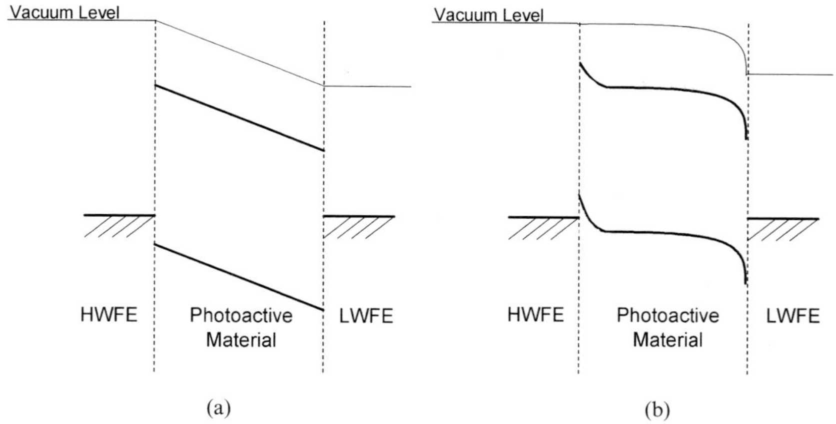

Fig. 8.4 MSM diode: (a) high purity limit, $W >> d$; (b) Schotky limit where $W < d$.

8.2 Semiconductors in which the Fermi energy can be controlled and shifted across the energy gap

In some ways, semiconducting polymers are simpler than their inorganic counterparts for use in device applications. In particular, whereas pinning of the Fermi energy by surface states is a major problem in inorganic semiconductors, the Fermi energy can be controlled and shifted all the way across the energy gap in semiconducting polymers.

In the bulk of a Si crystal, all the valence electrons participate in covalent bonds; there are no unpaired electrons. At a cleaved surface, however, the surface atoms have electrons in so-called "dangling bonds"; i.e. nonbonded electrons in states deep within the energy gap of the semiconductor (see Fig. 8.5a). Such nonbonded states within the gap would be expected to pin the Fermi level. Although this picture is much oversimplified because, for example, of surface reconstruction in inorganic semiconductors, the essential difficulty survives: in inorganic semioconductors, the Fermi level is typically pinned by states in the energy gap, as sketched in the band diagram of Fig. 8.5(b).

Semiconducting polymers as thin films in MSM structures do not have significant densities of dangling bonds. This was demonstrated, for example, through electroabsorption measurements[1] which were used to determine the built-in electric field in MSM structures of the kind described above. The results indicated that the maximum built-in potential that can be obtained by using two electrodes with very different work functions is nearly equal to the single particle energy gap; i.e. the energy required to remove an electron from the π–band and to then inject that electron into the π*–band.

Electroabsorption (EA) spectroscopy makes use of the Stark effect: an electric field acting on a semiconducting polymer sandwiched between two electrodes induces a change in the absorption spectrum of the film. In the EA experiment, a large dc field and a small modulation

[1] I. H. Campbell *et al.*, *Phys. Rev. Lett.*, **76**, 1900 (1996).

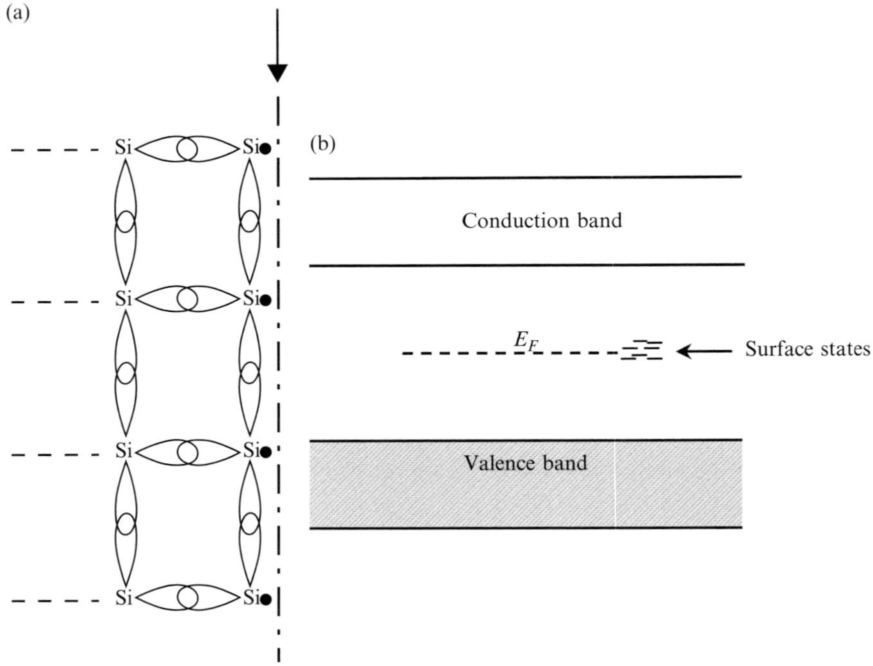

Fig. 8.5 (a) Schematic picture of cleaved Si; (b) band diagram showing surface states in the gap.

ac field (frequency f) are applied. At each frequency in the absorption spectrum, the fundamental frequency of the EA signal is proportional to the field within in the bulk of the polymer:

$$\Delta T/T \propto (V_{DC} - V_{bi})V_{AC}\sin(2\pi ft) \tag{8.3}$$

where T and ΔT are the transmission and the change in transmission, respectively, at photon energy $h\upsilon$, V_{dc} is the external applied field, V_{bi} is the built-in internal field and V_{ac} the amplitude of the ac modulation voltage at frequency f.

Figure 8.6 shows a plot of the magnitude of the electroabsorption signal as a function of dc bias for metal/MEH-PPV/Al structure where the second electrode is Au, Al or Ca. When $V_{dc} = V_{int}$, the EA signal strength goes to zero. From the difference in the voltage required for EA = 0 with the different metal electrodes, one concludes that the built-in field is determined by the difference in work functions. The observation that the built-in potential can approach the magnitude of the energy gap (for MEH-PPV, $E_g = 2.1$ eV) implies that the Fermi level can be moved all the way across the gap (see Chapter 9 on LEDs). The Fermi level is not pinned by defect or surface states in the gap. Conceptually, the ability to shift the chemical potential all the way across the energy gap greatly simplifies the device physics, and technologically, it enables the fabrication of polymer LEDs, polymer LECs, and polymer photodiodes with the simple MSM architecture.

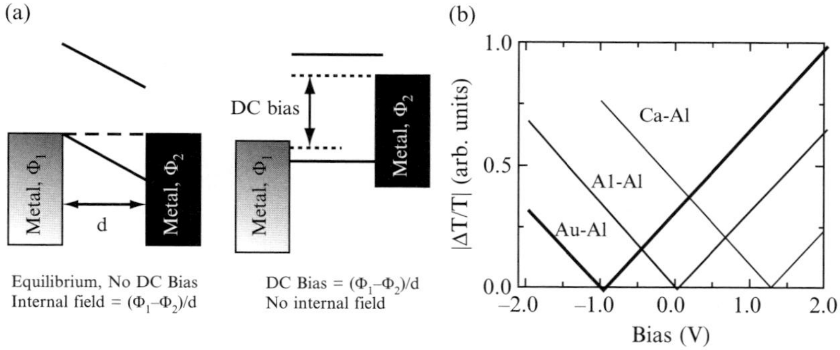

Fig. 8.6 (a) Schematic electron energy diagram of a metal/semiconducting polymer/metal structure in equilibrium and under dc, bias that cancels the internal electric field. Φ_1 and Φ_2 are the work function of two different metals. (b) Magnitude of the electroabsorption response at 2.1 eV as a function of bias for three metal/MEH–PPV/Al structures. [Reproduced from I. H. Campbell, *Phys. Rev. Lett.*, **76**, 1900 (1996) with permission from American Physical Society.]

8.3 Optical properties of semiconducting polymers

In the elementary energy band picture of semiconductor physics, the absorption of photons leads to the promotion of electrons from the valence band (π-band) to the conduction band (π*-band) resulting in mobile electrons and holes, as sketched in Fig. 8.7. Because the photogenerated electrons and holes are mobile, one expects to observe photoconductivity.

Even in the classical inorganic semiconductors such as Si, Ge and GaAs, where the band widths are larger than the separation of atomic levels, the electron–hole Coulomb attraction results in the formation of bound excited states, neutral excitons, with a binding energy that is sensitive to the dielectric constant of the material. In inorganic semiconductors for which the crystal structure and the electronic structure are three-dimensional, the binding energy and radius of the exciton can be described in direct analogy with the hydrogen atom.

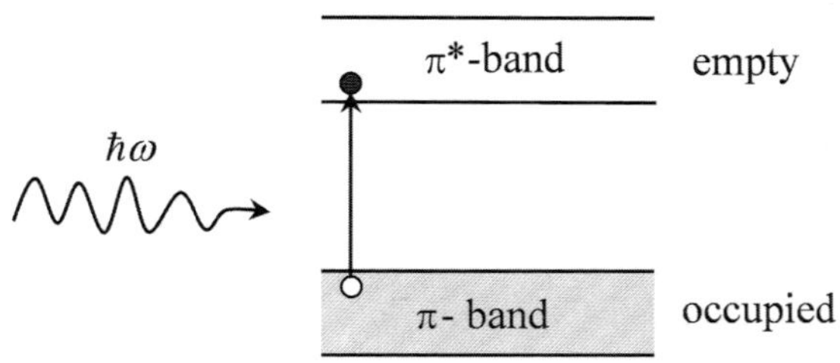

Fig. 8.7 π-π* interband transition.

$$\mathrm{E_B(exciton)} = E_b{}^* = \frac{m^* e^4}{\varepsilon^2 \hbar^2} \tag{8.4}$$

$$\mathrm{R_B(exciton)} = a_b{}^* = \frac{\hbar^2 \varepsilon}{m^* e^2} \tag{8.5}$$

where m^* is the effective mass, e is the electron charges, ε is the dielectric constant of the semiconductor and $\hbar$ is Planck's constant. For GaAs, E_B(exciton) = 3 meV, and R_B(exciton) is much larger than the interatomic spacing. The bound state in these traditional inorganic semiconductors, where the electron and hole in the bound state are delocalized over many atoms, is known as a Wannier exciton.

In a molecular solid such as anthracene, however, the band widths are much smaller than the separation of the molecular energy levels. As a result, the electron–hole bound state, the exciton, corresponds, in a first approximation, to an excited state of the molecule with an electron promoted from the HOMO energy level to the LUMO energy level. In a molecular solid, therefore, the exciton binding energy is large, typically larger than 1 eV. Although the so-called Frenkel excitons in molecular solids are localized primarily on a single molecule, they can hop from molecule to molecule as a result of the weak interactions between molecules in the solid state.

For semiconducting polymers, the intrachain bandwidth is large because of the π-electron overlap between closely spaced neighboring atoms in the repeat units, comparable to the bandwidths in inorganic semiconductors (several eV, see Chapter 1 for detailed discussion). However, the interchain interactions are weak, comparable to those in molecular crystals (a few tenths of an eV). Because of this intrinsic anisotropy, the exciton binding energies are intermediate between those of Wannier excitons and Frenkel excitons. In semiconducting polymers, the exciton binding energies range from E_B(exciton) < 0.1 eV in the PPVs and polythiophenes to a few tenths of an eV in the larger band gap systems such as the polyfluorenes. For these same reasons, the exciton wavefunction extends over a number of repeat units along the chain. Depending upon the detailed interchain stacking, the wavefunction of the excited state can spread onto neighboring chains; the multichain exciton is known as an "exciplex". Note, however, that the disorder within semiconducting polymers cast from solution, tends to localize the neutral excited states onto a single chain. Moreover, because of the disorder, the exciton diffusion range is typically less than 10 nm in semiconducting polymers.

Structural relaxation around the electrons and holes generated by the interband transition causes the formation of self localized polarons (Chapter 6). In addition to the photogenerated charge carriers and the singlet excitons, triplet excitons are also stable excitations in semiconducting polymers. The triplet excitons have significantly larger binding energies with wavefunctions that are much more localized than the singlet exciton wavefunctions.

The branching ratio between the intrachain excitations and interchain excitations is a function of the ratio between the interchain transfer integral and the intrachain transfer integral. Experiments indicate that upon photoexcitation, 10–30% of the initial excitations are charged polarons, presumably separated onto different chains. These charged polarons are responsible for the photoconductivity.

The electron–hole pairs created on a single chain at energies $h\upsilon > E_g$ rapidly relax (typically < 1 ps) to the band edge and form neutral excitons. As described in Chapter 6, the excitons are dressed with a local chain distortion that further lowers the total energy of the bound state. The electron–hole relaxation and self localization to form polarons, and, subsequently, neutral excitons is shown schematically in Fig. 8.8:

$$e^- + e^+ \rightarrow P^- + P^+ \rightarrow BP^0 \rightarrow \text{exciton} \tag{8.6}$$

The recombination of on-chain polaron pairs can be described in terms of these simple band diagrams.

The diagram on the right describes a neutral bipolaron. The process shown in Fig. 8.8 will go to completion only if the total energy decreases in each step:

$$E(P^-) + E(P^+) > E(\text{exciton}) \tag{8.7}$$

Defining$[E(P^-) + E(P^+)] \equiv E_g - 2E(\text{pol})_b$, where $E(\text{pol})_b$ is the polaron binding energy with respect to an electron at the bottom of the π^*-band (or with respect to a hole at the top of the π^*-band),

$$2E(\text{pol})_b < E_B(\text{exciton}) \tag{8.8}$$

Thus, the exciton binding energy (resulting from the Coulomb attraction in a rigid lattice) can be reduced because of the partial loss of the polaron binding energy by forming a neutral exciton (Fig. 8.8), Note that $E(\text{pol})_b$ cannot be inferred from the polaron energy level diagrams in Fig. 8.8 because these electronic diagrams do not include the positive energy contribution from the local structural distortion on the polymer chain (see Chapter 6).

Intrachain excitons can migrate, undergo intersystem crossing to the triplet state, dissociate at impurities, dissociate to create polaron pairs on neighboring chains and, of course, radiatively and nonradiatvely decay to the ground state.

Polarons on different chains can create weakly bound polaron pairs (Frankevich pairs). Their recombination can lead again to excitons and thereby to delayed photoluminescence. Polarons can be trapped, recombine with oppositely charged polarons, undergo charge transfer onto impurities or charge transport to the electrodes.

8.4 Optical properties of oriented and aligned semiconducting polymer chains

The intrinsic photophysics of semiconducting polymers has been explored using highly oriented chains of MEH-PPV in a matrix of ultrahigh molecular weight (UHMW) polyethylene (PE). Nearly perfect alignment of the PE and the MEH-PPV chains can be attained through gel processing followed by stretch orientation with draw ratio (final length divided by initial length) equal to approximately 50. The optical absorption and photoluminescence of these

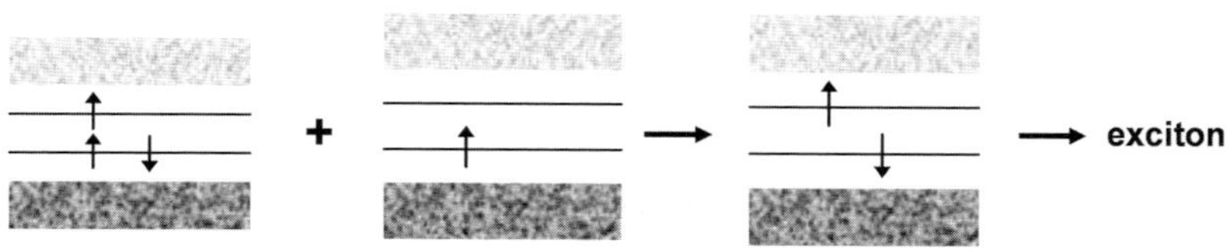

Fig. 8.8 Energy diagrams showing the $P^- + P^+ \rightarrow BP^0 \rightarrow$ exciton recombination process.

highly oriented chains is therefore characteristic of semiconducting polymers in which the role of structural disorder has been minimized.

As shown in Fig. 8.9, the alignment of the MEH-PPV chains can be inferred from the anisotropy in the absorption and the luminescence. The maximum absorption is obtained with light polarized parallel to the chain, consistent with the interband π-π* absorption sketched in Fig. 8.7. Moreover, the perpendicular absorption spectrum is completely different from that obtained with the light polarized parallel to the chains. Thus, the perpendicular spectrum arises from interchain absorption; i.e. promotion of an electron from the π-band on one chain to the π*-band on a neighboring chain. Because the parallel and perpendicular absorption spectra are different, the absorption data can set a lower limit for the orientational order parameter ($f > 50{:}1$).

The solid curve in Fig. 8.9(Right) is the luminescence obtained with a polarizer oriented to transmit light with polarization vector parallel to the orientation direction, and the dotted curve is the luminescence obtained with a polarizer oriented to transmit light with polarization perpendicular to the orientation direction. Note that the perpendicular data are multiplied by a factor of 80; the dotted curve at the bottom are the actual data obtained with polarization vector perpendicular to the chain. The orientational order parameter is approximately 80:1. The inset to Fig. 8.9(Right) shows that the angular dependence, $I(\theta)$, is accurately described by $I(\theta) = \alpha\cos^2\theta + \beta$ where θ is the angle between the chain orientation direction and the polarization vector. The orientational order parameter is defined as $f = \alpha/\beta$.

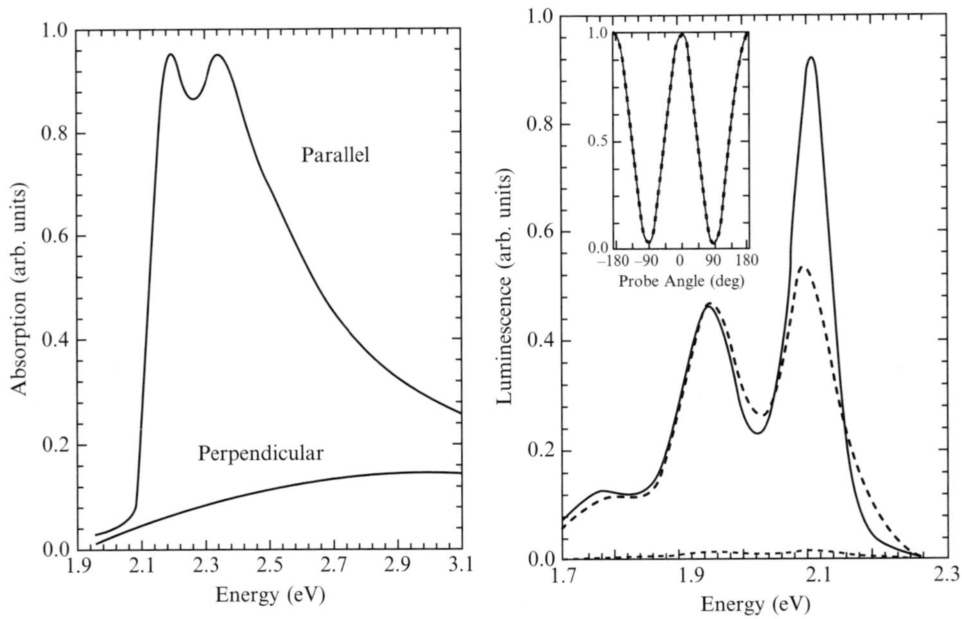

Fig. 8.9 The anisotropic absorption and emission in stretch oriented MEH-PPV blends in UHMW PE. Left: anisotropic absorption by aligned MEH-PPV. Right: anisotropic emission from aligned MEH-PPV. [Reproduced from T.W. Hagler *et al.*, *Phys. Rev. B*, **44**, 8652 (1991) with permission from American Physical Society.]

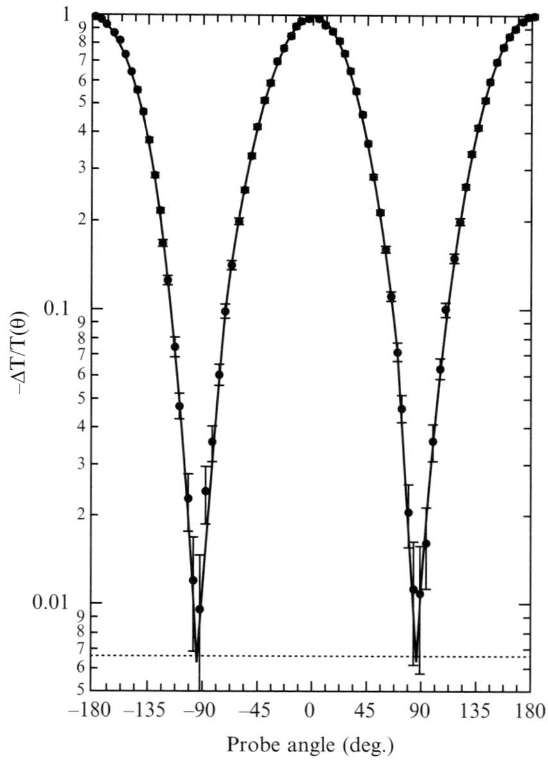

Fig. 8.10 Anisotropy in the electroabsorption with respect to the optical field and the chain orientation axis for oriented MEH-PPV in PE. [Reproduced from T.W. Hagler *et al.*, *Phys. Rev. B*, **49**, 10968 (1994) with permission from American Physical Society.]

Electroabsorption data obtained from these highly oriented films are shown in Fig. 8.10. The measured anisotropy of the electroabsorption for the oriented MEH-PPV blends in UHMW PE is larger than 100. The remarkably high orientational order parameter and the clear separation of the parallel and perpendicular absorption spectra give confidence that the intrinsic photophysics of semiconducting polymers can be explored through studies of these highly oriented polymer chains.

8.4.1 Vibronic structure

The vibronic structure in the absorption and emission can be understood in qualitative terms using a Jablonski diagram (Fig. 8.11) in which one plots the ground state and first excited state energies as a function of the amplitude of a structural distortion, Q; for example, the amplitude of a distortion with normal coordinates of a specific Raman mode of the repeat unit of the polymer. A more general distortion can be described as a superposition of several Raman modes.

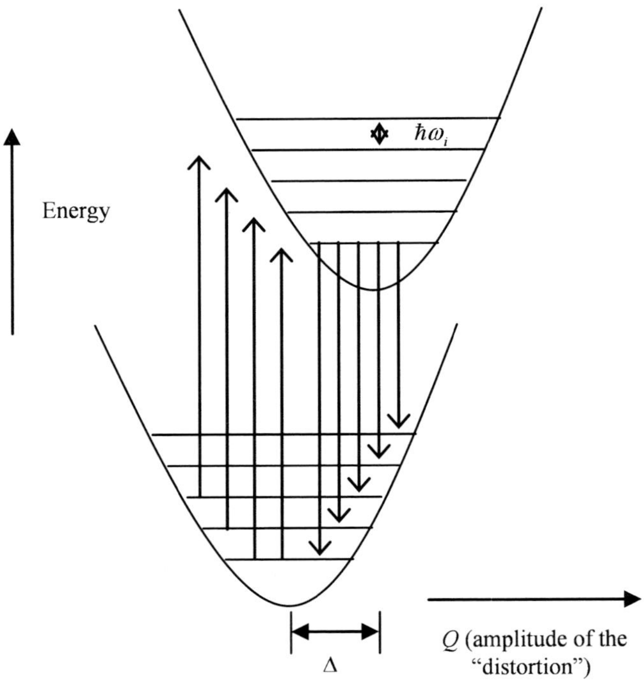

Fig. 8.11 Jablonski diagram showing the ground state and first excited state energies as a function of the amplitude of a structural distortion, Q.

The minimum in the excited electronic energy represented by the upper parabola occurs at a different value of Q than the minimum of the ground state energy (lower parabola), implying changes in the bond lengths in the excited state. This shift (ΔQ) arises from structural relaxation in the backbone geometry; the larger the structural relaxation, the larger the shift. The horizontal lines in Fig. 8.11 represent the energies of the vibrational normal mode (in this example, the specific Raman mode) in the ground state and the excited state.

If the electron in the excited π^*-state were delocalized over the entire chain of N repeat units, the bond length relaxation would be of order $1/N$ and thus not observable. The vibronic features observed in the absorption and emission spectra are therefore indicative of self-localization in the excited state. That there is structural relaxation and localization in the excited state is not a surprising result; self-localization with the formation of polarons is expected following photoexcitation (Chapter 6).

The total wavefunction is a product of the electronic wavefunction, $\psi_{\mathrm{electronic}}$, and the lattice wavefunction, $\psi_{\mathrm{structure}}$:

$$\psi = \psi_{electronic}\psi_{structure} \tag{8.9}$$

When there is no structural relaxation in the excited state, the normal mode vibrational wavefunctions are orthornormal to one another and $\Delta Q = 0$. In this limit, electronic transitions between different vibrational states in the ground and excited states are forbidden because of

(a)

(b)

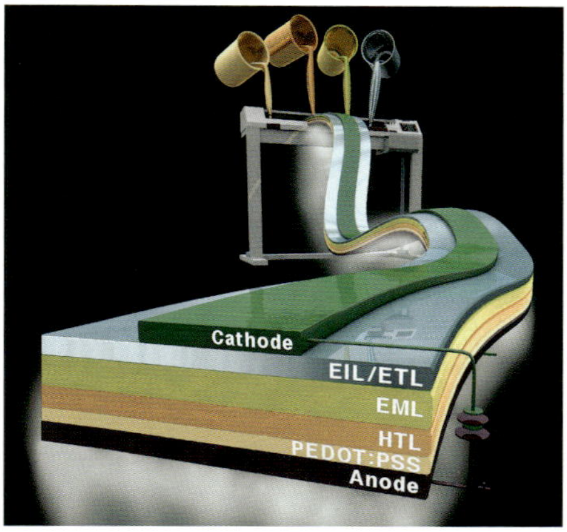

Figure 3.1: (a) Vials containing semiconducting and metallic polymers in solution. The colored solutions are semiconducting polymers with different band gaps (the colorless solution contains a blue-emitting polymer with absorption in the ultraviolet). The dark solution in the center contains conducting poly(ethylenedioxythiophene. (b) The organic semiconductors in solution form can be used as "inks" and printed onto surfaces with any printing technology. [Reproduced from *Convergence*, the *Magazine of Engineering and the Sciences* at UC Santa Barbara, 9, 2007.]

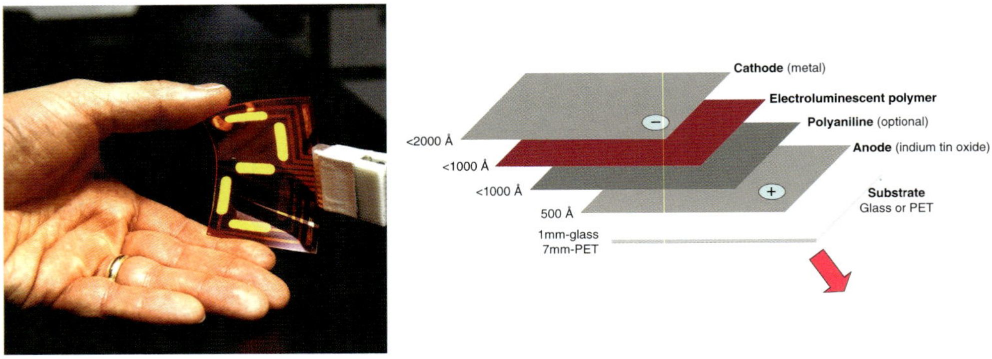

Figure 9.1: Device structure of an LED fabricated from semiconducting and metallic polymers and photo of a thin film flexible seven-segment display fabricated from semiconducting and metallic polymers.

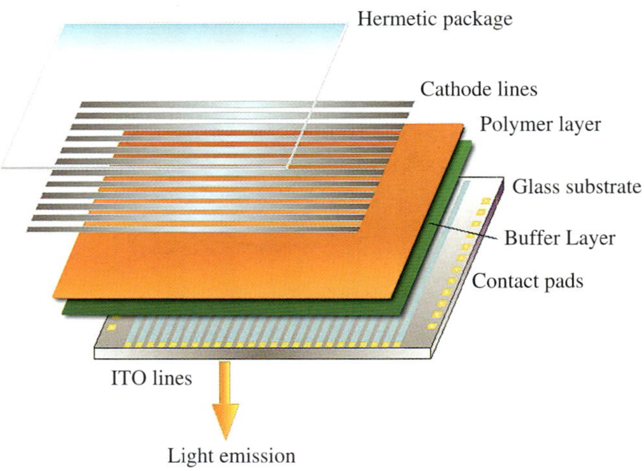

Shematic of Pixellated Display

Figure 9.17: Diagram of the structure of a passively addressed display.

Figure 9.18: Sketch of a printed array of pixels is shown on the top left. Top right is a photo of the colored pixels in an early device fabricated by ink-jet printing. Bottom image is a full color polymer emissive active matrix display showing a video. For active matrix display, a switching circuit is included into each pixel. [Courtesy of Dupont Displays.]

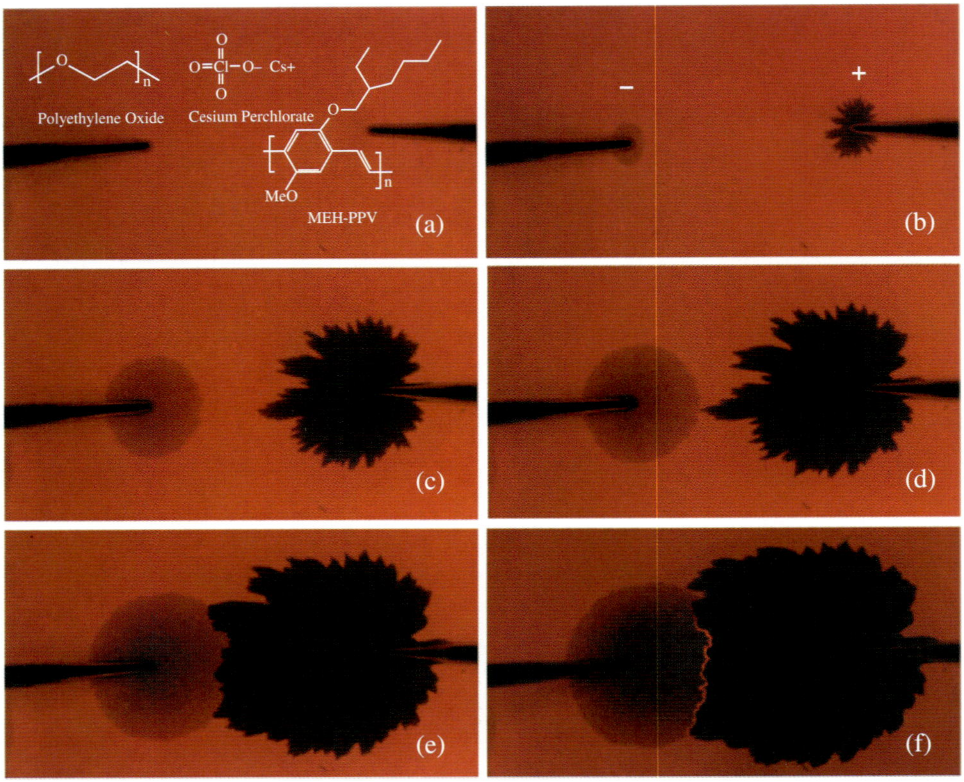

Figure 10.4: MEH-PPV:PEO:$CsClO_4$ film in contact with two tungsten probes separated by 4.8 mm. The film is at 330 K and under UV illumination. Time elapsed following application of 400V bias; (a) no bias, (b) 10 s, (c) 90 s, (d) 120 s, (e) 220 s, and (f) 270 s. [Reproduced from Y. Hu and J. Gao, J. Am. Chem. Soc. **131**, 18236 (2009) with permission from the American Chemical Society.]]

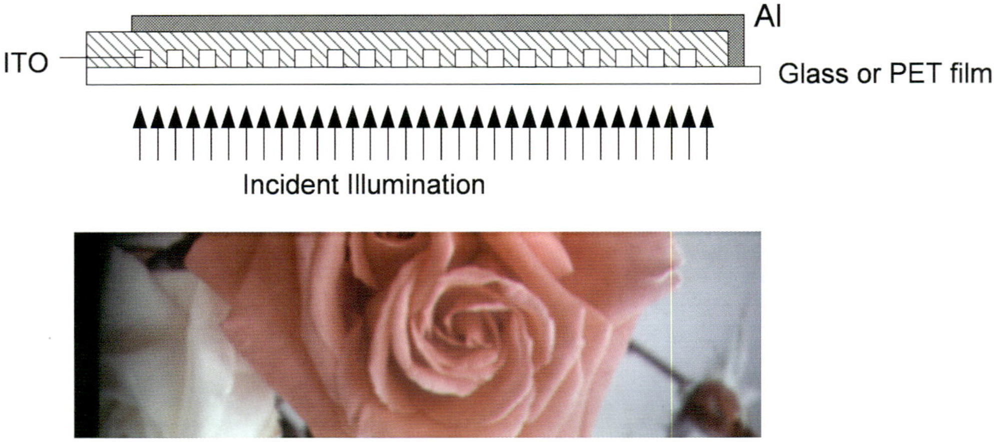

Figure 13.31: Top: a linear array of photodiodes for use as a scanner. The BHJ material was P3HT: PCBM. Bottom: color image recorded by the scanner described above.

the requirement that $\Delta q = 0$ for an optical transition (the wavelength of light is much larger than the lattice constant). Thus, for $\Delta Q = 0$, there are no vibrational features in the spectra; only direct transitions are allowed; e.g. 0-phonon to 0-phonon, 1-phonon to 1-phonon, etc. When $\Delta Q \neq 0$, the different vibrational states in the ground and excited electronic states are no longer orthogonal; the resulting overlap enables vibrational sideband transitions such as those shown in the Jablonski diagram in Fig. 8.11; e.g. 0-phonon to 1-phonon, etc.

The finite overlap of vibrational levels in the ground state and excited states can be expressed as

$$|\langle \psi_{(v_i)} | \psi_{(0)} \rangle|^2 = \frac{e^{-S} S_i^{v_i}}{v_i!} \tag{8.10}$$

where the S_i are defined as the Huang–Rhys factors

$$S_i = \frac{1}{2} \frac{m_i v_i^2}{\hbar\omega_i} (\Delta Q_i)^2 \tag{8.11}$$

$\hbar\omega_i$ is the vibrational level spacing, ΔQ_i are the normal coordinate displacements and v_i are the number of phonons in the excited state (since the temperature is small compared to the optical phonon energies, the ground state has $v_i = 0$).

Because the various $\Delta Q_i \neq 0$, the absorption coefficient exhibits vibronic sidebands

$$\alpha(\hbar\omega) = \sum_{\{v_i\}} \alpha_0 (\hbar\omega - \sum v_i \hbar\omega_i) \prod_i |\langle \chi(v_i) | \chi(0) \rangle|^2 \tag{8.12}$$

where $\alpha_0(\hbar\omega)$ is the absorption coefficient in the rigid lattice approximation ($\Delta Q_i = 0$). If a single optical phonon (Raman mode) dominates, only a single S and v are required in these equations.

The zeroth-order line shape, $\alpha_o(\omega)$, is assumed to be that derived for a one-dimensional semiconductor (e.g. as in SSH theory), but assuming a Gaussian distribution of the energy gaps (with width γ to account for residual disorder) centered about the mean value, E_g.

Figure 8.12 shows the excellent agreement between the theory (solid curve) and the absorption data (dotted curve) obtained from oriented MEH-PPV in PE; the data were obtained with the sample at 80 K to minimize the thermal disorder. The best-fit parameters were the following: $E_g = 2.136$ eV, $\hbar\omega_{ph} = 0.172$ eV, $\gamma = 0.040$ and $S = 0.41$. The value for $\hbar\omega_{ph} = 0.172$ eV is in excellent agreement with the strongest Raman line observed in the PPVs. Only a single Raman mode is required to fit the data. Using these values for Δ_0 and γ, one can compare $\alpha_0(\hbar\omega)$ obtained from experiment with that given by Eqn 8.13:

$$\alpha_0(\hbar\omega) = \frac{1}{(\hbar\omega)^2} \int d\Delta \frac{e^{-(\Delta-\Delta_0)^2/\gamma^2}}{\sqrt{(\hbar\omega)^2 - (2\Delta)^2}} \tag{8.13}$$

where $E_g = 2\Delta_o$,

A similar Huang–Rhys analysis was carried out for the emission data.

$$I(\hbar\omega) = \sum_{\{v_j\}} \frac{(\Gamma_{v_j}/2)^2}{[\hbar\omega - \hbar(2\omega_0) + \Sigma v_j \hbar\omega_j]^2 + (\Gamma_{v_j}/2)^2} \times \prod_j |\langle \chi(v_i) | \chi(0) \rangle|^2 \tag{8.14}$$

where Γ is the width of the emission band at $2\hbar\omega_0$ with vibrational side bands displaced by $\hbar\omega_j$.

The conclusion is that electrons and holes are initially generated via the interband π-π^* transition. These electrons and holes rapidly thermalize to the band edges in at most a few ps, and if the electron–hole pair is on the same chain, combine to form neutral excitons. The formation of polarons on separate chains is discussed above in terms of the yield of mobile carriers as measured by the photoconductivity.

The luminescence originates from the radiative decay of excitons to the ground state. The energy difference between the exciton excited state and the ground state is given by the measured luminescence emission. The best fit shown in Fig. 8.13 is obtained with $\hbar(2\omega_0) = 2.088$ eV, $\Gamma_0 = 0.066$ eV and $S = 0.47$. Again, only one phonon frequency is needed to obtain the excellent fit, $\hbar\omega_1 = 0.162$ eV implying that the strongest Raman line is at a slightly lower energy in the excited state (the corresponding optical phonon energy is 0.172 eV in the ground state as obtained from the analysis of the absorption data).

The data in Fig. 8.12 demonstrate that the absorption is dominated by the quasi-1D joint density of states associated with the interband transition. In contrast, the luminescence (Fig. 8.13) is from the relaxed exciton to the ground state. The Stokes shift obtained from the difference in energies of the 0-phonon absorption band edge and the 0-phonon emission line is quite small, 0.05 eV.

This difference in energies of the 0-phonon absorption band edge and the 0-phonon emission line *provides a quantitative measurement of the maximum possible exciton binding energy in well ordered MEH-PPV, E_B(exciton) $\approx$ 0.05 eV. Note that this value of the exciton*

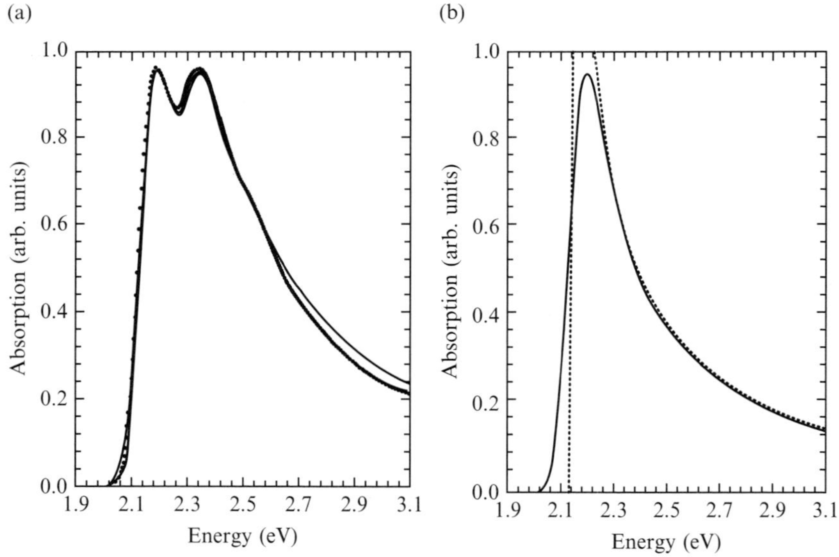

Fig. 8.12 Comparison of the optical absorption spectrum of MEH-PPV blends in UHMW PE with the 1D electronic density of states. (a) $\alpha(\hbar\omega)$; theory (solid curve) and experimental (dotted curve); (b) comparison of $\alpha_0(\hbar\omega)$ obtained from experiment with theoretical results for an perfect single chain. [Reproduced from T.W. Hagler *et al.*, *Phys. Rev. B*, **44**, 8652 (1991) with permission from American Physical Society.]

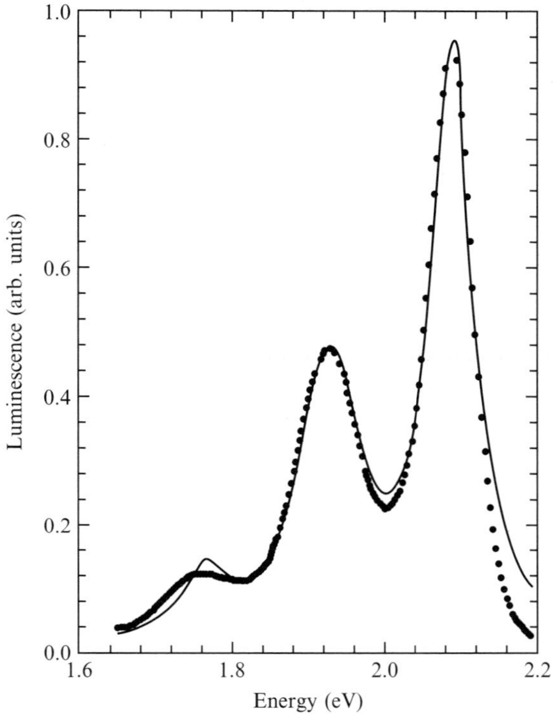

Fig. 8.13 Emission: comparison of theory (solid curve) and experiment (dotted curve) obtained from oriented MEH-PPV in PE. [Reproduced from T.W. Hagler *et al.*, *Phys. Rev. B*, **44**, 8652 (1991) with permission from American Physical Society.]

binding energy is completely independent of the photoconductivity measurements. The value obtained from photoconductivity measurements in MEH-PPV is in excellent agreement with E_B*(exciton)* $\approx$ *0.05 eV (see Section 8.8).*

The analysis summarized in this section leads to the conclusion that the π-π* absorption band is the interband transition between the π- and π*-bands separated by the energy gap E_g. As noted above, however, the exciton binding energies in semiconducting polymers are expected to be intermediate between those of Wannier excitons and Frenkel excitons. In the Frenkel exciton limit, the oscillator strength is almost completely removed from the higher energy interband transition and shifted into the exciton absorption. Thus one might attempt to explain the π-π* absorption band as resulting from a transition from the ground state to the exciton excited state (and indeed this point of view has been expressed repeatedly in the literature). There are, however, serious problems with this alternative point of view. For example, the lineshape for the $\Delta q = 0$ optical transition between the ground state and the lowest lying exciton would be symmetric, in contrast to the asymmetric lineshape obtained from theory and experiment; see Fig. 8.12(b).

The excellent agreement between theory and experiment for the most highly ordered system available, ordered MEH-PPV chains in aligned PE, provides confidence in and confirmation of the assumption that the so-called π-π^* absorption band is indeed the interband transition between the π- and π^*-bands separated by the energy gap, E_g, and that the exciton binding energy is small.

8.5 Structural disorder in semiconducting polymers

Disorder broadens the absorption and luminescence spectra and creates tail states in the gap near the edges of the bands. Such tail states are localized states described by the disorder theories of Mott and Anderson (disorder-induced localized states are discussed in Chapter 5). Transport through such localized states is dominated by variable range hopping (VRH) with occasional thermal excitation back into delocalized band states with a probability of $\exp(-\Delta E/kT)$ where ΔE corresponds to the energy from a localized state in the band tail to the mobility edge.

The absorption spectrum obtained from ordered and aligned MEH-PPV is compared with that obtained from spin-cast MEH-PPV in Fig. 8.14. It is clear from this comparison that one should be somewhat cautious about drawing conclusions regarding the intrinsic photophysics from such spin-cast disordered materials. On the other hand, the differences between the absorption spectra

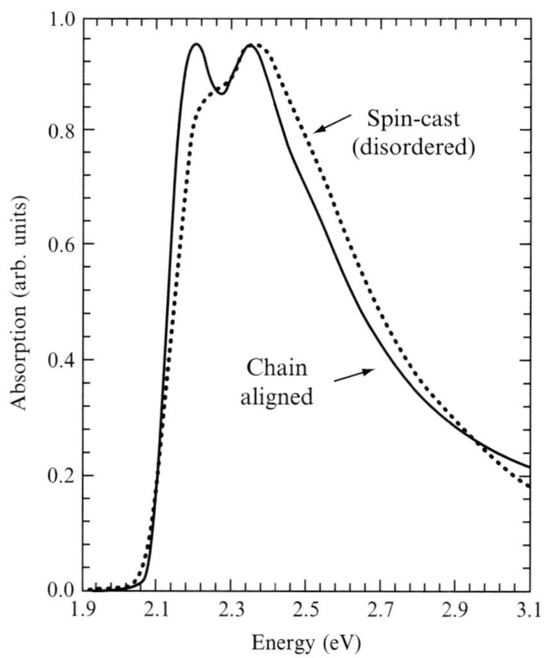

Fig. 8.14 Comparison of the π-π^* absorption from chain aligned MEH-PPV in PE and from spin cast films of MEH-PPV. [Reproduced from T.W. Hagler *et al.*, *Phys. Rev. B*, 44, 8652 (1991) with permission from American Physical Society.]

obtained from the highly ordered and oriented chains of MEH-PPV in PE and from spin-cast films are surprisingly small. The broadening on the low energy side implies a very weak band tailing into the gap as a result of the disorder (see Chapter 4 and in particular Fig. 4.3).

MEH-PPV is a good example because the molecular weight is high compared to that of many conjugated polymers. Thus, more severe broadening (and correspondingly more pronounced band tails) are observed in other systems. Nevertheless, the comparison shown in Fig. 8.14 is particularly important because MEH-PPV (and related soluble derivatives of PPV) have been heavily studied. In particular, since the lowest energy charged excitations are polarons, the transport (photoconductivity in thin films and gate induced conductivity in FETs) results from polaron hopping through localized states near the band edges in spin-cast materials.

8.6 Photoconductivity measurement of the onset of the interband absorption

Measurement of the onset of photoconductivity is the traditional method for determining the onset of the interband (band-to-band) transition in a semiconductor. If the exciton binding energy is large, one expects to observe the threshold for photogeneration of charge carriers at photon energies well above the onset of optical absorption; i.e. the onset of photoconductivity would be at $\hbar\omega_{\mathrm{ch}} = E(\mathrm{exciton})_{\pi\text{-}\pi^*} + E_{\mathrm{B}}$ where $E(\mathrm{exciton})_{\pi\text{-}\pi^*}$ is the photon energy required to create a neutral exciton, and E_{B} is the exciton binding energy; i.e. $\hbar\omega_{\mathrm{ch}} >> E(\mathrm{exciton})_{\pi\text{-}\pi^*}$ whereas the onset of optical absorption would occur at $\hbar\omega_{\mathrm{opt}} = E(\mathrm{exciton})_{\pi\text{-}\pi^*}$.

Photoconductivity action spectra obtained from PPV derivatives have been presented in the literature that show a major increase at approximately 3–4 eV. These data were interpreted as identifying the "real" onset of the π-π* interband absorption. If correct, this would imply that the main optical absorption band (see for example Fig. 8.14) observed at lower energies results from the photogeneration of neutral excitons and that the exciton binding energy is significantly larger than 1 eV.

Subsequent experiments demonstrated, however, that the apparent increase in photoconductivity at such relatively high photon energies resulted from photoemission of electrons from the semiconducting polymer into the vacuum rather than from true photoconductivity from photogenerated carriers in the polymer film.[2]

Measurements of the excitation profile (the photoconductivity as a function of the photon energy) demonstrate that there is only one threshold for the onset of photoconductivity (see Fig. 8.15); that single threshold must be associated with the onset of interband absorption. Moreover, in every case studied, that onset coincides with the onset of absorption. There is no evidence of a second threshold at higher energies as would be expected if the π-π* absorption were the result of photogeneration of excitons with a relatively large binding energy.

[2] D. Moses *et al.*, *Chem. Phys. Lett.*, **350**, 531 (2001).

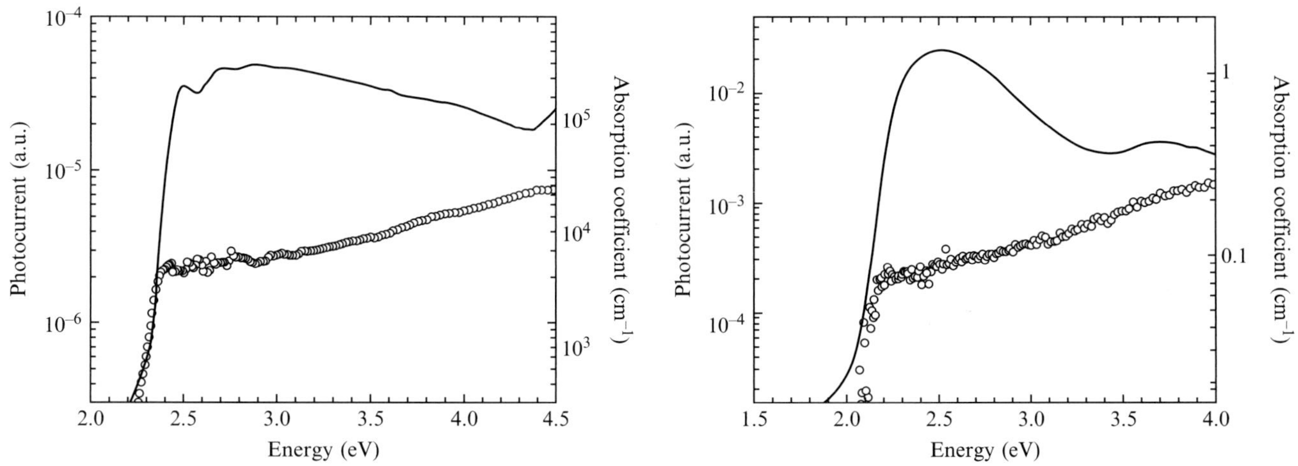

Fig. 8.15 Steady state photoconductivity and optical absorption in highly oriented PPV (left) and MEH-PPV (right) film. [Reproduced from Cesar Soci, Ph.D. thesis, *Elementary photoexcitations in conjugated polymers studied by optical spectroscopy and photoconductivity*, UCSB 2004.]

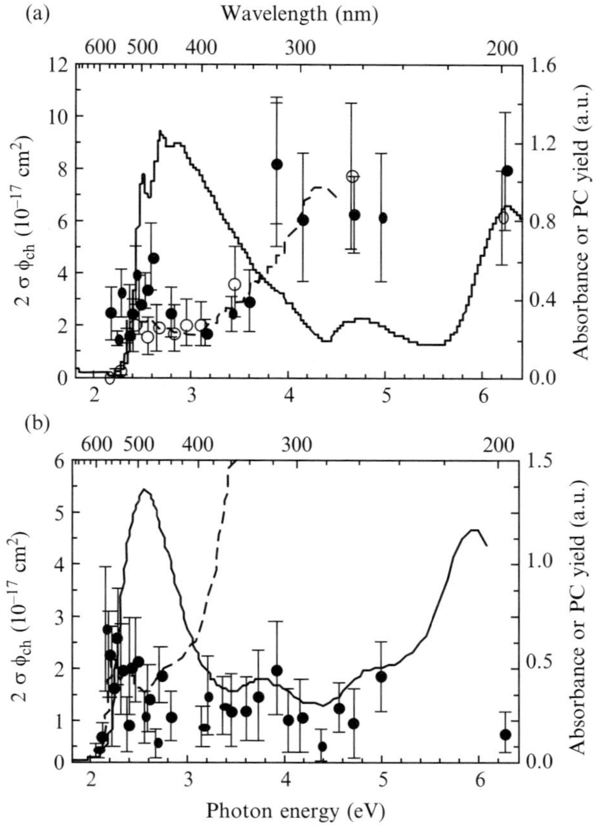

Fig. 8.16 Charge photogeneration excitation spectrum for (a) PPV and (b) pristine MEH-PPV. Solid circles are the ultrafast measurements (IR probe at 1100 cm^{-1}) and solid lines are the sample absorption spectrum. The empty circles in (a) are transient PC measurements. [Reproduced from P. B. Miranda, *et al.*, *Phys. Rev. B*, **64**, 081201 (2001) with permission from American Physical Society.]

8.7 Photon energy required for the photogeneration of charged polarons

The photogeneration of charge carriers can be directly monitored through measurements of the infrared active vibrational (IRAV) modes. As described in Chapter 7, the strength of the IRAV bands is directly proportional to the number of charged polarons. Thus, measurements of the excitation profile for the IRAV mode intensity are even better than photocurrent action spectra for determining the onset of the photogeneration of charge carriers because the photoconductivity is proportional to the product $n\mu$ (where n is the density of carriers and μ is the mobility of the carriers) whereas the IRAV mode strength is insensitive to the mobility of the charged carrier.

Ultrafast pump/and probe measurements of the IRAV intensity demonstrated that the polaron distortions and the signature IRAV modes appear at subpicosecond times following photoexcitation. These data again confirm the prediction of Su and Schrieffer (see Chapter 7, Section 7.6) that self-localization would occur within a few optical phonon periods (less than 100 femtoseconds).[3]

The excitation profile of the IRAV signal (the strength of the IRAV mode signal as a function of the photon energy of the pump pulse) provides the most reliable measurement of the onset of the photogeneration of charged polarons.

The excitation profile shown in Fig. 8.16 demonstrates that the onset of photogeneration of charged carriers coincides with the onset of absorption; i.e. the π-π^* absorption results from the interband transition as sketched in Fig. 8.7. This important result is fully consistent with the analysis of the absorption as summarized in Section 8.5.

8.8 Measurement of the exciton binding energy

The exciton absorption in inorganic semiconductor crystals, such as in single crystals of GaAs, can be observed indirectly via the photoconductivity action spectrum. At low temperatures, a neutral exciton is dissociated into an electron and a hole by the applied electric field and by scattering from charged impurities. In this way, the photogeneration of neutral excitons can be observed in the photoconductivity action spectrum. Data obtained from GaAs are shown in Fig. 8.17 (left). Analogous results were obtained from highly oriented and crystalline samples of PPV. At electric fields larger than 10^5 V/cm a sharp peak is observed just below the absorption edge. The PPV photoconductivity data are shown at several different electric field strengths in Fig. 8.17 (right).

Note that the magnitude of the peak at 2.36 eV grows as the electric field is increased, consistent with field-induced ionization of excitons by the high electric field. Thus, although the photogenerated excitons are neutral, the electrons and holes created by field ionization of these excitons (and by exciton dissociation via scattering from impurities) contribute to the photocurrent.

Measurements of the photocurrent action spectra with light polarized parallel and perpendicular to the chain axis demonstrate that the peak from field-ionized excitons appears only when the PPV film is exposed to light with polarization parallel to the chain orientation direction. Detailed analysis of these experiments indicated that the exciton peak is 0.06 eV below the onset of the interband transition at 2.42 eV. Thus, the exciton binding energy was determined as $E_\mathrm{B} = 0.06$ eV. This value agrees with the Stokes shift obtained from highly oriented MEH-PPV in PE as described in Section 8.5.

If indeed, $E_\mathrm{B} = 0.06$ eV, then E_B is only a few times $k_\mathrm{B}T$, and the photocurrent should be thermally activated. At 10^4 V/cm, where the applied field is smaller than that required for field ionization, an Arrhenius fit yields $E_\mathrm{A} \approx 70 \pm 11$ meV in agreement with the value obtained from the peak in the photocurrent excitation profile.

[3] W.-P. Su *et al., Proc. Natl. Acad. Sci. USA*, **77**, 5626 (1980).

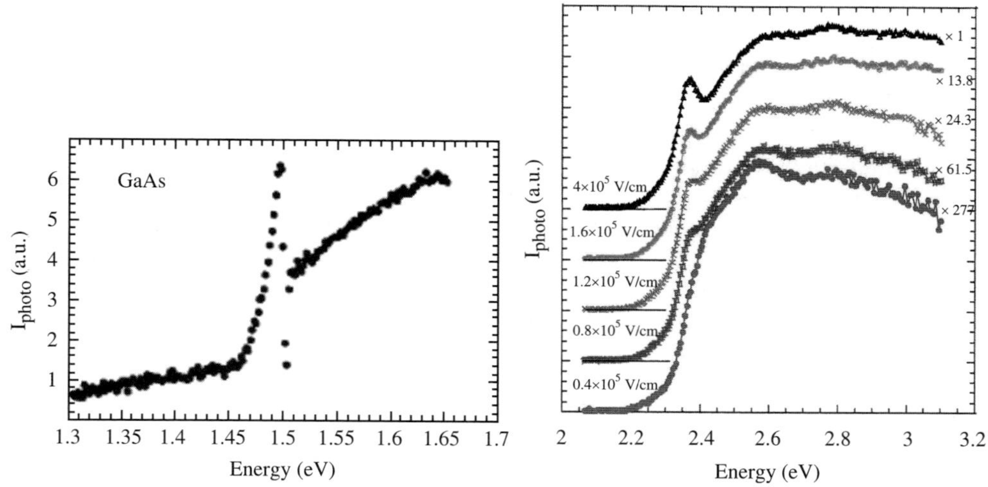

Fig. 8.17 Left: photoconductivity of GaAs at low temperatures; the exciton peak is observed at 3 meV below the interband absorption edge. Right: photoconductivity from highly oriented and crystalline samples of PPV; at high electric fields a sharp peak is observed just below the absorption edge. [Reproduced from D. Moses *et al.*, *Proc. Natl. Acad. Sci.*, **98**, 13496 (2001) with permission from National Academy of Sciences.]

8.9 Recombination of charge carriers (monomolecular and bimolecular)

The photoinduced charge carrier density decreases as a function of time because of recombination and the relaxation of the electronic system toward the ground state. When recombination occurs, energy is lost. Therefore, for example, all recombination processes need to be minimized to improve the conversion efficiency of solar cells.

There are two important general types recombination mechanisms in semiconductors: (a) monomolecular and (b) bimolecular.

(a) Monomolecular recombination:
Monomolecular recombination involves a single carrier and it occurs when one mobile charge carrier recombines with an immobile trapped charge. For example, when an electron encounters a recombination site (defect, impurity site, etc.) within the band gap, the electron is trapped and remains in the trap site until a hole comes along to recombine with the electron, thereby returning the system to the ground state. The rate of monomolecular recombination is given as:

$$\frac{dn}{dt} = \frac{dp}{dt} = -n/\tau_m \tag{8.15}$$

$$n(t) = p(t) = n_o.exp(-t/\tau_m) \tag{8.16}$$

where τ_m is an monomolecular lifetime (relaxation time). The monomolecular recombination rate and the corresponding carrier lifetime are independent of the excitation light intensity.

(b) Bimolecular recombination:

Bimolecular recombination involves both electrons and holes. The greater the number of electrons and holes in a finite volume, the greater the probability they will collide and recombine. For bimolecular recombination, the number of recombination is proportional to the product of number of electrons and number of holes (*np*). This process is dependent on excitation intensity and the recombination decay becomes faster as the excitation density increases. The rate of bimolecular recombination is given as:

$$\frac{dn}{dt} = \frac{dp}{dt} = -\beta(t)np \tag{8.17}$$

Assuming charge neutrality $n = p$, the solutions for *n(t)* and *p(t)* is given by:

$$n(t) = p(t) = \frac{n(0)}{1 + n(0) \int_0^t \beta(t)dt} \tag{8.18}$$

The time-dependence of $\beta(t)$ follows a power law decay at all temperature as $\beta(t) = \beta_0 t^{-(1-\gamma)}$, where β_0 and γ are temperature dependent parameters. Using this functional form of $\beta(t)$, the integral in Eqn 13 can be evaluated yielding

$$n(t) = p(t) = \frac{n(0)}{1 + (t/TB)^{\gamma}} \tag{8.19}$$

where $\tau_B = (\gamma/(n(0)\beta_0))^{1/\gamma}$ is an "effective" bimolecular lifetime.

8.10 Transport of charge carriers in disordered semiconducting polymers

In devices fabricated from semiconducting polymers, charge injection either by photon absorption or by injection at the electrodes is followed by the transport of charges. Minimal loss of carriers requires that the transit time t_r of a charge drifting through the interelectrode distance *d* under the influence of an electric field *E* is shorter than the lifetime of the charge carrier τ. The transit time can be calculated from the charge mobility μ as $t_{tr}=d/E\mu$. The product $\mu\tau$ is, therefore, an important parameter in LEDs, solar cells, and OFETs.

The mobility is the proportionality constant between the drift current density $j_d=en\mu E$ and the electric field *E*, where *n* is the charge carrier concentration and *e* is the elementary charge. In an ordered material, the mobility can be expressed as, $\mu=e\ \tau_{sc}/m^*$, and depends only on the scattering time between collisions τ_{sc} and the effective mass of the electron m^*. In disordered semiconducting polymers, however, the mean free path of charge carriers is comparable to the intersite distance, and the charges are localized (see Chapter 4). Their motion over macroscopic distances is viewed as a sequence of incoherent, thermally assisted hopping events that occur over a wide range of time scales. The VRH theory described in Chapter 4 is the standard approach to this type of problem. Alternative approaches to calculate the hopping rate of polarons between sites *i* and *j* with energy ε_i and ε_j utilize either the Marcus theory of electron transfer or the Holstein theory of polaron transport.

In disordered semiconducting materials, the structural disorder due to variations in the positions and local orientations of the polymer chains (positional disorder) and variations of the conjugation lengths and differences in the local environment (energetic disorder) lead to localization of states, especially in the band tails. Photoexcitations generated initially at energies well above the band edges for $h\nu > E_g$ rapidly relax toward the locaized band-tail states, because on average, the hops downward in energy have higher rate than the thermally activated upward hops. Thus, the direct consequence of the disorder and the distribution of hopping rates is that the mobility is no longer a material constant, but depends on the electric field, the charge concentration, and the temperature. In nonequilibrium conditions the transport will depend on the arrival time of the migrating carriers resulting in thickness dependent dispersive transport.

The application of an electric field results in the reduction of potential barriers in the direction of the field, thereby facilitating motion in the direction of the field. Thus, the mobility associated with transport through localized states has a strong dependence on both temperature and applied electric field. For example, VRH hopping is field dependent. Historically, the Poole–Frenkel field dependence has often been observed.

Normally the mobility increases in the direction of the field. However, the counter-intuitive negative field dependence is also possible in the presence of strong positional disorder. Some

Table 8.1: Experimental methods used to determine charge carrier mobility in organic materials.

		Transient	References
Nature of charge generation		Time of Flight (ToF)	M.A. Lampert and P. Mark, *Current Injection in Solids*, Academic Press, New York (1970)
	Photo-induced	Photo-CELIV	J. Mort and D.M. Pai (eds.), *Photoconductivity and Related Phenomena*, Elsevier, New York (1976)
		FP-TRMC	A.J. Mozer *et al.*, *Appl. Phys. Lett.* 86, 112104 (2005)
		Auston-Switch	D. Moses *et al.*, *Phys. Rev. B* 53, 4462 (1996)
	Doping-induced	CELIV	G. Juska, *et al.*, *Phys. Rev. Lett.* 84, 4946 (2000) A.J. Mozer *et al.*, *Phys. Rev. B* 71, 035214 (2005)
	Electrostatic	FET	S.M. Sze, *Physics of Semiconductor Devices*, 2nd edn., Wiley, New York (1981)
	Injected (from contacts)	DI SCLC	M.A. Lampert and P. Mark, *Current Injection in Solids*, Academic Press, New York (1970) C. Hosokawa *et al.*, *Appl. Phys. Lett.* 63, 1322 (1993)
		tr-EL	T.C. Wong *et al.*, *Chem. Phys. Lett.* 334, 61 (2001)

"energetically favorable" hops can be opposite to or perpendicular to the electric field. When this occurs, the electric field will inhibit such pathways and thereby lower the mobility.

As a consequence of the disorder, the experimental preparation conditions (nature of charge generation, electric field range, and charge concentration, nanomorphology, etc.) are expected to influence the measured mobility values even in nominally identical materials. A good example of this variation is the orders of magnitude higher charge carrier mobility of P3HT observed in FETs compared to that obtained from measurements of space charge limited current (SCLC) measurements or from time-of-flight (ToF) measurements. The higher mobility in the FET structure has been attributed to the higher charge carrier concentration regime at which FETs normally operate. An additional complication arises from the fact that the field effect mobility is sensitive to the interface chemistry associated with the material used for the gate dielectric. FETs fabricated with semiconducting polymers in the channel are discussed in detail in Chapter 14.

As shown in Table 8.1, a number of methods have been developed to measure the charge carrier mobility and to monitor the charge density decay. A detailed description of each of these methods is included in the references within Table 8.1. The acronyms used in Table 8.1 are defined as follows:

CELIV	Charge Extraction by Linearly Increasing Voltage;
FP-TRMC	Flash Photolysis – Time Resolved Microwave Conductivity;
DI SCLC	Dark Injection Space Charge Limited Current;
tr-EL	transient ElectroLuminescence.

8.11 Summary

To a first approximation, conjugated polymers can be treated as semiconductors with well defined band structures. Many of the concepts developed in the study of inorganic semiconductors can be carried over directly. Disorder plays an important role; the ordered MEH-PPV in oriented PE example discussed in this chapter is a notable exception. Although the chemical purity is certainly not at the level obtained in inorganic semiconductors, the density of electronically active impurities can be remarkably small. Promotion of electrons to antibonding states by the absorption of photons causes bond relaxation. This geometrical relaxation in the excited state from polaron formation leads to vibronic structure in the absorption and the emission spectra. These vibronic sidebands would not be observed in rigid band systems; they are a direct result of the structural relaxation in the excited state. Although the intrachain bandwidths are large, comparable to the bandwidths in inorganic semiconductors, the interchain interactions are weak, comparable to those in molecular crystals. Because of this intrinsic anisotropy, the exciton binding energies are intermediatebetween those of Wannier excitons and Frenkel excitons. In semiconducting polymers, the exciton binding energies range from E_B(exciton) < 0.1 eV in the PPVs and polythiophenes to a few tenths of an eV in the larger band gap systems such as the polyfluorenes.

The transport of polarons (e.g. following photoexcitiation) through the localized states in the band tails is complex. Nevertheless, a number of experimental methods have been successfully used to characterize the transport in the band tails in the presence of disorder.

Despite the apparent complexity of semiconducting polymers cast from solution, their fundamental optical and electronic properties and, importantly, the device physics of polymer LEDs, FETs, and LEFETs can be understood using the concepts introduced in this chapter.

Further reading

V. M. Agronovich and G. F. Bassani (eds.), *Electronic Excitations in Organic Nanostructures*, Elsevier, Amsterdam, 2003.

G. Hadziioannou and P. F. Hutten (eds.), *Semiconducting Polymers*, Wiley-VCH, Weinheim, 2000.

G. Hadziioannou and G. G. Mallarias (eds.), *Semiconducting Polymers*, Vols. 1–2, Wiley-VCH, Weinheim, 2007.

T. Holstein, *Ann. Phys. (N.Y.)*, **8**, 325 (1959).

T. Holstein, *Ann. Phys. (N.Y.)*, **8**, 343 (1959).

H. Klauk (ed.), *Organic Electronics*, Wiley-VCH, Weinheim, 2006.

R. A. Marcus, *Ann. Rev. Phys. Chem.*, **15**, 155 (1964).

A. Moliton, *Optoelectronics of Molecules and Polymers*, Springer, Berlin, 2006.

D. Moses, J. Wang, A. J. Heeger, N. Kirova, S. Brazovski, *Proc.Natl Acad. Sci. USA*, **98**, 13496 (2001).

H. S. Nalwa (ed.), *Handbook of Organic and Conductive Molecules and Polymers*, John Wiley, Chichester, 1997.

M. C. Petty, M. R. Bryce and D. Bloor, *Introduction to Molecular Electronics*, Edward Arnold, London, 1995.

M. Pope and Ch. S. Swenberg, *Electronic Processes in Organic Crystals and Polymers*, Oxford University Press, Oxford, 1999.

N. S. Sariciftci (ed.), *Primary Photoexcitations in Conjugated Polymers: Molecular Exciton versus Semiconductor Band Model*, World Scientific, Singapore, 1997).

M. Schwörer and H. C. Wolf, *Organische Molekulare Festkörper*, Wiley-VCH, Weinheim, 2005.

J. Simon and J. J. Andre, *Molecular Semiconductors*, Springer, Berlin, 1985.

T. A. Skotheim (ed.), *Handbook of Conducting Polymers*, Marcel Dekker, New York, 1986.

T. A. Skotheim and J. R. Reynolds (eds.), *Handbook of Conducting Polymers*, 3rd edition, CRC Press, Taylor and Francis, Boca Raton, 2007.

9

Polymer based light-emitting diodes (PLEDs) and displays fabricated from arrays of PLEDs

There has been remarkable progress in organic display technology during the last 20 years. The field has evolved from the early work by C. W. Tang[1] in 1987 to viable products that have a wide range of applications spanning across consumer markets, general lighting, industrial products and displays for the military. The light emission from polymer LED was first reported by Burroughes et. al[1b]. The active matrix polymer LED display can open totally new applications such as thin film foldable, bendable and transparent and large area displays. The efficiency of organic light-emitting diodes (OLED) has approached nearly 100% using phosphorescent materials. Phosphorescent OLEDs are promising candidates for the next generation energy saving solid state light sources to replace conventional incandescent bulbs and fluorescent lamps. For use in solid state lighting applications, large areas can be fabricated by low-cost roll-to-roll printing methods.

The goal of this chapter is to introduce OLED devices based on luminescent semiconducting polymer materials, to describe the fundamental physics of electroluminescent emission and to outline the device physics associated with high information content displays and solid state lightning.

9.1 LEDs fabricated from semiconducting polymers

Inorganic light-emitting diodes (LEDs) are made of direct band gap inorganic crystals (e.g. GaAs or GaN). Although LEDs made from inorganic crystals work remarkably well for many applications, they are not well suited for applications that require large areas, arrays of different colored LEDs, or mechanical flexibility; crystals are brittle and must be grown by expensive epitaxial methods such as molecular beam epitaxy or chemical vapor deposition. LEDs fabricated from semiconducting polymers are therefore attractive because they can be easily patterned in large areas onto almost any type of substrate, they can emit a range of colors that span the entire visible spectrum (each color resulting from a semiconducting polymer with a specific molecular repeat unit), and they are flexible.

Figure 9.1 shows a schematic of the device structure and a photograph of a thin film flexible seven-segment display fabricated from a semiconducting polymer. The bottom electrode

[1] C. W. Tang, *Appl. Phys. Lett.*, **51**, 913 (1987); (b) J. H. Burroughes *et al. Nature* **347**, 539 (1990).

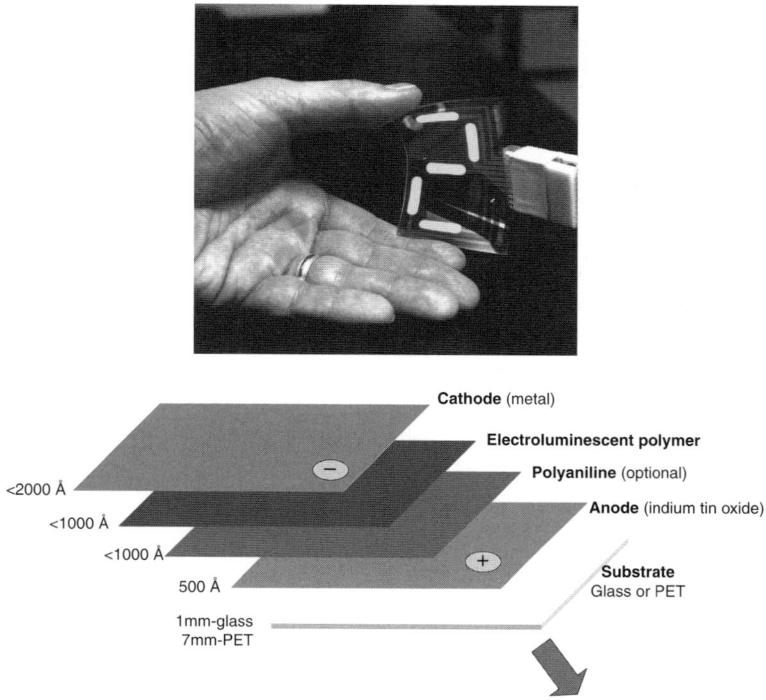

Fig. 9.1 Device structure of an LED fabricated from semiconducting and metallic polymers and photo of a thin film flexible seven-segment display fabricated from semiconducting and metallic polymers. (This figure is reproduced in color in the color plate section.)

(the anode) of this flexible display was made by spin-casting a layer of metallic polyaniline onto a flexible plastic substrate. More generally, however, a bilayer electrode comprising a thin layer electrode comprising a thin layer of metallic polymer cast onto ITO is used for the anode; the ITO carries the current and the metallic polymer layer serves to planarize the ITO and to improve the injection of holes into the luminescent semiconducting polymer. The emissive layer of the display shown in Fig. 9.1 was formed by spin casting a thin layer (approx. 100 nm) of MEH-PPV (poly(2-methoxy-5-(2-ethylhexyloxy)-1.4-phenylenevinylene) onto the polyaniline. The top electrodes that define the seven-segment display were formed by evaporating calcium through a patterned shadow mask. Because of the low conductivity of undoped emissive polymers and the very thin MEH-PPV layer, it is not necessary to pattern the polymer or the bottom electrode to prevent current spreading between neighboring pixels.

Semiconducting polymers have been synthesized with different molecular structures and with an associated wide range of energy gaps. Consequently, luminescent semiconducting polymers can be obtained with emission colors that span the full range of the visible spectrum. Functionalization of the same main chain with different side-chains (e.g. with alkoxy or alkyl groups) can be used to shift the color of the emitted light over a substantial portion of the visible spectrum. The synthesis of homopolymers and copolymers with different molecular structures is the "band gap engineering" methodology for semiconducting polymers. Figure 9.2 shows a few examples that demonstrate the range of different band gaps and

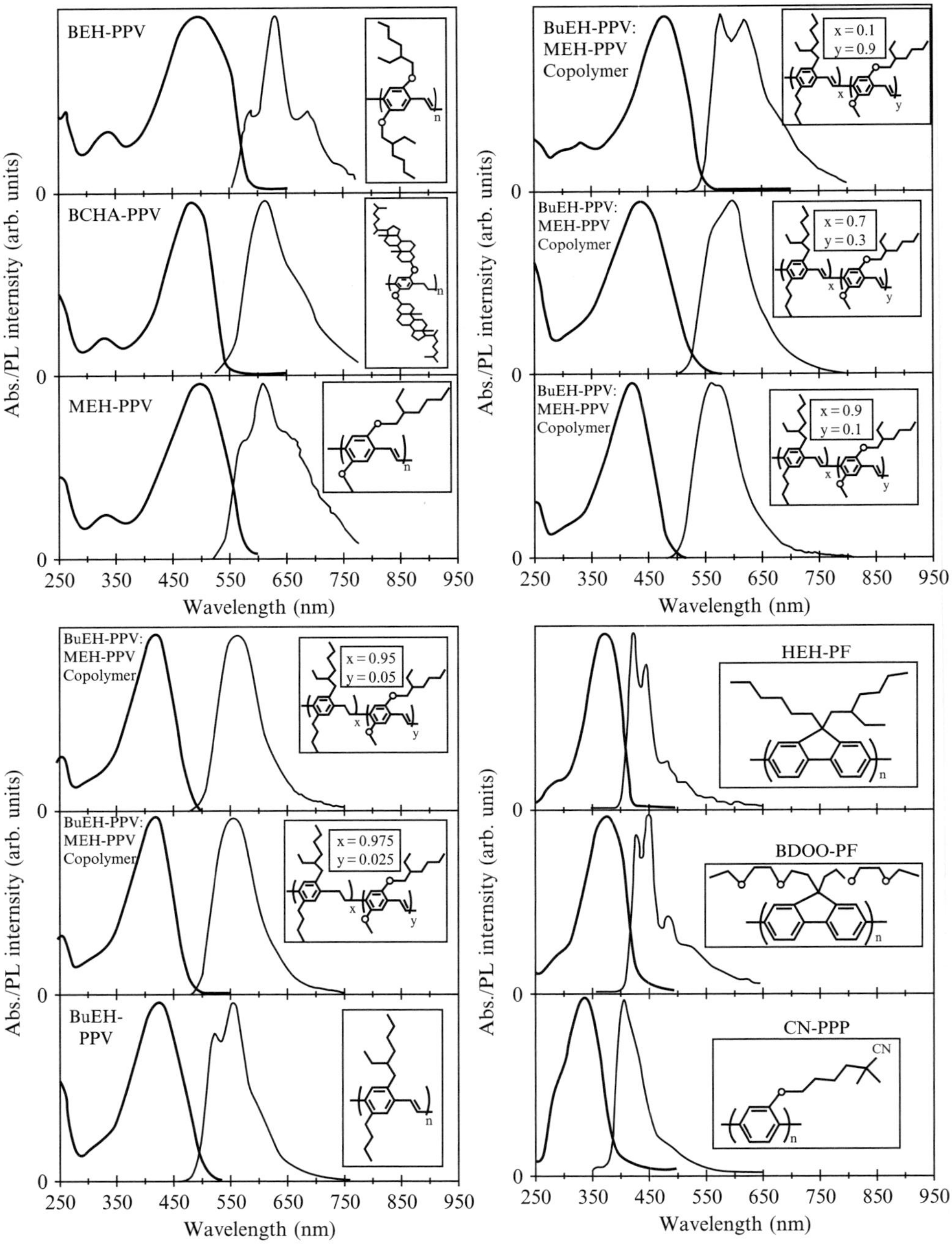

Fig. 9.2 Absorption and emission spectra of a number of semiconducting polymers with different molecular structures. The emission colors span the visible spectrum.

emission colors that can be obtained through band gap engineering. The use of copolymers is particularly interesting. By using block-copolymers, well-defined quantum-well structures can be created.

Concentration quenching is not a major problem in semiconducting polymers. Typically, the quantum efficiency for photoluminescence from thin films is comparable to that from dilute solutions. The absence of strong concentration quenching is the result of the spatial delocalization of the excited states. Because the weakly bound excitons are spread over manyrepeat units, the Davidov splitting that results from interchain interactions is small. As a result, the disorder that is present in films cast from solution is sufficiently large to mix the "dark" and emissive Davidov-split states. Thus, quantum efficiencies in excess of 60–70% can be obtained from thin solid films of luminescent semiconducting polymers.

9.2 Accurate measurement of OLED/PLED device parameters

In addition to transmittance, absorbance, and color temperature, light is described by the following:

flux – the total luminous power (measured in lumens);
intensity – the angular concentration of flux (candelas);
illuminance – the surface density of incident flux (lux or lumen/m^2);
luminance – the intensity emitted per unit area (candelas/m^2).

Table 9.1 summarizes the various quantities, their units and the related measurement techniques.

Table 9.1. The quantity and unit of measurement.

Technique of measurement	Type of measurement	Quantity	Unit of measurement
Photometry	Total flux	Total luminous flux	lumens
Radiometry	Total flux	Total radiant flux	Watts
Spectroradiometry	Total flux	Total spectral flux	Watts/nm
Photometry	Angular intensity	Luminous intensity	candelas=lumen/sr
Radiometry	Angular intensity	Radiant intensity	Watts/sr
Spectroradiometry	Angular intensity	Spectroradiometric intensity	Watts/(sr nm)
Photometry	At a surface	Illuminance	Lux=lumen/m^2
Radiometry	At a surface	Irradiance	Watts/m^2
Spectroradiometry	At a surface	Spectral irradiance	Watts/(m^2 nm)
Photometry	At a surface	Luminance	candelas/m^2
Photometry	At a surface	Radiance	Watts/(sr m^2)
Spectroradiometry	At a surface	Spectral radiance	Watts/(sr m^2 nm)

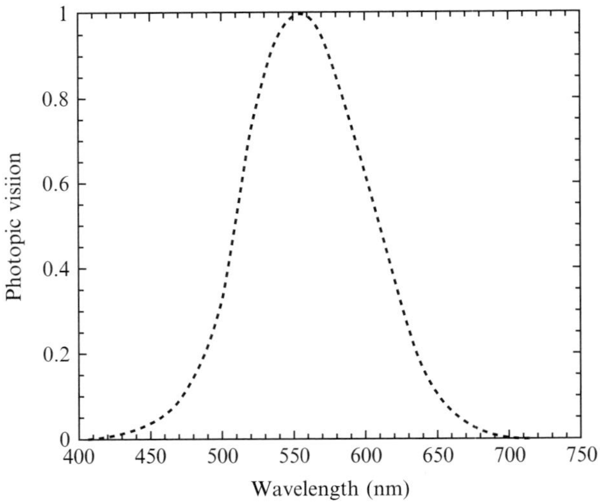

Fig. 9.3 The photopic luminosity function.

Because the most important applications of OLEDs/PLEDs are for displays and for lighting, the response of the human eye, described by the photopic luminosity function, must be taken into account. The photopic luminosity function is shown in Fig. 9.3. By using the photopic luminosity function, the radiance [watts/(sr m^2)] from display devices is converted into the luminance [candela/m^2, cd/m^2 or lumen/(sr m^2)]. Therefore, photometry is used to measure the forward viewing luminance at the surface of an OLED/PLED.

The luminous intensity (luminance) can be determined by measuring the flux in any given solid angle, Ω (the ratio of the size of aperture divided by the square of the distance between the light and the aperture). Consider a flat emitting surface, each point of which emits light equally in all directions, i.e. a "Lambertian" source. For a Lambertian source, the light intensity follows a cosine law decreasing from maximum when viewed along the normal to the surface to zero at 90° to the normal. An OLED/PLED is a Lambertian source if the luminous intensity follows the cosine law when measured with a very small area detector placed far away from the surface (very small Ω, see below).[2] Although the light emitted from OLEDs/PLEDs typically cannot be precisely described as the emission from a flat radiating surface for which the light intensity falls off as the cosine of the observation angle with respect to the surface normal, we will assume that these deviations from Lambertian emission are relatively small.

Therefore, to measure the luminous intensity one must first choose a reference direction for the measurements then determine the solid angle to be used in the measurement. For display applications, the reference direction should be chosen as the forward viewing direction (along the direction perpendicular to the surface of the OLED/PLED). The luminous intensity is defined as the emission in cd/m^2 from the emitting surface.

[2] N. C. Greenham *et al.*, *Adv. Mater.*, **6**, 491 (1994).

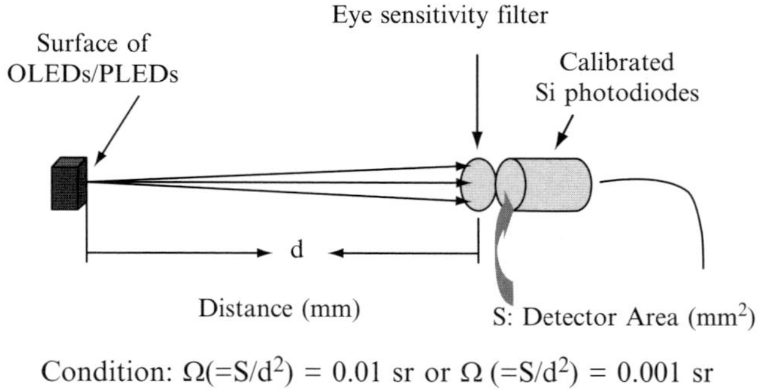

Fig. 9.4 The configuration for measuring OLED/PLED luminous intensity.

An accurate and convenient configuration for measuring luminous efficiency from OLEDs/PLEDs is shown in Fig. 9.4. Because the luminous efficiency strongly depends on the spectrum of the OLED/PLED spectrum (even for constant quantum efficiency), an eye-sensitivity filter is mounted directly onto the surface of a calibrated photodiode (see Fig. 9.4)

Assuming a Lambertian intensity profile and a disk-shaped source with radius *r*, the light intensity on a point detector placed a distance *d* away (see Fig. 9.5) can be expressed as follows:

$$I = I_o(r^2/(r^2 + d^2)) \tag{9.1}$$

For a detector with a finite area, this expression can be used if the detector subtends a sufficiently small solid angle. The conditions recommended by CIE correspond to $\Omega \leq 0.01$

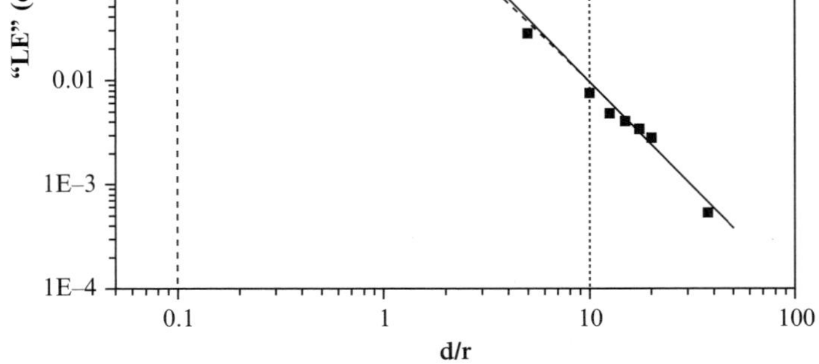

Fig. 9.5 Inverse square law approximation of measured luminous efficiency.

sr. Under these conditions, the luminous intensity (, in candelas) can be obtained by comparing the signal from the PLED to that obtained from a calibrated lamp. The brightness is then given by $L = \Lambda/\pi r^2$.

As expected, the signal falls off as d^{-2}. There is no serious error introduced by the finite area of the detector even for $d/r \approx 1$. For the data IN Fig. 9.5, $r = 2$ mm.

Once the luminance, L (cd/m^2), is accurately measured, the luminous efficiency, LE (cd/A), and luminous power efficiency, PE (lm/W) can be determined using the following equations:

$$LE(cd/A) = \frac{L}{j}$$

$$PE(lm/W) = \frac{\pi^* L}{jV} \tag{9.2}$$

where j is the current density, and V is the applied voltage.

Another option is to place the OLED/PLED into an integrating sphere containing a calibrated detector, and then to measure the total light output from the LED. In an integrating sphere, all the emitted photons are collected, including those that are guided to the edge of the substrate. However, because photons emitted from the edge of the substrate are not useful in display applications, the integrating sphere approach must be used with caution. If the edges of the OLED/PLED are sealed to absorb the waveguided and scattered light, the external quantum efficiency can be accurately obtained with this measurement technique.

Typical conditions for measurement of PLEDs are the following: the surface area (S) of the calibrated photodiode with eye-sensitivity filter is 16 mm^2 (4 mm 4 mm); the distance (d) between the photodiode and PLEDs is 4 cm; i.e. $\Omega = 0.01$ sr ($\Omega = S/d^2$).

Note that the response of a diffuser/fiber combination that is often used is far from uniform or ideal. Not only does the fiber/diffuser combination fail to provide the correct measurement area, but also the nonuniform spatial response makes the result highly sensitive to alignment. Therefore, the fiber/diffuser combination should be avoided if the OLED/PLED intensity is to be correctly measured.

9.3 Fowler–Nordheim tunneling in semiconducting polymer MSM diodes

The operating mechanism of polymer LEDs is quite different from conventional p-n junction LEDs. In a polymer LED, a pure undoped film of luminescent semiconducting polymer is sandwiched between a high work function metal anode and a low work function metal cathode. The charge carrier concentration in such pure semiconducting films is sufficiently low ($\approx 10^{14}$–10^{15} cm^{-3}) that any residual carriers introduced by impurities, etc. are swept out by the built-in field that arises from the difference in work functions of the two electrodes. As noted in Chapter 8, the depletion depth of luminescent semiconducting polymers is, therefore, larger than the thickness of the polymer layer in an LED (typically < 100 nm). Consequently, the electronic structure of the LED can be approximated by the simple band model displayed in Fig. 9.6. The built-in field causes the uniform slope; there is negligible band bending.

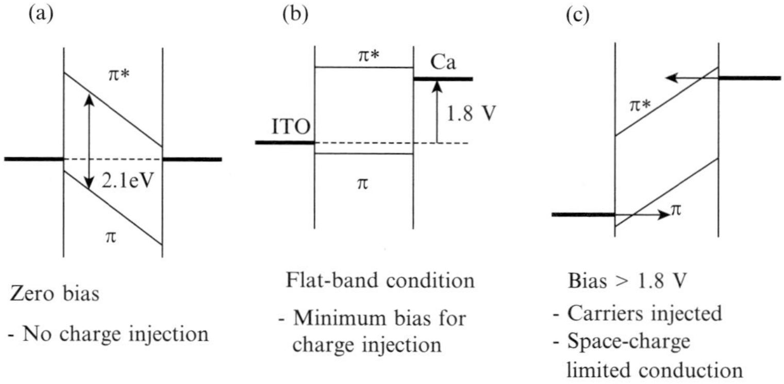

Fig. 9.6 The electronic structure of the LED.
(a) Zero bias with a common Fermi level across the device (note the within the semiconducting polymer layer, the Fermi level moves across the energy gap on going from the anode to the cathode).
(b) Flat band condition occurs when the applied voltage equals the difference in the work functions of the anode and the cathode. This is the minimum voltage required for injection of electrons and holes. Ideally, the electroluminescent emission should turn on at this voltage.
(c) Forward bias; carriers are injected through the triangular barriers at the anode (holes) and cathode (electrons) and meet within the polymer film where they radiatively recombine (electroluminescence).

The height of the barrier for hole injection is determined by the difference between the work function of the anode and the energy level of the π (valence) band; the height of the barrier for electron injection is determined by the energy difference between the work function of the cathode and the energy level of the π* (conduction) band. The barrier height can be determined in this simple way because conjugated polymers do not have surface states that would pin the Fermi level (see discussion in Chapter 8).

When a positive bias is applied to the LED, the Fermi level of the cathode is raised relative to that of the anode, as shown in Fig. 9.6(c). Thus the thickness of the triangular tunneling barrier is a function of the applied voltage; the barrier thickness decreases as the voltage is increased. Carriers tunnel through the barrier primarily by Fowler–Nordheim field emission tunneling from the anode and cathode into the π-band (holes from the anode) and the π*-band (electrons from the cathode) of the semiconducting polymer.

Thermionic emission over the barriers can also play a role if the barriers are small and the temperature is relatively high. Since the rate of injection by Fowler–Nordheim tunneling is determined by the strength of the electric field, it is important for the polymer layer to be thin so that high electric fields can be obtained at low voltages.

Note that in Fig. 9.6(a) and (b), the Fermi level moves all the way across the energy gap of the polymer (from one side of the device to the opposite side). The ability to freely move the Fermi level across the gap is an important feature of the device physics of semiconducting polymers (see Chapter 8, Section 8.2). If the Fermi level were pinned by gap state, as is typically the case for inorganic semiconductors, the thin film LED would not function.

To optimize the performance of polymer LEDs, it is important to minimize the barriers for charge injection by choosing electrodes with work functions that are well matched to the bands of the polymer. ITO, polyaniline, polypyrrole, and poly (3,4-ethyendioxthiophene) (PEDOT) are the most commonly used anode materials. They are also semitransparent and therefore allow the emitted light to escape from the device. Calcium, barium, and magnesium are commonly chosen as the cathode because of their low work functions. Unfortunately, low work function metals are highly reactive. Polymer LEDs must, therefore, be hermetically sealed for long life. Improved electron injection from stable metals such as aluminum can be achieved by coating the electrode with a polar self-assembled monolayer. The dipole layer effectively shifts the electrode work function.

When the electrodes are well matched to the energy bands of the semiconducting polymer, the barrier for charge injection is small. After injection of electrons and holes into the π^*- and π-bands, they must come together and recombine radiatively to give off light. In this context, the low mobility of the charge carriers (polarons) in semiconducting polymers is helpful since the slow drift of the charge carriers across the thickness of the semiconducting polymer will allow sufficient time for the carriers to meet and recombine.

If one carrier type is injected much more efficiently and drifts in the applied electric field with higher mobility than the other, then many of the majority carriers will traverse the entire polymer layer without recombining with a minority carrier. Thus, maximum efficiency can be achieved only through balanced electron and hole currents.

As demonstrated in Fig. 9.7, diodes fabricated as described above show excellent rectification ratios and strong electroluminescent light emission in forward bias. Light emission turns on close to the flat band condition, when the applied voltage is greater than the difference between the work functions of the two electrodes (see Fig. 9.6). The current at voltages below 2 V in Fig. 9.6(b) is the residual leakage current resulting from microshorts and imperfections.

Parker observed that the turn-on voltages for devices with different thicknesses of the semiconducting polymer layer are roughly the same and equal to the voltage required to reach the flat band condition.[3] As shown in Fig. 9.8, however, the current density is a strong function of the thickness. By re-plotting the data in Fig. 9.8 as a function of the electric field strength (rather than the voltage), all the data collapse onto a single curve (see Fig. 9.9). Thus polymer LEDs are electric field driven rather than voltage driven.

9.3.1 Single carrier devices

In order to study the details of the carrier injection, single carrier devices can be fabricated in which the current density of one of the carriers is reduced to negligible levels.

"Hole-only" devices can be fabricated by replacing the low work function Ca cathode (2.9 eV) with higher work function metals such as In (4.2 eV), Al (4.4 eV), Ag (4.6 eV), Cu (4.7 eV) or Au (5.2 eV); see Fig. 9.9.

Increasing the work function of the cathode increases the offset between the Fermi energy of the cathode and the bottom of the π^*-band (for example, in MEH-PPV the bottom of the π^*-band is at –2.8 eV with respect to the vacuum). This reduces the number of injected electrons to levels below which the injected holes dominate. Despite the fact that the work function of the cathode increases by more than 1 eV (from In to Au), the *I-V* curves of the

[3] I. D. Parker, *J. Appl. Phys.*, **75**, 1656 (1994).

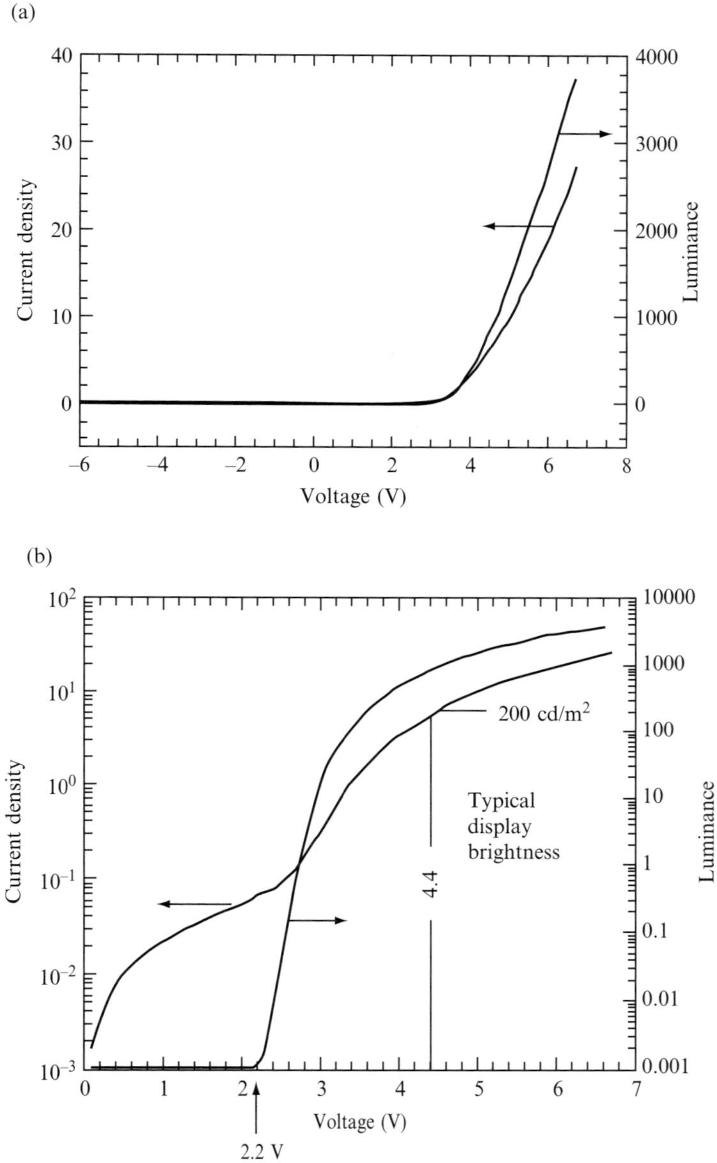

Fig. 9.7 Current density and luminance versus voltage for devices fabricated with MEH-PPV using polyaniline as the anode and calcium as the cathode.

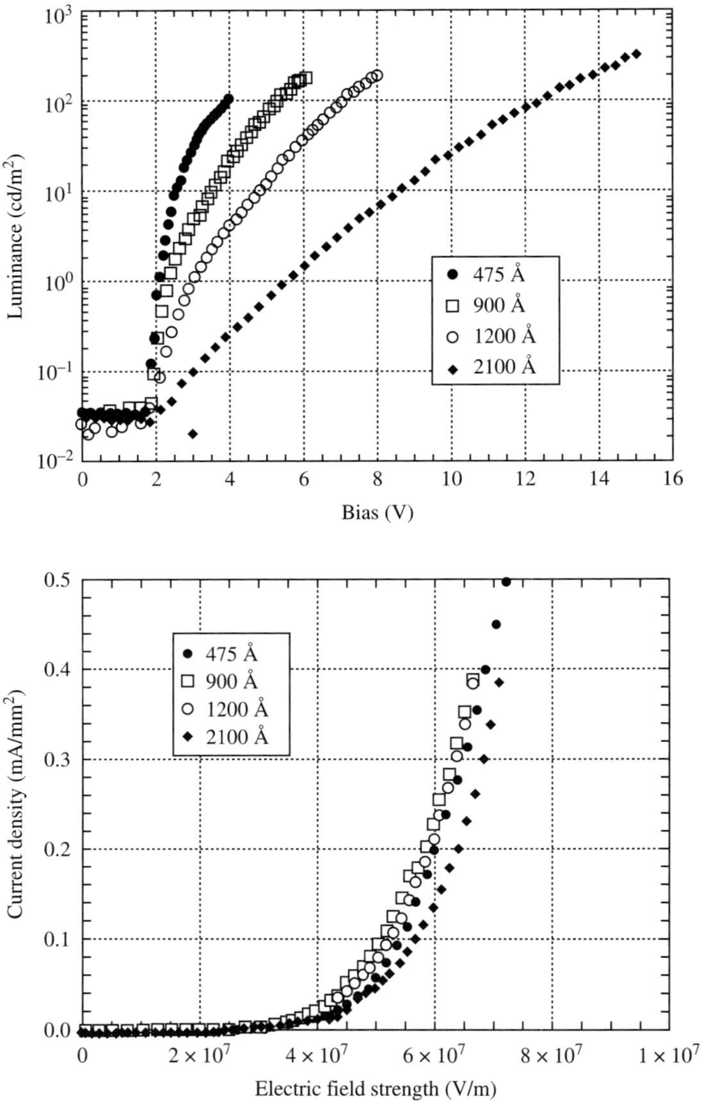

Fig. 9.8 Top: luminance versus voltage for ITO/MEH-PPV/Ca LEDs with semiconducting polymer layers of different thicknesses. Bottom: current versus electric field for devices fabricated with MEH-PPV layers of different thicknesses. [Reproduced form A. J. Heeger, *et al.*, *Synthetic Metals*, **67**, 23 (1994), with permission from Elsevier.]

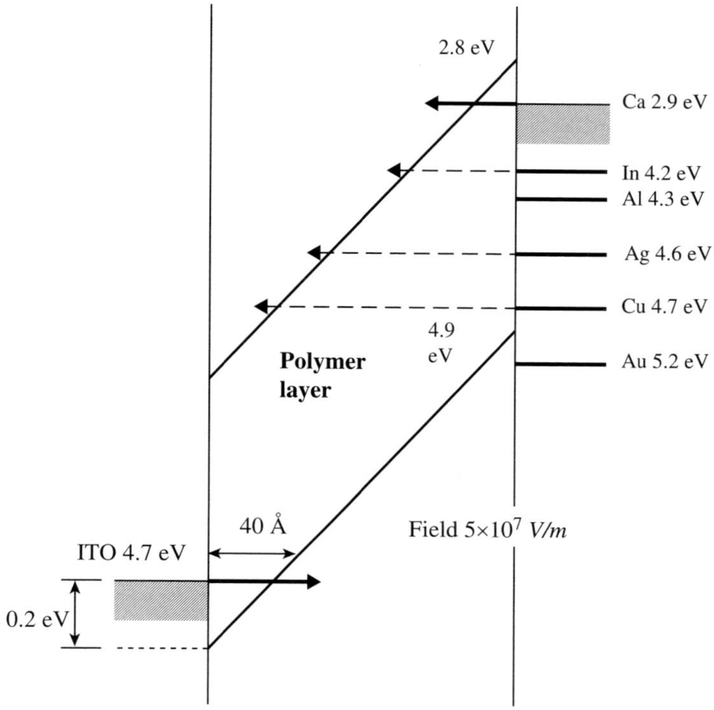

Fig. 9.9 Band diagram (in forward bias) indicating the magnitudes of the triangular tunneling barrier for various cathode materials.

devices made with In, Al, Ag, Cu and Au cathodes are almost identical, indicating that electron injection has been shut off.

Fowler–Nordheim tunneling theory predicts that the tunneling current is an exponential function of $1/F$

$$I \propto F^2 \exp(-\kappa/F) \tag{9.3}$$

where I is the current, F is the electric field strength and κ is a parameter that depends on the barrier shape. For a triangular barrier,

$$\kappa = 8\pi(2m^*)^{1/2}\varphi^{3/2}/3qh. \tag{9.4}$$

Here, ϕ is the barrier height, m^* is the effective mass of holes in the semiconducting polymer, q is the electron charge and h is Planck's constant.

The *I-V* data from a hole-only device fabricated from MEH-PPV with Au as the cathode and ITO as the anode are plotted as I/F^2 versus $1/F$ in Fig. 9.9. As predicted, the plot is close to linear particularly at high fields. The literal assumption of tunneling through a triangular barrier appears to be an excellent approximation. The deviation from linearity at lower fields probably indicates an additional contribution to the current from thermionic emission.

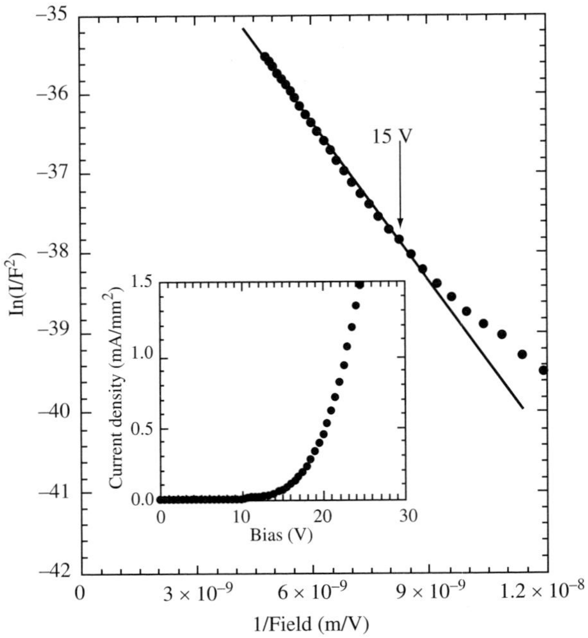

Fig. 9.10 Fowler–Nordheim plot of $\ln[I/F^2]$ versus $1/F$ for a hole-only device fabricated using MEH-PPV with ITO as the anode and Au as the cathode. [Reproduced form I. D. Parker, *J. Appl. Phys.*, **75**, 1656 (1994) with permission from American Physical Society.]

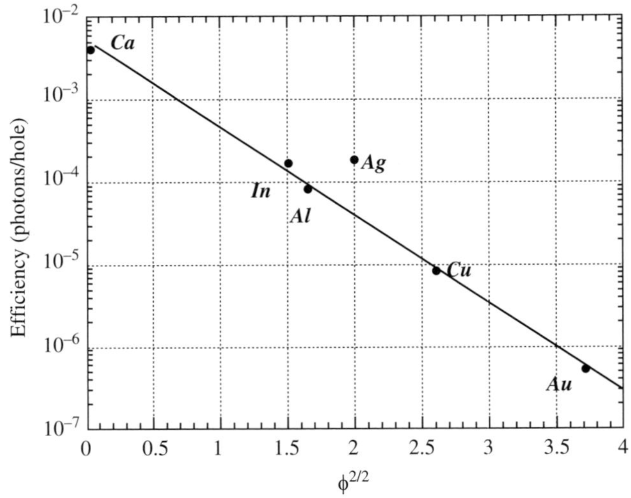

Fig. 9.11 Semilog plot of quantum efficiency (photons/electron) versus $\varphi^{3/2}$. [Reproduced from A. J. Heeger *et al.*, *Synthetic Metals*, **67**, 23 (1994), with permission from Elsevier.]

Assuming that the electric field is constant across the semiconducting polymer and that $m^* = m$ (the free electron mass), the calculated barrier heights are 0.2–0.3 eV for all the hole-only devices indicated in Fig. 9.10. The fact that the inferred barrier height is unchanged despite the large variation in cathode work function indicates that the barrier must be at the ITO/polymer interface. Indeed, $\varphi = 0.2$–0.3 eV is in agreement with the energy diagram in Fig. 9.9.

A similar approach can be taken to study electron injection. Replacing ITO with a lower work function metal (e.g. Nd or Mg) yields devices in which the carriers are almost exclusively electrons. A similar analysis of the data from a range of "electron-only" devices indicates that electrons tunnel into the π^*-band through a triangular barrier at the polymer/cathode interface. With Ca as the cathode, this barrier is small (approx. 0.1 eV for MEH-PPV), again in agreement with the energy diagram in Fig. 9.10.

9.3.2 PLED operating voltage and efficiency

The operating voltage is very sensitive to the barrier height as predicted by the Fowler–Nordheim tunneling model. Equations 9.3 and 9.4 indicate that for the same tunneling current, the ratio $\varphi^{3/2} / V$ must be the same. Thus, increasing the barrier height from 0.1 eV (PANI) to 0.6 eV (Cr) should increase the operating voltage by a factor of 11. The experimentally determined increase was a factor of 9.

It is obvious that the device efficiency, η, must be very sensitive to the barrier height, since the efficiency is limited by the minority carrier density. As indicated by Eqns 9.3 and 9.4 the efficiency is a function of the barrier height: $\ln(\eta)$ versus $\varphi^{3/2}$. The excellent agreement between the theory and the data in Fig. 9.11 confirms the use of the Fowler–Nordheim tunneling model for describing the carrier injection into the band structure of the semiconducting polymer.

There is evidence of interface modification (chemical reactions between the metal atoms and certain groups on the polymer) when metal films are deposited onto semiconducting polymers. Nevertheless, the data in Fig. 9.12 demonstrate a clear systematic variation in the device efficiency as a function of the barrier height, implying that such local chemical reactions at the surface are not of major importance to the device operation.

9.3.3 Limits of the MIM model

When the barrier heights are small, the simple comparison of work functions for the prediction of barrier heights is inappropriate; thermionic emission becomes important for barriers less than a few tenths of an eV. Moreover, the barrier will certainly not be perfectly triangular as the model assumes, but must be somewhat ill-defined as a result of disorder-induced band-tailing near the polymer band edges. In addition, the evaporated electrode materials are likely to be full of defects (possibly even amorphous) and are therefore not expected to have precisely the work function values listed in reference books. Other factors, such as the mobility of each charge species and space-charge limiting effects, may also have increasing influence over the performance for low barrier-height devices. Nevertheless, it is clear that an energy band picture for the semiconducting polymer supplemented by Fowler–Nordheim tunneling theory provides an excellent starting point for understanding the operating mechanism of LEDs fabricated from semiconducting polymers.

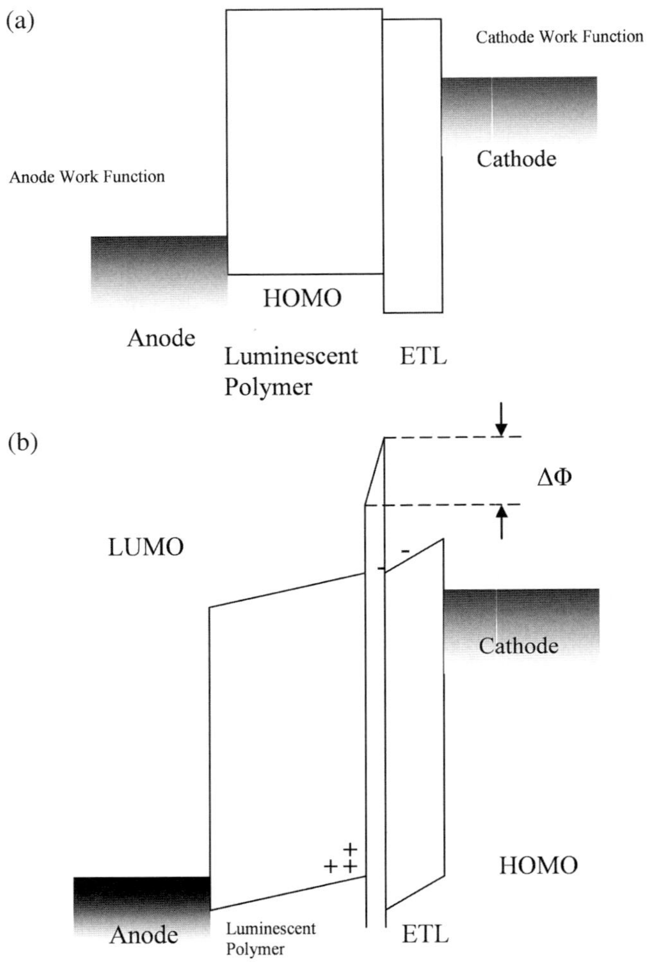

Fig. 9.12 (a) Flat band diagram of the electronic structure of an LED with a hole blocking layer. (b) Electronic structure of an LED with a hole blocking layer in forward bias. The electric field is primarily across the ETL; the field in the luminescent polymer is nearly zero.

9.3.4 Approaches to improved carrier injection: Hole-blocking layers, electron–acceptors and hole–acceptors

It is not always possible to find electrode materials that are well matched to the electronic structure of an electroluminescent polymer. For example, for large band-gap blue-emitting polymers such as the polyfluorenes, the energy barrier at the cathode is typically too large for efficient injection of electrons via Fowler–Nordheim tunneling. Similarly, for more electronegative, stable semiconducting polymers, the barrier at the anode will be too large for efficient

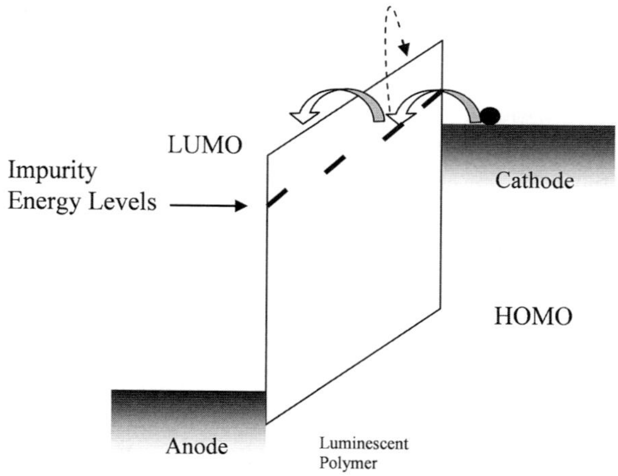

Fig. 9.13 End-capping with electron accepting groups and/or hole accepting groups leads to the formation of localized states near the band edges; electron accepting states are at energies just below the π^*-band, and hole accepting states are at energies just above the π-band. Alternatively, such acceptor units can be inserted into the main chain as "dopants". Small molecule acceptors can also be used.

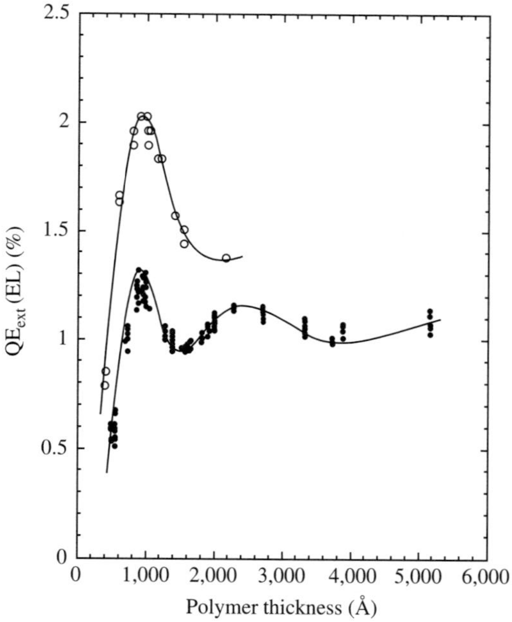

Fig. 9.14 Thickness dependence of the external quantum efficiency of PLEDs fabricated from OC1C10-PPV and MEH-PPV. [Reproduced from Y. Cao, *Nature*, **357**, 477 (1992) with permission from Nature Publishing Group.]

hole injection. In such cases, hole and/or electron blocking layers can be added to improve the balance of electron and hole currents.

Figure 9.12 shows the electronic structure of an LED with a hole blocking layer. The blocking layer creates a barrier at the interface of two polymers that blocks the flow of the majority carrier. As the density of the majority carrier increases at the blocking interface, the electric field at the minority carrier injecting electrode increases, thereby enhancing the minority carrier injection. As a result, the electron and hole currents become more nearly balanced. An analogous diagram can be drawn for a device with an electron blocking layer.

End-capping with electron accepting groups and/or hole accepting groups is also a useful example.

As shown in Fig. 9.13, such groups lead to the formation of localized states near the band edges. Thus in forward bias an electron can tunnel into the π^*-band in two steps; first from the cathode into a localized state on an electron accepting site, and then from that site into the π^*-band. Since the tunneling probability decreases exponentially as $\varphi^{3/2}$, the two-step injection can be more efficient than direct tunneling through the barrier into the π^*-band. As sketched in Fig. 9.13 (dotted arrow), a two-step process involving tunneling followed by thermal excitation from the impurity acceptor state into the band can also help.

Electron accepting and/or hole accepting groups can also be incorporated into the chain (copolymers) or by addition of appropriate small molecules into the film. In the latter approach, however, clustering and phase separation are likely to be problems.

9.4 Thickness dependence of electroluminescence efficiency

Figure 9.14 shows the thickness dependence of the external quantum efficiency for electroluminescence, $QE_{ext}(EL)$ as a function of the film thickness for two closely related PPV derivatives, OC1C10-PPV and MEH-PPV. The quantum efficiencies of these devices were measured using an integrating sphere.

The oscillatory dependence on the thickness (d) of the polymer layer arises from the proximity of the metallic mirror electrode (the cathode); the emitting oscillator interacts with the virtual image oscillator "behind the mirror". Because the radiation from the emitter and the retarded radiation from the image oscillator interfere, the PL decay rate is an oscillatory function of the distance from the mirror. Consequently, for a thin film, the quantum yields for photoluminescence and electroluminescence are oscillatory functions of d. Such thin film interference effects are also important for solar cells as discussed in Chapter 13.

When the average distance from the mirror to the oscillator is too small ($d < 100$ nm), the losses from the metallic electrode quench the luminescence. Although thinner devices would have lower operating voltages (the polymer LED is electric field driven), the efficiency decreases rapidly for thicknesses below 100 nm. As shown in Fig. 9.14, the optimum thickness is approximately 80 nm.

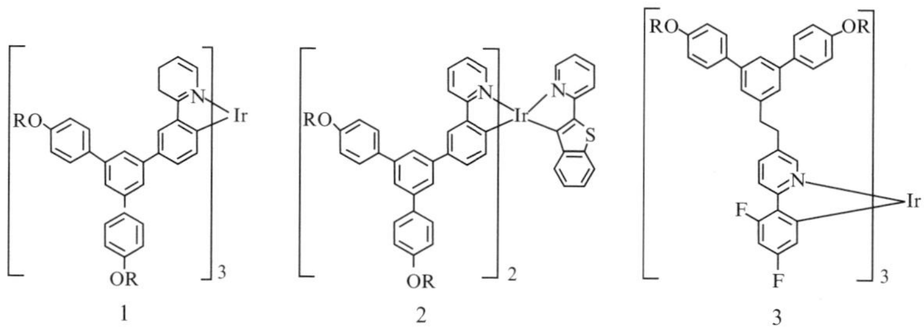

Fig. 9.15 Phosphorescent light-emitting dendrimers (R = 2-ethylhexyl); 1) Green emitter, 2) red emitter and 3) sky blue emitter. Phosphorescent dendrimers consist of core dendrons, and surface groups. This modular construction enables solution processing and the electronic and optical properties to be tuned independently.

9.5 Limits on the electroluminescence efficiency

9.5.1 Phosphorescent LEDs

In organic LED, electrons and the holes are injected from the contacts, and the charges move through the device under the influence of the electric field. Some of the charges recombine forming singlet state and triplet state excitons. Based on spin statistical arguments, the singlet to triplet excitons formation on recombination would be 1:4. The efficiency of light emission is given by,

$$EQE = \Phi_{\text{capture}} \times \Phi_{\text{spin}} \times \Phi_{\text{rad}} \times \Phi_{\text{escape}} \tag{9.5}$$

where Φ_{capture} is the fraction of excitations which recombine, Φ_{spin} is a factor to take account of spin statistics ($\Phi_{\text{spin}} = 1$ in phosphorescent materials and $\Phi_{\text{spin}} = 1/4$ for fluorescent materials), Φ_{rad} is the fraction of excitons which once formed decay radiatively (approximated by the photoluminescence quantum yield) and Φ_{escape} is the fraction of photons which once generated escape from the device (approximately $1/2n^2$ where n is the refractive index).

In early reports on polymer LEDs, the Φ_{capture}, Φ_{rad} and Φ_{spin} factors were far from optimized. The polymers were mainly hole transporting materials so charge balance and Φ_{capture} was poor. To improve charge capture, one or more extra charge transporting layers were incorporated between the light-emitting layer and the cathode and/or the anode. By using extra layers the charge capture efficiency was improved to almost 100%. If the light-emitting material is fluorescent, then only the singlets can emit light contributing a maximum of 25% to emission process. This sets the upper limit of electroluminescence efficiency to ~5% (assuming $\Phi_{\text{escape}} \sim 20\%$ and $\Phi_{\text{rad}} \sim 100\%$). In contrast, for phosphorescent materials which harvest singlet and triplet excitons, the external electroluminescence efficiency can approach ~20%. This implies that the internal quantum efficiency of phosphorescent LED can be achieve values close to 100%. Such a high efficiency was first demonstrated using phosphorescent emitter, fac-tris(phenylpyridyl)iridium(III), in a multilayer device architecture using thermal deposition of the layers.[4]

For solution processed phosphorescent dendrimer material (see Fig. 9.15), 16% external efficiency was achieved; i.e. very close to the theoretical limit of 20%. A key point of the

[4] Adachi *et al.*, *J. Appl. Phys.*, **90**, 5038 (2001).

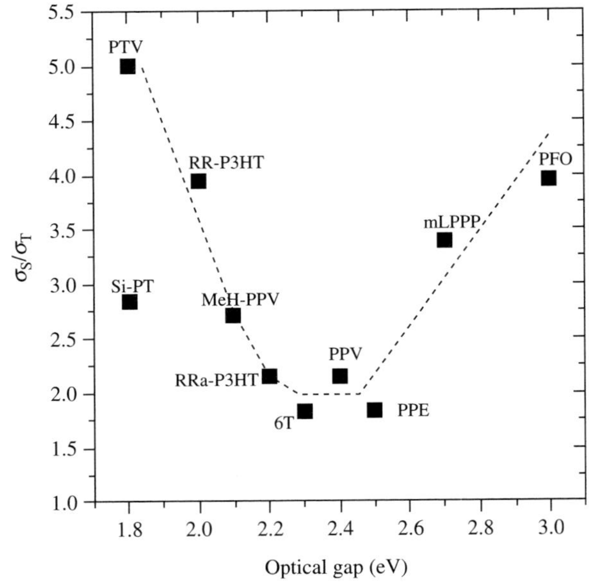

Fig. 9.16 Ratio of the singlet to triplet cross-sections for a series of luminescent semconducting polymers with different energy gaps (and thus different emission colors). [Reproduced from M. Wohlgenannt, *et al.*, *Nature*, **409**, 494 (2001) with permission from Nature Publishing Group.]

phosphorescent dendrimer was that the efficiencies were achieved in a simple device structure consisting of only three organic materials rather than the more complicated multilayer device structure based on evaporated small molecules.

Very high efficiencies for green and red emission were demonstrated with phosphorescent dendrimers. However phosphorescent deep blue emitters still represent a significant challenge. The deeper blue phosphorescent materials are generally less luminescent complexes. This is due to strong coupling of vibrational mode to the excited state. Emission from aggregates or excimers is also a problem. Excimer formation leads to a change of emission color (generally red shift) and a reduction in photoluminescence. Therefore, the ability to engineer the intermolecular interactions at a molecular level for deep blue phosphorescent emitter is critical.

9.5.2 Fluorescent LED (singlet–triplet ratio)

The performance of polymer LEDs is ultimately limited by the quantum efficiency for luminescence. If, however, the lowest energy excited states are strongly bound excitons (electron–hole pairs in singlet or triplet states), the theoretical limit for the quantum efficiency of electroluminescence (EL) is 25% of the corresponding quantum efficiency for photoluminescence (PL): an electron in the π^*-band and a hole in the π-band can form a triplet with multiplicity of three or a singlet with multiplicity of one, but only the singlet can decay radiatively. However, if the cross-section for an electron–hole pair to form a singlet bound state is significantly higher than the cross-section to form a triplet bound state, the ratio of EL quantum efficiency to PL quantum efficiency would be correspondingly larger than that predicted by the simple statistical argument given above (i.e. > 25%). Finally, if the electron–hole binding energy is sufficiently weak, the maximum quantum efficiencies can theoretically approach that of the PL. In the weak interaction limit, electron–hole pairs in a triplet configuration would simply scatter but not recombine. Eventually, the spin–orbit interaction would cause a spin-flip after which singlet formation and radiative recombination could occur.

Values for $QE_{ext}(EL) / QE_{ext}(PL) > 0.5$ indicate that either the exciton binding energy is small (of order k_BT) or that the cross-section for an electron–hole pair to form a singlet bound state is significantly higher than the cross-section to form a triplet. There is experimental evidence that both of these alternatives are true. The exciton binding energy in the soluble PPVs, as inferred from photoconductivity measurements, is approximately 50–70 meV; i.e. only 2–3 k_BT at room temperature (see Chapter 8). Wohlgenannt *et al.*[5] were able to directly measure the ratio of singlet to triplet cross-sections; their data are reproduced in Fig. 9.16. The singlet cross-section is 3–4 times larger than the triplet cross-section for all the luminescent semiconducting polymers; i.e. consistent with $QE_{ext}(EL) / QE_{ext}(PL) > 0.5$.

9.6 Pixelated displays

There are two general classes of pixilated displays that differ in the manner in which the individual pixels are switched on and off: passive matrix displays and active matrix displays.

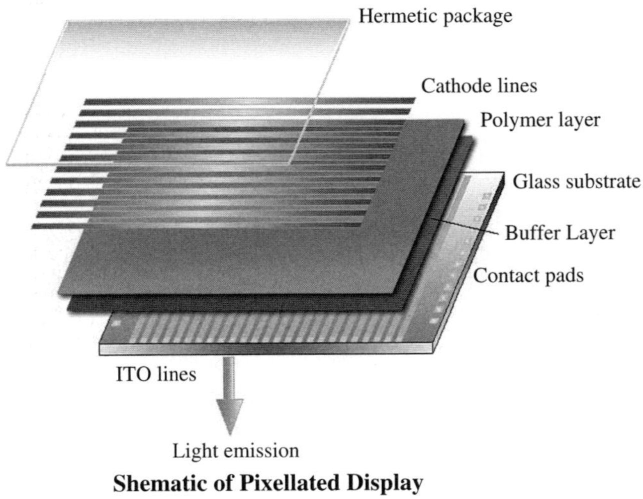

Fig. 9.17 Diagram of the structure of a passively addressed display. (This figure is reproduced in color in the color plate section.)

[5] M. Wohlgenannt *et al.*, *Nature*, **409**, 494 (2001).

Fig. 9.18 Sketch of a printed array of pixels is shown on the top left. Top right is a photo of the colored pixels in an early device fabricated by ink-jet printing. Bottom image is a full color polymer emissive active matrix display showing a video. For active matrix display, a switching circuit is included into each pixel. [Courtesy of Dupont Displays.] (This figure is reproduced in color in the color plate section.)

Active matrix displays have a thin film transistor (TFT) switching circuit embedded in the area of each individual pixel. Although the TFT backplanes needed for polymer emissive displays are similar to those developed for liquid crystal displays (LCDs), the TFT circuits must be capable of switching much higher currents than are required for LCDs. For active matrix OLED displays, the luminescent semiconducting polymer and cathode, etc. are deposited directly onto the pre-manufactured TFT backplane (the anode for the LED pixel is built onto the TFT circuit). In an active matrix display, the pixels are held at constant brightness by the TFT circuit and the image is refreshed at video rates (e.g. 60 Hz).

Passively addressed displays utilize electrodes that are patterned into columns and rows as sketched in Fig. 9.17. In a passively addressed display, only the pixel defined by the intersection of a specific column of the anode and a specific row of the cathode that are simultaneously addressed will give off light. The image is produced by scanning the columns and rows with appropriate driver circuits. Because of the low duty cycle, each individual pixel is pulsed to high brightness when it is on; the viewer averages the intensity. For example, with a 64 column display, the duty cycle is 1/64. Thus to achieve an average brightness of 200 cd/m^2, the pixels must be pulsed on to a brightness of 12,800 cd/m^2.

For monochrome displays, a uniform film of the semiconducting polymer (as sketched in Fig. 9.17) can be spin-cast from solution. For full color displays, both active matrix and passively addressed, individual red, green and blue pixels must be patterned and individually addressed.

The colored liquids in Fig. 3.1(a) are examples of different semiconducting polymers in solution. These colored solutions can be thought of as "functional inks" and applied to substrates using the methods of modern printing technology (see the discussion in Chapter 3). Ink-jet printing appears to be a feasible way to deposit the red, green and blue pixels of a full color display (Fig. 9.18).

In display applications, fast (video rate) switching of the pixels is required. The intrinsic lifetime of the electroluminescence is the decay time of the photoluminescence; i.e. less than a nanosecond. Thus, for pixilated polymer emissive displays, the switching rate is limited only by the RC time constant of the device. For the small pixels of a full color display, C is sufficiently small that the devices can be switched in times that are in the nanosecond regime.

The possibility of using flexible substrates to fabricate fully flexible displays is particularly attractive (see Fig. 9.2). The critical limitation of the flexible LED technology is the permeability of most plastics to oxygen and water vapor. This permeability brings atmospheric oxygen and water vapor into the device and accelerates degradation. Water vapor is especially harmful since it will react with the low work function cathode. The development of technology for making plastic substrates capable of forming a hermetic seal will be a critical step in the evolution of the field of polymer LEDs. A photograph of a full color polymer emissive active matrix display is shown in Fig. 9.18.

9.7 White light emission

PLEDs which emit white light are of interest for use as back-lights in high efficiency active matrix displays (with color filters) and because they can be used for solid state lighting. Several approaches have been used to generate white light from PLEDs. As an illustrative example, we

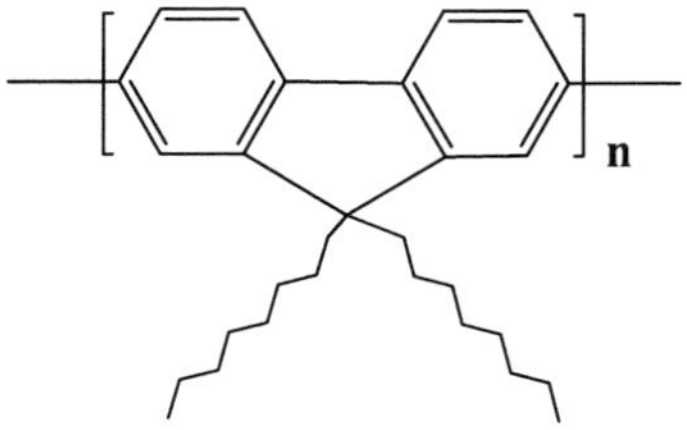

PFO

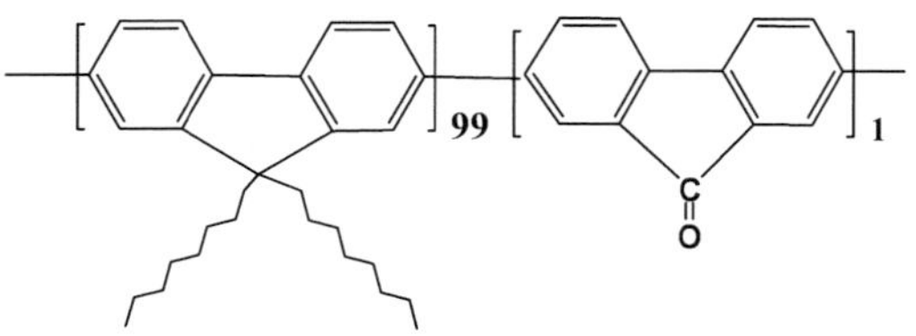

PFO-F (1%)

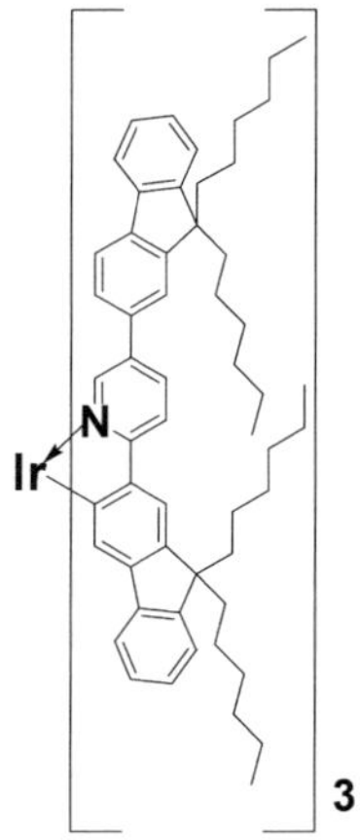

Ir(HFP)$_3$

Fig. 9.19 Molecular structures of polyfluorene (PFO, polyfluorene containing 1% fluorenone, and the Ir metal–organic complex, Ir(HFP)$_3$.

focus on the use of polyfluorene (PFO) as a host material containing small quantities of poly (9,9-dioctylfluorene-*co*-fluorenone) with 1% fluorenone, (PFO-F(1%)) and the Ir complex, tris [2,5-bis-2'-(9',9'-dihexylfluorene)pyridine-$\kappa^2NC_{3'}$]Iridium(III), Ir(HFP)$_3$. The PFO provides the blue component, the PFO-F(1%) provides the green component and the Ir(HFP)$_3$ provides the red component. The molecular structures of the various components are shown in Fig. 9.19. As host materials, the PFOs enable full color (blue, green and red) emission via energy transfer to longer wavelength emitters in blends with other conjugated polymers, phosphorescent dyes and organometallic emitters. Consequently, the PFs are interesting candidates as materials for use in the fabrication of PLEDs that emit white light. In such white emitting PLEDs, the PFs can function both as the host and as the blue emitter.

Typically, however, a low-energy (2.2–2.4 eV) green PL band appears in the emission from PFOs over time and degrades the blue emission. The green emission band originates from *fluorenone defects* (impurities) in the PFO chain. One can take advantage of the fluorenone defect for creating materials with stable green emission.

White-emitting PLEDs can be fabricated using PFO, PFO-F and Ir(HFP)$_3$. Figure 9.20(a) shows the EL spectra obtained from devices with a configuration of (ITO)/PEDOT:PSS/ emitting layer/Ba/Al. The emitting-layer comprises Ir(HFP)$_3$:PFO-F(1%):PFO. The same spectrum, renormalized to the sensitivity of the human eye, is shown in Fig. 9.20(b). The emission is well matched to the response of the human eye.

In the electrophosphorescent PLEDs made from the blends of Ir(HFP)$_3$:PFO-F(1%):PFO, injected holes and electrons recombine by two processes: direct recombination on the main chain (PFO) to produce blue emission in parallel with electron and hole trapping on the

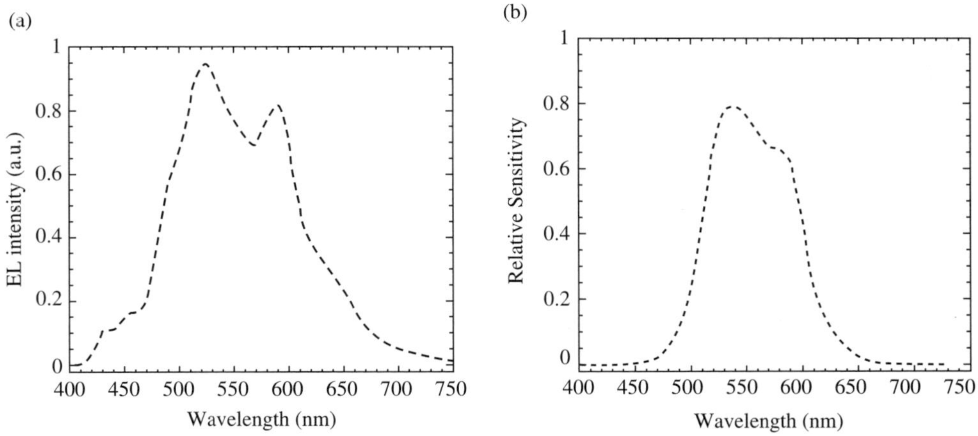

Fig. 9.20 (a) EL spectra obtained from white emitting electrophosphorescent PLEDs and (b) the same spectra as in (a) but renormalized to the sensitivity of the human eye. [Reproduced from X. Gong *et al.*, *Adv. Mater.*, **16**, 615 (2004) with permission from Wiley-VCH.]

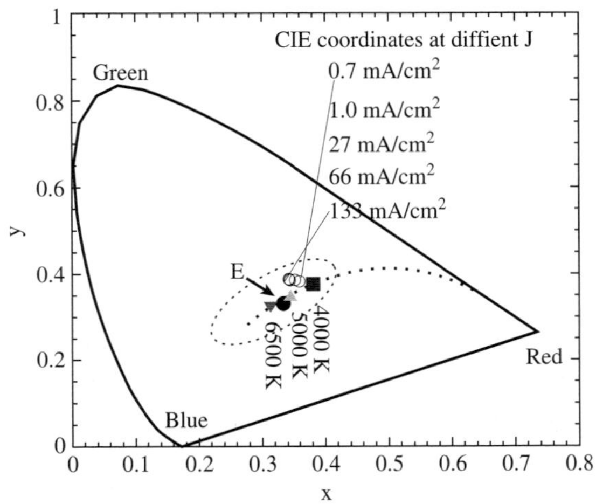

Fig. 9.21 CIE (1931) chromaticity diagram, with coordinates corresponding to the emission from white emitting devices (○ ○ ○) biased at different current densities. Also shown are the equal-energy point (E) for pure white light (0.333, 0.333) (●) and the coordinates corresponding to color temperatures of 4000 K (■), 5000 K (▲) and 6500 K (▼). The dotted line indicates different color temperatures; the dotted oval indicates the approximate area where the human eye perceives the color as white. [Reproduced from X. Gong *et al.*, *Adv. Mater.*, **16**, 615 (2004) with permission from Wiley-VCH.]

fluorenone units and on the $Ir(HFP)_3$ followed by radiative recombination, with green light from PFO-F (1%) and red light from the triplet excited state of $Ir(HFP)_3$.

The CIE (Commission International d'Eclarirage) coordinates, color temperature (CT) and color rendering index (CRI), were quantitatively evaluated from EL spectra. Figure 9.21 shows the 1931 CIE chromaticity diagram, with coordinates corresponding to the emission from electrophosphorescent PLEDs: data points are shown for devices (open circles) biased at different current densities. The dotted line indicates different color temperatures; the dotted oval indicates the approximate area where the human eye perceives the color as white. The CIE coordinates for the $Ir(HFP)_3$:PFO-F(1%):PFO devices are (0.352, 0.388) at J=1 mA/cm^2, very close to the CIE coordinates for pure white light, (0.333, 0.333) as shown in Fig 9.22.

The $Ir(HFP)_3$:PFO-F(1%):PFO devices had CT ~ 4600 K and CRI=86; the CT ~ 4600 K is very close to the CT of sunlight at 20° solar altitude (4700 K). The CTs and CRIs are insensitive to brightness and J.

The white emission turns on at approximately 5 V, with a luminance (L) of L = 6100 cd/m^2 at 17 V. A Lambertian intensity profile was assumed to calculate the LE (cd/A) and the power efficiency (lm/W) with the following results: LE = 3 cd/A and L= 255 cd/m^2 at J= 8.5 mA/cm^2. Unfortunately, the power efficiency was only approximately 1 lm/W; significant improvement will be required before such white light-emitting PLEDs can be competitive for use in solid state

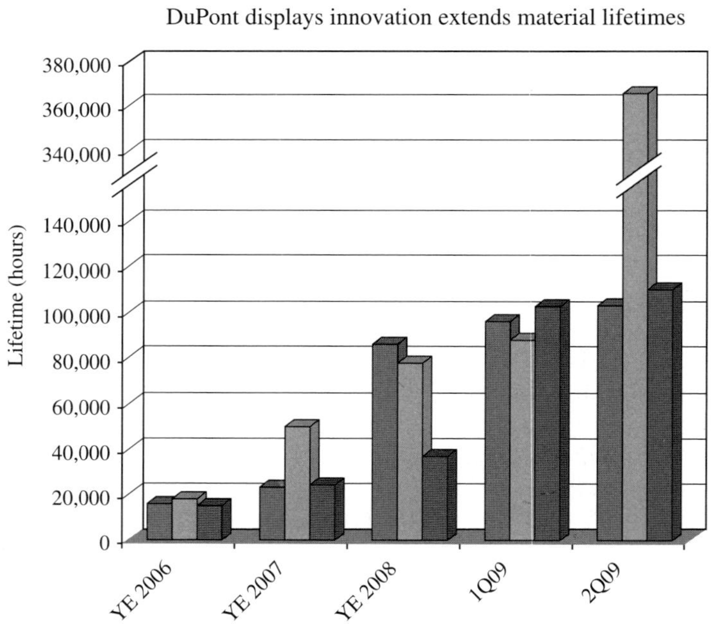

Fig. 9.22 Lifetime data of DuPont Displays' new "Gen 3" material: Progress for red (left), green (middle) and blue (right) materials is shown as simulated display lifetime. [Reproduced from http://www2.dupont.com/Displays/en_US/news_events/article20090527.html]

lighting. Higher power efficiencies have been reported using multilayer vapor depositied films of small molecules.

9.8 Lifetime of solution processed light-emitting diodes

Over the past few years the light operating lifetime of light-emitting LED has improved significantly. A recent report indicates (see Fig. 9.22) that DuPont proprietary "Gen-3" green solution-based materials can last a lifetime of more than 1 million hours (which equates to over 100 years of continuous usage). The performance of blue light-emitting materials has been the most challenging.

Again, however, DuPont Gen 3 solution blue materials are demonstrating significant performance gains. One of the blue materials has demonstrated a lifetime of 38,000 hours, which is one of the longest blue OLED material lifetimes publicly reported. These results indicate that with solution based materials, it will be possible to begin to penetrate the television and general lighting markets using lower-cost printing approaches to manufacture full color displays.

9.9 Conclusion

Research on PLEDs has resulted in significant progress during the last decade. There are, however, significant challenges including achieving high material purity, low-cost, high brightness, and long operating lifetimes.

The use of white-emitting PLEDs for solid state lighting will require the development of low-cost manufacturing of large-area devices with high efficiency and long operating lifetimes. Because roll-to-roll printing offers the promise of low-cost mass production, manufacturing by implementing printing technology may eventually make the cost of producing white PLEDs competitive.

Further reading

I. H. Campbell, B. K. Crone and D. L. Smith, in: *Semiconducting Polymers*, edited by G. Hadziioannou and G. G. Mallarias, Wiley-VCH, Weinheim, 2007, Chapter 10.
Commission International de I'Éclariage, *Measurement of LEDs*, CIE publication 127.
Z. Kafafi (ed.), *Organic Electroluminescence*, SPIE and Taylor and Francis, Boca Raton, 2005.
H. A. E. Keitz, *Light Calculations and Measurements*, 2nd edition, Macmillan, London, 1971.
Z. Li and H. Meng (eds.), *Organic Light Emitting Materials and Devices*, Taylor and Francis, Boca Raton, 2007.
K. Müllen and U. Scherf (eds.), *Organic Light Emitting Devices*, Wiley-VCH, Weinheim, 2006.
D. Parker, *J. Appl. Phys.*, **75**, 1656 (1994).
D. Ryer, *Light Measurement Handbook*, International Light Inc., 1998.
J. Shinar (ed.), *Organic Light Emitting Devices*, Springer, New York, 2004.

10 Light-emitting electrochemical cells (LECs)

10.1 Introduction

Polymer LECs provide an alternative approach to electroluminescence (EL) from semiconducting polymers. Unlike polymer light-emitting diodes (LEDs), which are based on the intrinsic properties of undoped semiconducting polymers, the polymer LEC is a solid-state device in which the semiconducting polymer is doped to create a p-i-n junction *in situ*. The active materials used in LECs must be both electrical and ionic conductors. Two different approaches to materials for LECs have been demonstrated:

1. the active material is a phase-separated mixture of luminescent polymer and polymer electrolyte;
2. the active material is a luminescent semiconducting polymer into which a small concentration of an ionic liquid has been added.

When a sufficiently large voltage bias is applied to an LEC, the ions redistribute, redox occurs near the electrodes, and a p-i-n junction is formed *in situ* by electrochemical doping (see Chapter 2). The low resistance doped regions facilitate injection of electrons and holes, respectively, into the n-type and p-type regions. Light is emitted from an LEC through the radiative recombination of electrons and holes (or, more precisely, positive and negative polarons) in the undoped, intrinsic (i) region.

The LEC is the first semiconducting polymer device that is based on a doping-induced homojunction; an LEC is more like an inorganic p-n junction LED than a polymer LED. This chapter outlines the operating mechanism and device characteristics of LECs. Frozen-junction LECs, which offer many of the combined advantages of LECs and LEDs, are discussed in some detail.

10.2 LEC device architecture

The electrochemistry of conducting polymers is discussed in detail in Chapter 2. In particular, the concept of the all-polymer battery is presented in the diagram of Fig. 2.9. The LEC is a "polymer battery" in which the electrochemical redox material and the electrolyte are combined as a phase-separated mixture in a single composite film comprising semiconducting, luminescent polymer, positive and negative ions, and an ion transport medium The ion transport medium can be the semiconducting polymer as a polyelectrolyte (see Section

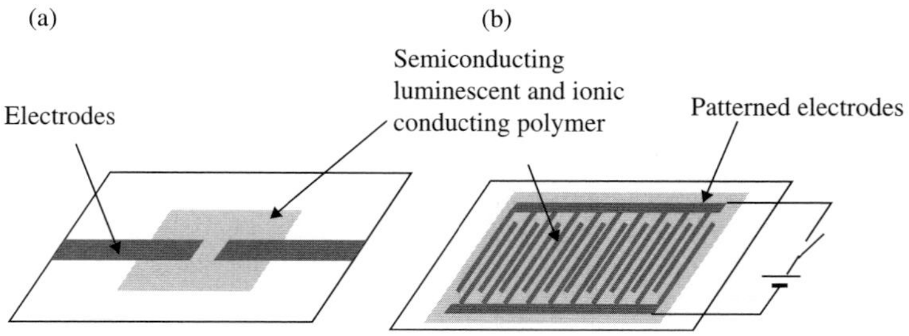

Fig. 10.1 Surface cell configuration for the LEC; in (a) the anode and cathode are two simple metallic pads whereas in (b) the electrodes are interdigitated.

10.10), but typically a third component, an ion transport polymer is added to facilitate ion transport.

An LEC can be made in both the "sandwich" configuration (the composite mixture thin film is sandwiched between two electrodes), similar in structure to a polymer LED, and in the surface cell configuration. Figure 10.1 shows two versions of the surface cell configuration.

The semiconducting luminescent polymer is chosen from the class of soluble conjugated polymers. The polymer electrolyte is a solid-state ionic conductor formed by the dissolution of a salt (e.g.$Li^+CF_3SO_3^-$) in an ion-solvating polymer (e.g. polyethylene oxide, PEO) or a crown ether, etc.

The processing of a sandwich LEC includes three steps. First, a blend solution is made by co-dissolving the semiconducting polymer and polymer electrolyte in a common solvent. Then, the blend solution is spin-cast onto an ITO glass substrate to form a polymer thin film. Finally, top electrodes (e.g. Al) are thermally evaporated on top of the polymer film through a shadow mask. A surface cell LEC can be made by casting a tin film of the emitting material onto a substrate with pre-patterned electrodes. Therefore, LECs are fabricated by relatively simple processing steps, an advantage which is a more general feature of polymer semiconducting devices.

10.3 LEC operating mechanism

The operating mechanism of an LEC is shown schematically in Fig. 10.2. At zero bias, the semiconducting polymer is in the undoped state. When a voltage greater than the energy gap of the polymer is applied to the metallic contacts, the polymer is reduced near the negative electrode, the polymer is oxidized near the positive electrode, and the ions move (e.g. the Li^+ ions in Fig. 10.2), both toward the anode and out of the transport medium into the PPV phase, such that the excess electronic charge (positive or negative) is locally compensated by redistribution of the ionic charge. In this process, the polymer is doped n-type near the negative electrode and doped p-type near the positive electrode with a nondoped intrinsic region in between, and a p-i-n junction is formed *in situ* within the semiconducting, luminescent polymer.

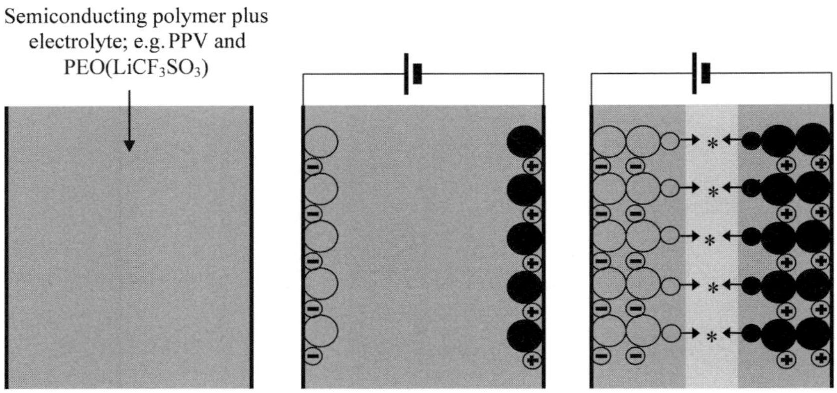

Fig. 10.2 Diagram showing the LEC operating mechanism; when the applied voltage is greater than the energy gap of the polymer, the p-i-n junction is formed *in situ*. The small circles which enclose a "+" sign represent the positive ions (Li^+), and the small circles which enclose a "−" sign represent the negative ions (SO_3^-); the larger open circles represent holes injected into the π band, and the larger filled circles are electrons injected into the π*-band. Light is emitted in the intrinsic region where the electrons and holes meet and recombine.

Anode reaction:

$$(\pi - \text{polymer}) + [\text{D}^+\text{A}^-]_{\text{solv}} \rightarrow [(\pi - \text{polymer})^{x+}(\text{A}^-)_x] + x\text{D}^+ + xe^- \qquad (10.1a)$$

Cathode reaction

$$(\pi - \text{polymer}) + xe^- + [\text{D}^+\text{A}^-]_{\text{solv}} \rightarrow [(\pi - \text{polymer})^{x-}(\text{D}^+)_x] + x(\text{A}^-) \qquad (10.1b)$$

where D^+A^- is the salt in the electrolyte. The doping near the metallic contacts facilitates charge injection, holes into p-type doped material and electrons into n-type doped material. As sketched in Fig. 10.2 (right) the electrons and holes meet and recombine in the intrinsic region, and light is emitted. The emission spectrum of an LEC is typically the same as that of an LED made with the same luminescent polymer.

In general, two regimes are expected corresponding to $eV < E_g$ and $eV > E_g$, respectively, where $E_g$is the energy gap between the π- and π*-bands of the semiconducting polymer and V is the voltage applied across the polymer:

1. $eV < E_g$: This is the double-layer regime with carriers in the semiconductor polymer layers near the electrodes of the semiconductor and a corresponding increase in the local density of negative or positive ions in the semiconducting polymer.
2. $eV > E_g$: This is the electrochemically doped regime with n-type doping on the cathode side and p-type doping on the anode side and the formation of a p-i-n junction.

When the voltage applied to the semiconducting polymer is increased to a value sufficiently large to remove an electron from one side of the polymer and inject an electron into the opposite side of the polymer ($eV > E_g$), the oxidized semiconductor becomes conducting (near the anode side) the reduced semiconductor becomes conducting (near the cathode side) with associated redistribution of the ions.

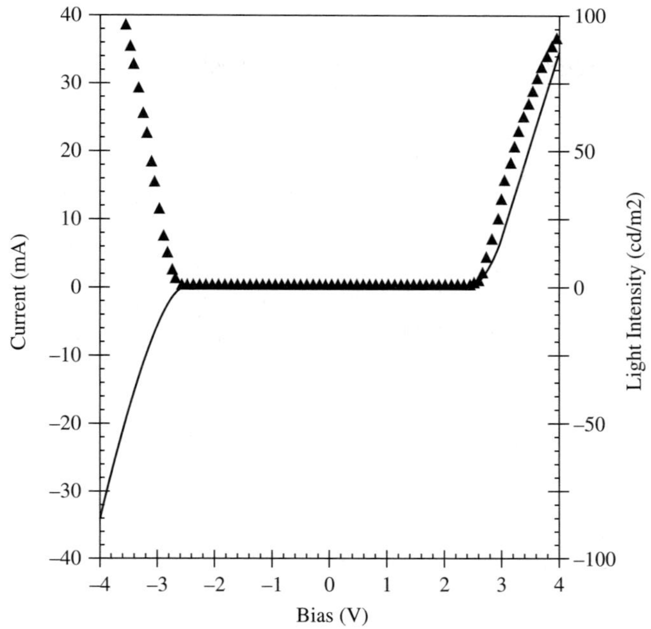

Fig. 10.3 Typical current–light–voltage characteristics of an ITO/PPV + PEO(Li)/Al cell. The voltage scans from 0 to 4 V and from 0 to −4 V, respectively. [Reproduced from Q. Pei *et al., J. Am. Chem. Soc.*, **118**, 3922 (1996) with permission from the American Chemical Society.]

The electrochemical doping in LECs is reversible, and the p-i-n junction formed *in situ* is dynamic. When the external bias is removed, the LEC relaxes by discharging, and the semiconducting polymer returns to the undoped semiconductor state. In addition, changing the polarity of the applied bias can reverse the polarity of the p-i-n junction. Thus, polymer LECs show bipolar current–voltage (*I-V*) and light emission–voltage (*L-V*) characteristics, as shown in Fig. 10.3.

The *in situ* doping that creates the p-i-n junction within an LEC does not require that both ionic species are mobile. In the example of Fig. 10.2, the mobility of Li^+ ions in PEO is much greater than that of the $CF_3SO_3^-$ ions. Even if only one ionic species is mobile (e.g. Li^+), the ionic re-distribution will enable n-type doping on one side with a local excess of Li^+ ions and p-type doping on the opposite side with a local excess of $CF_3SO_3^-$ ions.

Because ion motion and transport are involved, however, the detailed performance of an LEC is strongly dependent on the rate at which the voltage is increased or decreased; ionic transport with associated redistribution of the ionic species is the rate limiting step.

The LEC operating mechanism is elegantly demonstrated in Fig. 10.4 which shows a series of time-lapse images of an LEC film under ultraviolet (UV) illumination. The MIEC[1] film

[1] The composite of a polymer electrolyte and a luminescent conjugated polymer is a mixed ionic/electronic conductor (MIEC).

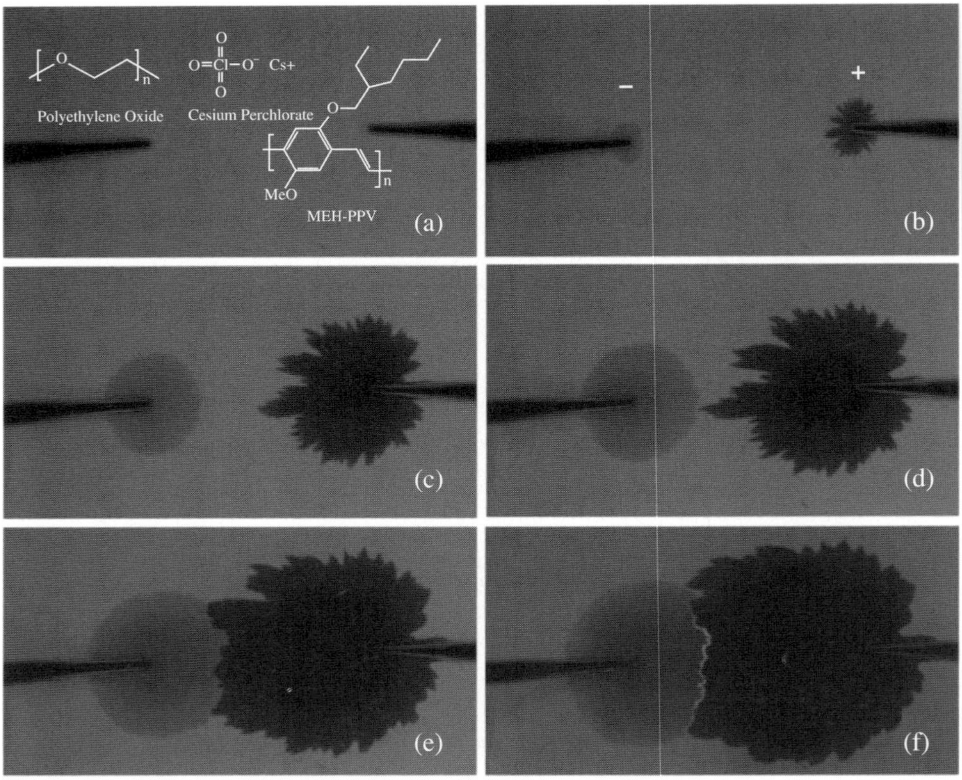

Fig. 10.4 MEH-PPV:PEO:$CsClO_4$ film in contact with two tungsten probes separated by 4.8 mm. The film is at 330 K and under UV illumination. Time elapsed following application of 400V bias; (a) no bias, (b) 10 s, (c) 90 s, (d) 120 s, (e) 220 s, and (f) 270 s. [Reproduced from Y. Hu and J. Gao, J. Am. Chem. Soc. **131**, 18236 (2009) with permission from the American Chemical Society.]] (This figure is reproduced in color in the color plate section.)

contains MEH-PPV, a red-emitting luminescent conjugated polymer. The red background color comes from the photoluminescence of MEH-PPV.

The LEC material also contains PEO and cesium perchlorate. Two tungsten probes, tip radius of 10 μm and separated by approximately 4.8 mm, were brought into contact with the polymer film. This probe separation was large enough for imaging with a digital SLR camera without additional optical magnification. Upon the application of a 400 V bias, darkened profiles emerge from the probe tips and grow in size with time (Fig. 10.4b–d). The darkening of the polymer film under UV illumination indicates fluorescence quenching, and is caused by the electrochemical doping of the luminescent polymer. The point-like metal–polymer contacts used in Fig. 10.4 give rise to an n-doping profile that is nearly circular and smooth at the doping front. The p-doping front is uneven, although the overall shape of the p-doped region also resembles a circle. The p- and n-doped regions eventually make full contact to form a p-n junction (Fig. 10.4e) in a region of the MEH-PPV that is far from either of the metal contacts.

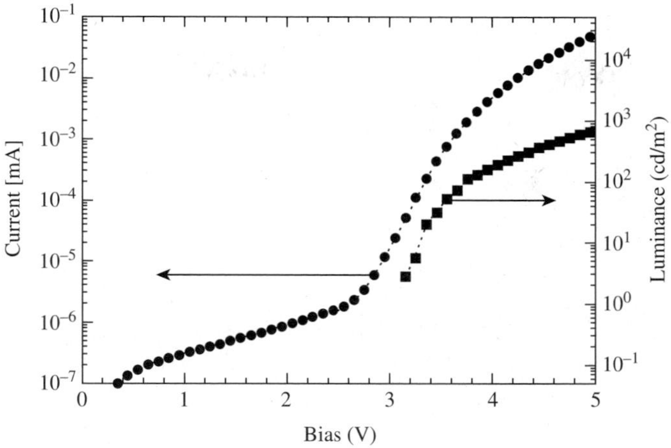

Fig. 10.5 Current versus voltage and luminance versus voltage for an LEC with maximum brightness of approximately 1000 cd/m^2. [Reproduced from G. Yu *et al.*, *Appl. Phys. Lett.*, **70**, 934 (1997) with permission from the American Physical Society.]

Electroluminescence can be observed at the p-n interface when it becomes stronger than the background fluorescence (Fig. 10.4f).

As shown in Fig. 10.5, LEC devices can reach relatively high current densities and emit light at high luminance values. Because of the excellent contacts to the emitting intrinsic layer via the n- and p-type doped regions within the same material, injection is optimized and balanced.

10.4 LECs in the surface geometry configuration

Figure 10.6 shows photographs of a polymer LEC in the surface cell configuration with interdigitated gold electrodes (PPV plus PEO with $LiCF_3SO_3$). The photos were obtained during operation using an optical microscope. The spacing between electrodes (and between light-emitting regions) is 15 μm. Yellow-green light emission can be seen when a voltage bias of 4 V is applied; the width of the emitting region within the p-i-n junction is about 2 μm.

The "wiggles" in the emission "line" originate from inhomogeneities in the p-i-n junction. Larger scale "wiggles" and instability are inhibited by the capacitance of the junction. The energy stored in a capacitor is given by

$$E_{cap} = \frac{1}{2} C V^2$$

Since the capacitance, C, is proportional to the length of the intrinsic region in the p-i-n junction, any wiggles in the emission line will increase the total energy stored, and will therefore tend to straighten the emission line and prevent shorts (e.g. direct contact of the p-type region with the opposite Au contact). Moreover, shorts from p-type to n-type are

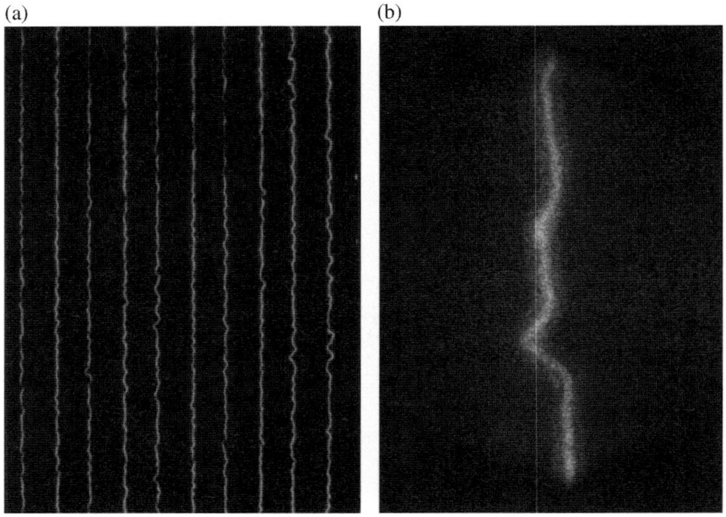

Fig. 10.6 Images of the LEC emission as viewed under a microscope. The distance between interdigitated electrodes (see Fig. 10.1) and the emission between successive emission zones is approx. 15 μm. The image is shown at higher magnification on the right; the width of the emission zone is approx. 2 μm.

naturally self-healing; every local p-to-n short circuit will result in recombination of the partially oxidized and reduced sections resulting in an intrinsic region with high electrical resistivity.

10.5 Direct confirmation of electrochemical junction formation in the LEC

Optical beam induced current (OBIC) microscopy was used to image the structure of the junction in an LEC. In OBIC measurements, the electric field profile across the LEC structure is measured by scanning a focused laser beam across a surface cell LEC. The surface cell LEC was prepared with a pair of gold strips as the contact electrodes (Fig. 10.1a), separated by a gap of approximately 20 μm. When the excitation beam is within the junction, photogenerated electrons and holes are separated by the built-in electric field, resulting in an external photovoltaic signal. When the photoexcited region is confined either to the p-type side or to the n-type side, there is no external signal.

Measured OBIC photovoltage versus position characteristic curves are shown in Fig. 10.7; the built-in electric field is zero near the two contacts and maximum in the i-region. The peak value of the built-in electric field and the spatial width of the field distribution change with the bias voltage. Spatially resolved PL and EL were measured on the same device. The built-in electric field profile measured by OBIC microscopy is consistent with the PL and EL profiles. Thus, the OBIC electrical data are consistent with the images in Fig. 10.4 and directly confirm the electrochemical doping and junction formation in an operating LEC.

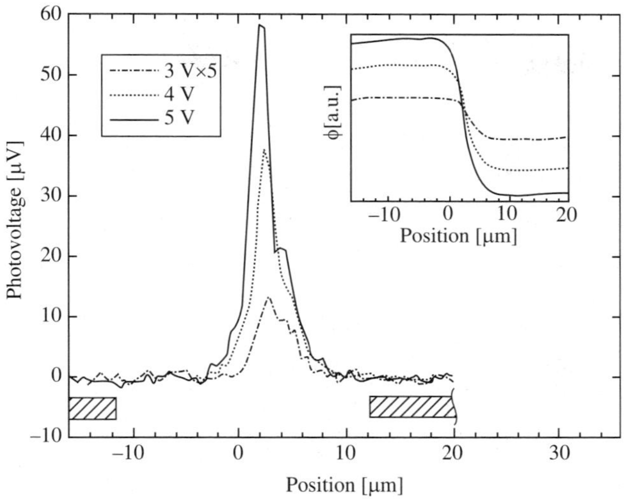

Fig. 10.7 Photovoltage (proportional to the local built-in electric field) versus position in the OBIC experiment; the built-in electric field is zero near the two contacts and maximum in the i-region. Inset: built-in potential obtained by integrating the electric field data. [Reproduced from D. J. Dick *et al.*, *Adv. Mater.*, **8**, 985, (1996) with permission from the VCH Verlagsgesellschaft Germany.]

In situ optical absorption studies (Fig. 10.8a) during the formation of the p-i-n junction demonstrate doping of the semiconducting polymer film upon applying an electrode potential $V > E_g/e$. Upon releasing the voltage, the electrochemically oxidized/reduced regions relax and after a time sufficient for back-migration of the ions, the original undoped state of the

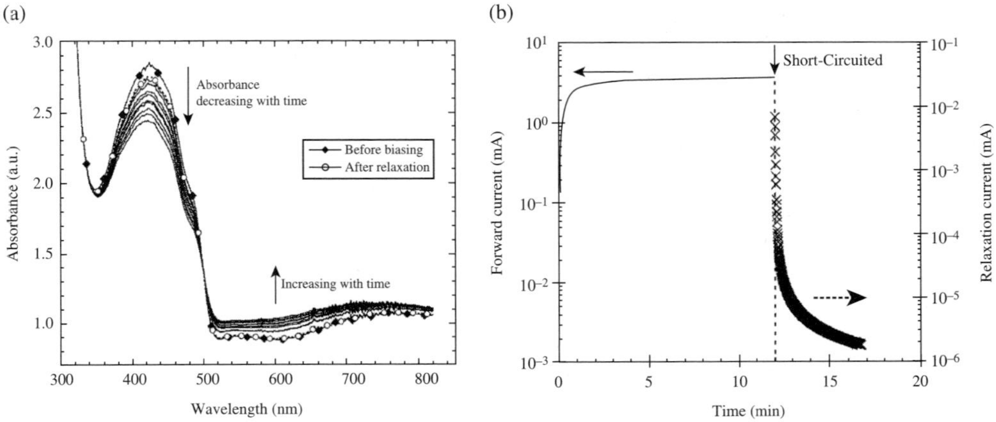

Fig. 10.8 (a) Optical absorption studies during the formation of the p-i-n junction; the decrease in absorption at wavelengths below 500 nm and the corresponding increase in the near-IR indicate electrochemical doping of the semiconducting polymer (PPV). (b) Build-up of the current in the external circuit during charging and decay during discharge. [Reproduced from Q. B. Pei *et. al.*, *J. Am. Chem. Soc.*, **118**, 3922 (1996) with permission from the American Chemical Society.]

device is re-established. This of course assumes negligible degradation of the active material during operation in the p-i-n junction mode. The kinetics of this relaxation are shown in detail in Fig. 10.8(b).

After shorting through an external ammeter, the "battery" discharges. The LEC is a very poor battery since the p- and n-doped regions are electronically in contact through the intrinsic region. Current through this internal electronic pathway does not contribute to the external current, but to internal discharge of the electrochemical cell. As shown in Fig. 10.7(b), the cell discharges over a period of several minutes. The relatively long time decay of the voltage and current is consistent with the need for ion transport during the discharge.

10.6 Control of phase separation: Conjugated polymer plus solid electrolyte

The time required for build-up and decay of the p-i-n junction is limited by the need for ionic transport. By creating a phase separated microstructure in the form of a bicontinuous interpenetrating network of the semiconducting polymer and the ion transport polymer, both electronic and ionic conductivity can be achieved. Figure 10.9 shows a TEM micrograph of an "optimized" morphology for the composite material in an LEC.

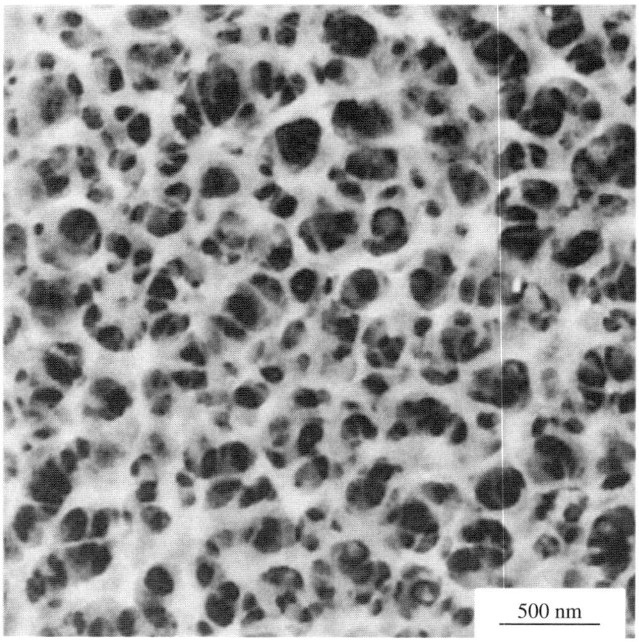

Fig. 10.9 TEM micrograph of the phase separated microstructure of the composite material in an LEC. The interpenetrating network was formed by using a bifunctional "surfactant" molecule that lowers the interfacial energy between the two components. [Reproduced from Y. Cao *et al. Appl. Phys. Lett.*, **68**, 3218 (1996) with permission from the American Physical Society.]

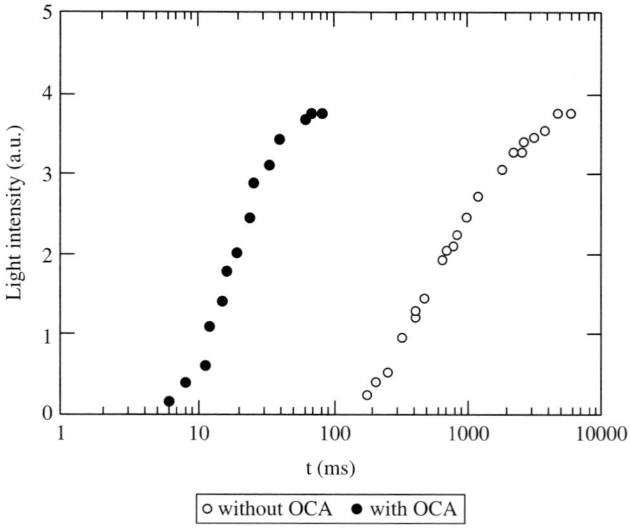

Fig. 10.10 Comparison of the turn-on of the light emission (and associated formation of the p-i-n junction) with and without the OCA additive that causes the formation of the interpenetrating network morphology shown in Fig. 10.9. [Reproduced from Y. Cao *et al.*, *Appl. Phys. Lett.*, **68**, 3218 (1996) with permission from the American Physical Society.]

The ability to manipulate the nanostructure with a bifunctional additive is the guiding strategy; one end of the molecular additive coordinates with one component and the other end with the other component in the phase separated composite material, in direct analogy with the "surfactant" counter-ions used to make polyaniline soluble and processible as discussed in Chapter 2.

By reducing the length scale which defines the interpenetrating network to the nanometer regime, the distances that ions must move will be correspondingly reduced, thereby improving the response time for the formation of the junction. As demonstrated in Fig. 10.10, the nm scale bicontinuous interpenetrating network material enables more rapid formation of the *in situ* p-i-n junction and therefore more rapid turn-on of the light emission.

10.7 LECs versus LEDs: Advantages and disadvantages

Polymer LEDs can be modeled as metal–semiconductor–metal (MSM) structures; they are field-driven devices in which the electron and hole currents are limited by carrier injection at the metal–semiconductor contacts. As a result, polymer LEDs are characterized by sensitivity to electrode materials (low work function metals are needed for injection of electrons) and device thickness. Because to the p-i-n junction nature of LEC operation, polymer LECs offer advantages when compared to polymer LEDs.

(i) *Balanced electron and hole currents.* Since the doped semiconducting polymer has low resistivity and tends to form ohmic contacts with metal electrodes, more nearly balanced electron and hole injection is expected.
(ii) *Low operating voltages.* The light turn-on voltage of an LEC is roughly equal to E_g/e (2–3 V). Red, green, blue and broadband white emission from polymer LECs have been demonstrated with low operating voltages (<4 V).
(iii) *Insensitivity to electrode materials.* The charge carrier injection in an LEC is not sensitive to the work function of the electrode material. Air stable metals such as aluminum and gold are used for electron injection in LECs in contrast to polymer LEDs where reactive low work function metals are required for good electron injection. Moreover, polymer LECs can be made in the surface cell configuration with symmetric contact electrodes (e.g. Au/polymer/Au).
(iv) *Insensitivity to film thickness.* LECs are voltage-driven devices. The applied bias voltage drops mostly in the more resistive undoped intrinsic region. As a result, the device characteristics of LECs are not sensitive to device thickness. Surface cell LECs have been demonstrated with an operating voltage of only few volts. For surface cell LEDs with similar dimensions, however, the operating voltage is few hundred volts!

These advantages of polymer LECs result from electrochemical doping and p-i-n junction formation during LEC operation. However, unlike an inorganic p-n junction, the p-i-n junction in an LEC is dynamic; i.e. it relaxes by discharging when the externally applied voltage is removed. A new operation requires that the junction is rebuilt before the maximum light output can be achieved. As a result, LECs also have serious disadvantages:

(i) *Slow response.* The light emission turn-on of an LEC is slow because seconds or even minutes are required for the redistribution of ions during the creation of the p-i-n junction.
(ii) *Poor stability at high operating voltage.* When the applied bias exceeds the electrochemical stability window of the system (semiconducting polymer, electrolyte and metal electrodes), the device degrades due to "overdoping". As a result, the operating voltage must be limited to below about 4 V to avoid device failure. Thus, high brightness cannot be achieved in dynamic-junction LECs.
(iii) *Hysteresis behavior.* The cyclic current–voltage (*I-V*) and voltage–light intensity (*V-L*) curves of polymer LECs exhibit hysteresis (see Fig. 10.13). The hysteresis is caused by the interplay of the slow ionic processes and the fast electronic processes involved in LEC operation.
(iv) *Bipolar light emission and I-V characteristics.* The bipolar characteristics of LECs will introduce cross-talk in column–row addressable displays.

Consequently, dynamic-junction LECs cannot be used in any serious display applications. In order to solve the problems of dynamic-junction LECs, a stabilized junction is needed so that after the junction is formed only fast electronic processes are involved in the operation. The dynamic p-i-n in an LEC can be stabilized by the "frozen-junction" approach, as discussed in Section 10.10.

10.8 Junction formation in the "sandwich" configuration

To characterize the p-i-n junction built *in situ* in the sandwich geometry, it is useful to recognize that the LEC is a capacitor. Before the p-i-n junction is formed, the device consists of a homogenous dielectric between two metal electrodes (see Fig 10.11). This will result in a capacitance, *C*,

$$C = \varepsilon A/d$$

where ε is the dielectric constant of the dielectric, *A* is the area of the LEC, and *d* is the thickness of the active layer.

After the *in situ* formation of the p-i-n junction, however, the p- and n-doped regions become highly conducting and thereby become part of the anode and cathode, respectively. The dielectric region is then confined to the intrinsic region in between with a thickness of d_i. This shrinking of the thickness from *d* to d_i will result in an increase of the capacitance from $C_{undoped} = \varepsilon A/d$ to $C_{p\text{-}i\text{-}n} = \varepsilon A/d_i$. Measurements of the complex impedance ($Z = R + i/\omega C$) have confirmed the expected change in capacitance. At low frequencies, the impedance must be analyzed as that of an electrolytic capacitor; the behavior of Z'' at low frequencies is dominated by ionic motion. At sufficiently high frequencies, however, the ions cannot follow the oscillating electric field, and the device can be modeled as a simple capacitor. As shown in Fig. 10.12, the imaginary part does exhibit the expected $Z'' = 1/\omega C$ dependence at high frequencies with a change in slope for applied voltages above (e.g. 3 V) and below (e.g. 1.5 V) the threshold for electrochemical doping.

Figure 10.12 demonstrates that in the sandwich geometry the cross-over from n-type to p-type doping occupies most of the film thickness (≈ 80%). The width of the p-i-n junction in the sandwich cell is relatively small (of order 0.2 μm) compared to the 2 μm junction width observed in Figs 10.6 and 10.7, implying that the width of the cross-over is dependent on the boundary conditions. In the surface cell, the metallic electrodes are widely spaced so that the electrochemical potentials are free to equilibrate across 20 μm. In the sandwich cell, the boundary conditions require that the cross-over distance be less that the thickness, *d*.

10.9 Frozen junction LECs

Ion transport in polymer electrolytes takes place only in the amorphous, elastomeric phase of the ion–polymer complexes. In PEO-Li^+ electrolytes, the motion of Li^+ ions is enabled by the

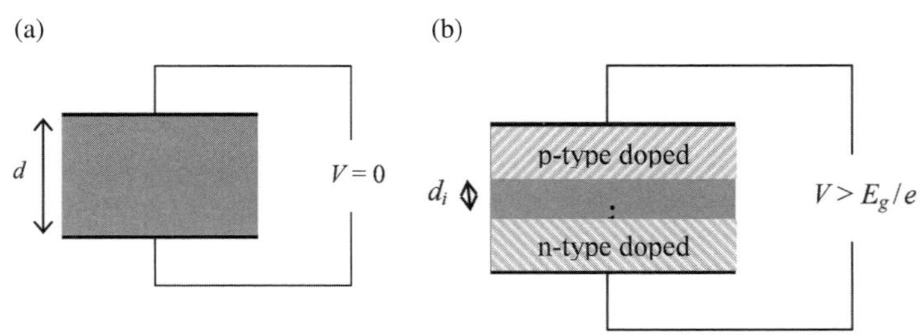

Fig. 10.11 The LEC as a capacitor (a) before junction formation and (b) after junction formation.

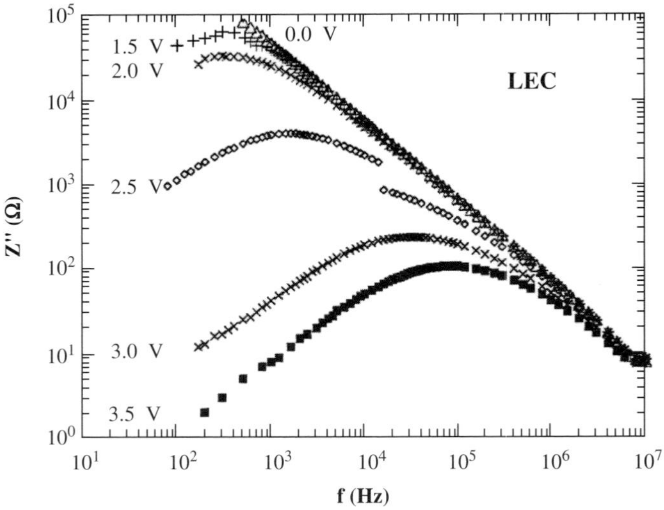

Fig. 10.12 Imaginary part of the complex impedance, Z'', as a function of frequency; the applied bias voltages are given on the left. From the high frequency slopes, $C = 2.1$ nF for $V<2.0$ V, $C = 2.2$ nF for $V = 2.0$ V, and $C = 2.6$ nF for $V = 3.0$ V. Thus, $d_i/d \approx 0.8$ where $d = 0.2$ μm. [Reproduced from Y. Li *et al.*, *Chem. Phys. Lett.*, **287**, 83 (1998) with permission from the Elsevier Science B.V.]

simultaneous making and breaking of bonds with the oxygen atoms of the PEO chain, and is assisted by segmental motion of the PEO chain. Therefore, ion transport in polymer electrolytes is thermally activated and is related to the glass transition phenomenon. The temperature dependence of the ionic conductivity can be expressed by the VTF (Vogel–Tamman–Fulcher) equation:

$$\sigma = AT^{-1/2}\exp\{-E_a/(T - T_0)\} \tag{10.2}$$

where A is a constant, T is the temperature, T_0 is the "ideal" glass transition temperature (about 50°C below the actual glass transition temperature, T_g) and E_a is the activation energy. The strong temperature dependence of ionic conductivity is a general feature. At temperatures below T_g, ion mobility is negligible because the chain segmental motion is quenched.

This "freeze-out" of the ion mobility allows the stabilization of the dynamic p-i-n junction in LECs by lowering the operating temperature. At temperatures below T_g of the ion-transport polymer ($\approx$ 208 K for PEO), the ions are immobilized because of the absence of polymer chain segmental motion. Consequently, both the doping profile and the p-i-n junction are stabilized, giving "frozen-junction" LECs.

In order to make a frozen-junction LEC, first, a voltage bias is applied to the LEC at a temperature above the glass transition temperature to initiate the electrochemical redox and to build the p-i-n junction. Then, the LEC is cooled below the glass transition temperature under bias. The voltage bias applied to the LEC during junction formation and during the subsequent cooling is called the *pre-bias*. Without a pre-bias, the cooled device is highly resistive and nonemitting, the p-i-n junction has not been created. This methodology is reminiscent of the "poling" of polymers for use in applications in nonlinear optics.

Frozen-junction LECs have unique characteristics which distinguish them from both polymer LEDs and dynamic-junction LECs; see Fig. 10.13. The LEC exhibits both rectifying diode behavior and unipolar light emission, with polarity the same as that is used to create the junction. Moreover, since the electrochemistry is "turned off", frozen junctions LECs can be operated at relatively high voltages; i.e. voltages that would be outside the window of electrochemical stability if ion motion were not frozen out. Thus, frozen-junction LECs can, in principle, overcome all the disadvantages of dynamic-junction LECs listed above.

Frozen-junction LECs respond to changes in the applied voltage much faster than dynamic-junction LECs. While seconds or even minutes are required for the dynamic junction to form and for the LEC to achieve maximum light emission, the light turn-on time of the frozen-junction LEC is less than 1 μs. The fast response of frozen-junction LECs implies that only electronic processes are involved in the operating mechanism.

The existence of the stabilized p-i-n junction in a frozen-junction LEC can be verified by measurement of the photovoltaic current/voltage that results from the built-in electric field of the junction. If there were no junction, there would be no photovoltaic effect! The photovoltaic response of an LEC is shown in Fig. 10.14 (see also Fig. 10.7 for evidence of the built-in potential in the surface cell configuration).

The rectification ratios of frozen-junction LECs are quite small compared to those characteristic of polymer LEDs. A frozen-junction LEC can be regarded as a diode in series with a resistor. Because the resistance arises mostly from the undoped region, it becomes larger at low temperatures, thereby reducing the rectification ratio.

Since the glass transition temperature for PEO is 208 K, frozen junction LECs fabricated with PEO as the ion solvating and ion transporting material must be operated well below room temperature. Frozen junction LECs that operate at room temperature have been demonstrated

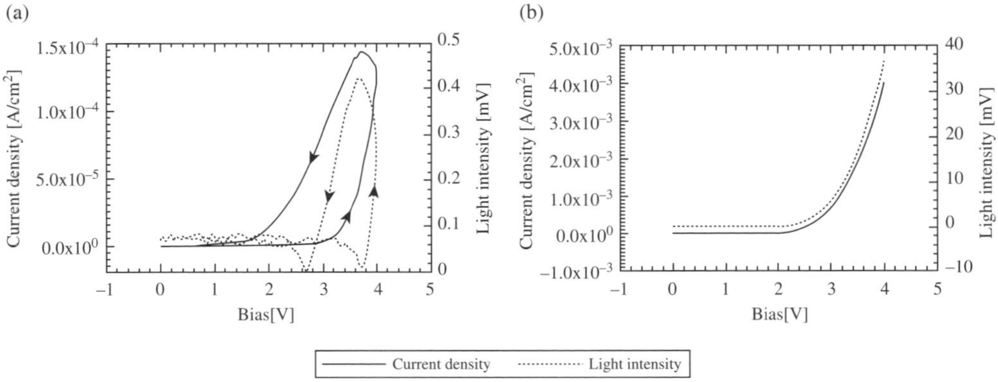

Fig. 10.13 Current versus voltage and light intensity versus voltage for LECs fabricated in the Al / MEH-PPV: PEO (Li-triflate) / ITO configuration with ratios MEH-PPV:PEO:Li = 12:5:2.
(a) Room temperature, with hysteresis indicating electrochemical doping/undoping above 2 V. (b) Unipolar emission without hysteresis; increasing and decreasing voltage curves are identical. [Reproduced from J. Gao *et al., J. Appl. Phys.*, **86**, 4594 (1999) with permission from the American Physical Society.]

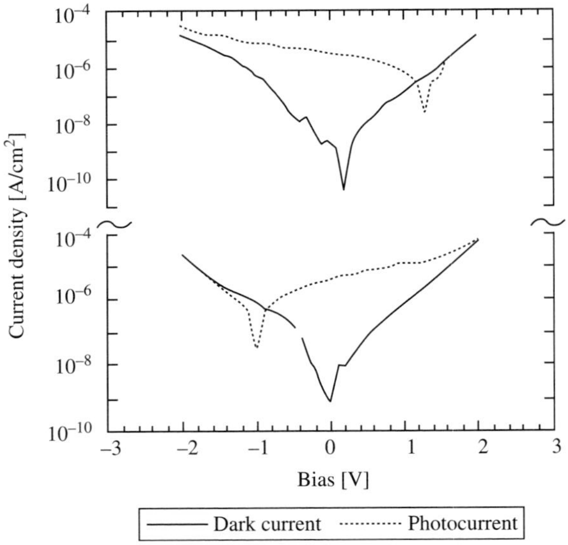

Fig. 10.14 The photovoltaic effects in an LEC fabricated with MEH-PPV:PEO:Li. The sign of the short circuit current as well as the sign of the open circuit voltage depend on the sign of the bias voltage applied to form the p-i-n junction above the glass transition. The photocurrent is reversed when the pre-bias is reversed (dark currents are shown for comparison). The open-circuit voltage (voltage at when the photocurrent is zero) results from the built-in potential across the p-i-n junction. [Reproduced from J. Gao *et. al., Appl. Phys. Lett.*, **71**, 1293 (1997) with permission from the American Physical Society.]

using crown ether molecules to solvate the Li^+ ions. Since crown ethers form a complex with Li^+ with a binding energy significantly greater than k_BT at room temperature, ionic motion of the Li^+ is inhibited. As a result, mobile ions become available for p-i-n doping and junction formation only by raising the temperature well above room temperature (to approximately 80 °C). After forming the junction, the crown-ether/luminescent polymer LECs can be operated at room temperature.

As shown in Fig. 10.15, relatively high performance light emission has been achieved at room temperature in LECs using Li^+– complexed crown ether as the electrolyte and BuEH-PPV as the luminescent polymer. Although relatively stable because the electrochemistry has been "turned off", even these devices last only a few days. Thus, although the principle of the room temperature frozen-junction LEC has been demonstrated, work needs to be done to achieve sufficiently long-term stability to enable commercial applications.

10.10 LECs fabricated with ionic luminescent polymers and with ionic liquid additives

Single-component polymer LECs have been demonstrated with conjugated polyelectrolytes as the active materials; e.g. poly[9,9′-bis(6″-(*N,N,N*-trimethylammonium)hexyl) fluorene-*alt-co*-

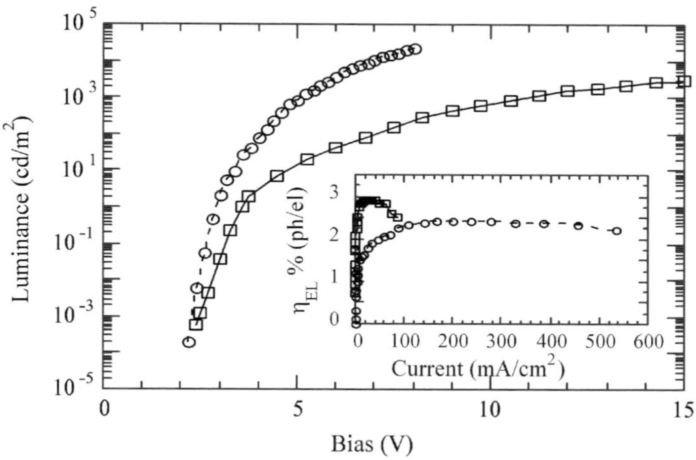

Fig. 10.15 *L-V* in two frozen junction LECs with different blend ratios (circles, BuEH-PPV:Crown Ether: Li-triflate:OCA = 1:1:0.2:0.1; squares, BuEH-PPV:CE:Li-triflate = 2:1:0.2). The corresponding external quantum efficiencies are shown in the inset. [Reproduced from G. Yu *et al.*, *Adv. Mater.*, **10**, 385 (1998) with permission from the VCH Verlagsgesellschaft, Germany.]

phenylene]bromide (PFN^+Br^-) as shown in Fig 10.16. Indium-tin-oxide/PFN^+Br^-/aluminum sandwich structures exhibit blue light-emission.

Thermophysical characterization shows that PFN^+Br^- is in a metastable amorphous phase after spin casting, but that crystallization takes place at elevated temperatures. Since ionic transport occurs only in the amorphous phase, this nonreversible phase transition provides a mechanism for creating frozen-junction LECs that are stable at room temperature.

Sandwich cell devices can be "charged" via ionic redistribution at elevated temperatures in the amorphous phase with the associated formation of a p-i-n junction, and then stabilized through crystallization. After forming the p-i-n junction, the onset of light emission occurred at 2.9 V independent of the device thickness. The brightness reached 150 cd/m^2 at approximately 3.7 V.

The lifetime of the PFN^+Br^- LEC is encouraging; the devices can be run for hours (see Fig 10.17). The significantly improved lifetimes and relatively fast turn-on times for these single component LECs operating at room temperature indicate that this is a promising direction.

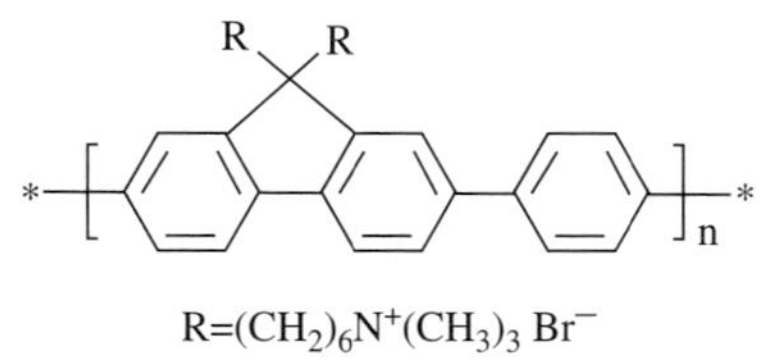

Fig. 10.16 Molecular structure of the cationic polymer, PFN^+Br^-.

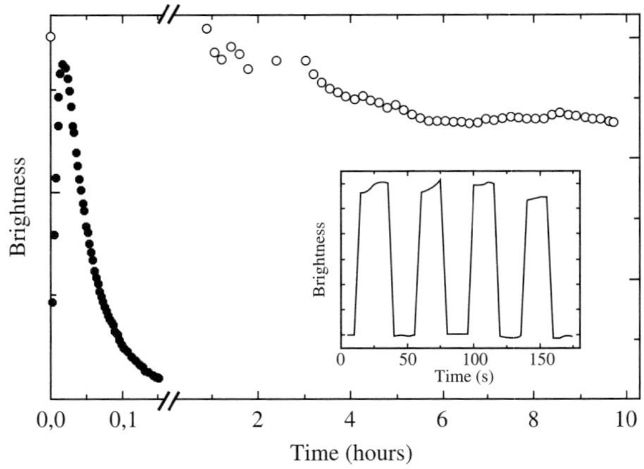

Fig. 10.17 Brightness for ITO/PFN$^+$Br$^-$(270 nm)/Al devices as a function of time at $V = 3.5$V. The devices were operated at room temperature (open circles) and allowed to self-heat (solid circles); heating significantly increased the rate of degradation. Inset: brightness response to a square wave function with 4.0 V amplitude is presented for a stabilized device operated at room temperature. [Reproduced from L. Edman *et al.*, *Appl. Phys. Lett.*, **82**, 3961 (2003) with permission from the American Physical Society.]

A number of water soluble, ionic semiconducting polymers have been synthesized, including both cationic and anionic materials.

10.11 LECs with long operating lifetimes: Ionic liquid additives

LECs with long operating lifetimes have been fabricated using an ionic liquid as the source of ions. These devices utilized the simple sandwich structure PEDOT-PSS coated indium-tin-oxide (ITO) glass as the anode and aluminum as the cathode. The well-known soluble phenyl-substituted PPV) copolymer, "superyellow", was selected as the host light-emitting polymer and an organic ionic liquid; methyltrioctylammonium trifluoromethanesulfonate (MATS) was used to introduce a dilute concentration of mobile ions into the emitting polymer layer. The molecular structure of MATS is shown in Fig. 10.18. The merits of MATS include its good solubility in common organic solvents, such as toluene, hexane, and acetonitrile, and its relatively high decomposition temperature (approximately 220 °C). Because MATS has a melting temperature of approximately 56°C, frozen junction devices can be prepared for operation at room temperature.

MATS shows excellent compatibility with superyellow as confirmed by atomic force microscope (AFM) studies. The single-phase nature of the solid solution of MATS in superyellow implies that the system is thermodynamically stable. In the superyelow:MATS emitting layer, the ionic liquid comprises only 2 weight percent in the solid thin films. As a result of the good solubility of MATS in superyellow, the films exhibited properties almost identical to those of pure superyellow films. The single-phase stability of this material system with the ionic liquid additive is the origin of the long operating lifetime of the devices. These

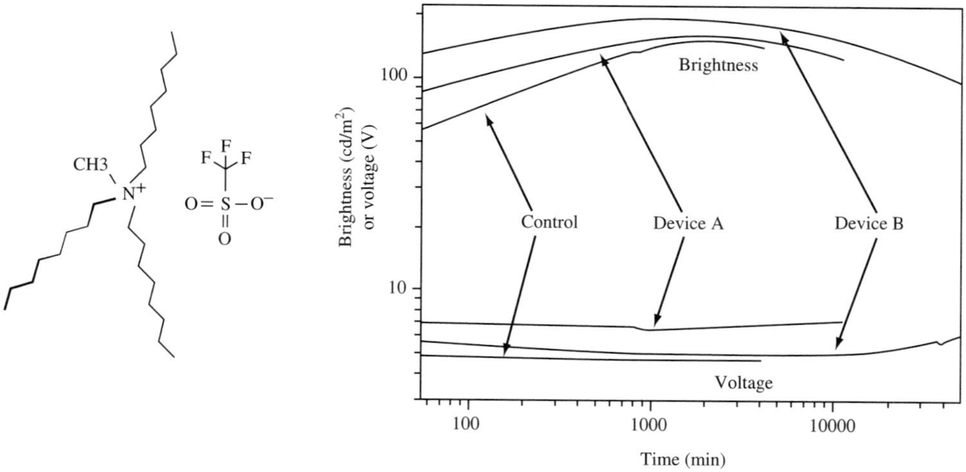

Fig. 10.18 Molecular structure of methyltrioctylammonium trifluoromethanesulfonate (MATS); brightness versus time continuous operation mode for device A (ITO/PEDOT/PS-TPD-PFCB/Superyellow:2% MATS/Al), device B (ITO/PEDOT/VB-TCTA/Superyellow:2%MATS/Al) and control device with structure (ITO/PEDOT-PSS/Superyellow:2%MATS/Al). [Reproduced from Y. Shao *et al.*, *Adv. Mater.*, **21**, 1972 (2009) with permission from the Wiley-VCH.]

ionic liquid based LECs, with aluminum metal as cathode, exhibit excellent current-rectification diode properties, a single-phase active layer, short response time, and continuous operating lifetimes of hundreds of hours. In one device configuration, the extrapolated lifetime to half brightness was 1500 hours (60 days). After sealing with epoxy and a glass cover slide, LECS with MATS can be operated continuously in air for weeks.

The use of ionic liquids as the source of ions in LECs is a promising direction. After the frozen junction is formed, the devices exhibit all the good characteristics of both PLEDs and PLECs including excellent current-rectification ($\sim 10^4$), fast response (<2 ms), high brightness (>10,000 cd/m^2), high efficiency ($\sim$3.3 cd/A), low turn-on voltage (2.2 V), stable cathodes, and long operating lifetimes.

10.12 LECs: Summary

LECs are novel polymer light-emitting devices, fundamentally distinct from polymer LEDs. The operation of an LEC involves the formation of a dynamic p-i-n junction by *in situ* electrochemical doping. The dynamic junction in an LEC can be frozen, or stabilized by lowering the temperature to below the T_g of the polymer electrolyte, by using crown ethers to complex the ions, or by crystallization after forming the junction in the amorphous phase. Frozen-junction LECs offer combined advantages of polymer LEDs and LECs. The remaining goal of LEC research is to make frozen-junction LECs that can be operated at room temperature with long lifetimes. Progress toward this goal using ionic liquids as the ion source is promising. Further enhancement of the lifetime should be possible using the techniques developed for stable second order nonlinear optical (NLO) polymers; i.e. by developing polymer blends with high T_g and/or by cross-linking the one component after junction formation.

The frozen-junction LEC is the first polymer device in which the operation is based on a stable, doping-induced homojunction. The frozen junction concept can be used in other device applications where electrochemically induced (*in situ*) changes in the electrical properties of materials are desired. For example, polymer frozen-junction photovoltaic cells have been demonstrated with unique polarity-reversible photovoltaic response.

Further reading

J. Gao, G.Yu, and A. J. Heeger, *Appl. Phys. Lett.*, **71**, 1293 (1997).

A. J. Heeger and J. Gao, Electrochemical light emitting cells from semiconducting polymers, in: *Encyclopedia of Materials: Science and Technology*, Vol. 6, edited by G. Hadziiannou, (2001), pp. 2525–2531.

P. Matyba, K. Maturova, M. Kemerink, N. D. Robinson and L. Edman, *Nature Materials*, **8**, 672 (2009).

Q. Pei and A. J. Heeger, *Nature Materials*, **7**, 167 (2008).

G. Yu, Y. Cao, C. Zhang, Y.-F. Li, J. Gao and A. J. Heeger, *Appl. Phys. Lett.*, **73**, 111 (1998).

11
Semiconducting polymers as laser materials

Lasers are used extensively – they can weld, cut or drill, transmit phone calls through optical fibers, act as chemical sensors, produce computer printouts and much more. Following the discoveries of luminescence and electroluminescence from semiconducting polymers, the development of lasers using semiconducting polymers became a major research activity around the world. Semiconducting polymers combine novel optoelectronic properties, with simple fabrication and processing making them attractive candidates for laser materials. The use of these materials opens up the prospect of compact, low-cost, disposable lasers suitable for a wide range of applications. Lasing in the full range of the visible spectrum has been demonstrated using a number of semiconducting polymers with different molecular structures.

Almost all semiconducting polymer lasers reported have needed another laser to optically pump them to reach threshold. This is a cumbersome and expensive configuration which has limited their usefulness. There is therefore great interest in developing an electrically pumped laser, as it would provide a new class of laser that can be made very simply. Although indirect pumping of a polymer laser by a nitride LED was proposed a decade ago,[1] it has only recently been demonstrated,[2] opening the way to compact and convenient polymer lasers. Direct injection of charge into a polymer laser would be even more convenient and is a major challenge in current research.

The goal of this chapter is to introduce lasers based on luminescent semiconducting polymer materials, to describe the distinct characteristics of polymer lasers and to outline the device physics and challenges associated with electrical pumped lasing.

11.1 Optical gain in semiconducting polymers

In order to understand light amplification, one must focus on the concept of optical gain. Optical gain arises from two physical phenomena:

(1) stimulated emission;
(2) population inversion.

Consider first a two-level system as sketched in Fig. 11.1(a).

[1] F. Hide *et al.*, *Appl. Phys. Lett.*, **70**, 2664 (1997).
[2] Y. Yang *et al.*, *Appl. Phys. Lett.*, **92**, 163306 (2008).

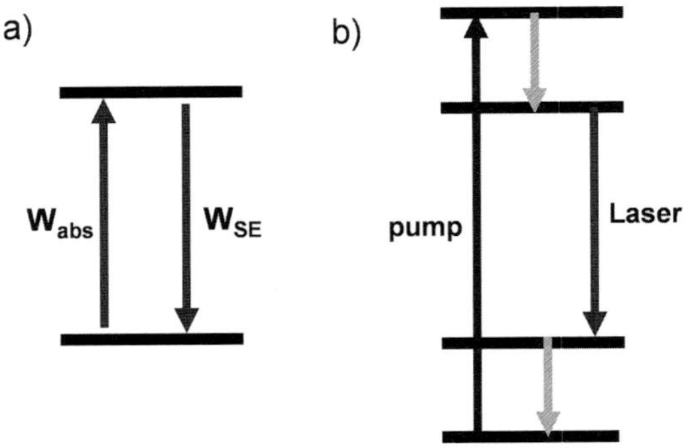

Fig. 11.1 Two-level system with transition probabilities W_{ABS} for absorption and W_{SE} for stimulated emission; (b) four-level laser system.

In his famous 1917 paper, Einstein predicted stimulated emission; an incident photon with energy $\hbar\omega$ would stimulate the W_{SE} transition provided that $\hbar\omega = \Delta E$ where E is the energy difference between the two levels in, for example, an atom or a molecule. Consequently, if there are more "molecules" in the excited state than in the ground state, i.e. if the population is inverted from what one would expect in thermal equilibrium, light with photon energy $\hbar\omega = \Delta E$ would be amplified.

In the same paper, however, Einstein showed that for such a two-level system the transition probabilities of absorption and stimulated emission are equal

$$W_{abs} = W_{SE} \tag{11.1}$$

and proportional to the intensity of the incident radiation. In the high intensity limit where the transition rates greatly exceed the relaxation time (τ) for the two-level system to return to thermal equilibrium, $W_{abs} = W_{SE} >> 1/\tau$, the populations of the two levels become equal. Thus, for a two-level system, one cannot achieve population inversion through optical pumping.

Stimulated emission (SE) is essential for gain and, since $W_{abs} = W_{SE}$, the strength of the SE is directly related to the transition probability for absorption (or the cross-section for absorption). Thus, a good laser material must have strong absorption.

In a two-level system, population inversion must be achieved indirectly, e.g. by transfer of an excited state from another species (for example by Frster transfer or collision transfer) or, in a semiconductor, by injection of an electron and a hole into the lowest unoccupied band and highest occupied band, respectively. The latter, which is the origin of the population inversion in a semiconductor diode laser (injection laser), is sketched in Fig. 11.2.

When a new class of materials is under investigation for possible use in diode lasers, it is beneficial to start by excitation through photopumping so that stimulated emission and gain can be studied without the complications that are associated with current injection, charge transport and electrode incorporation.

For optically pumped amplification of light, it is desirable that the material have the electronic structure of a "four-level" system (see Fig. 11.1b). In a four-level system, the stimulated emission spectrum does not overlap with the ground state absorption spectrum because photoexcitation is rapidly followed by structural relaxation (bond length relaxation) so

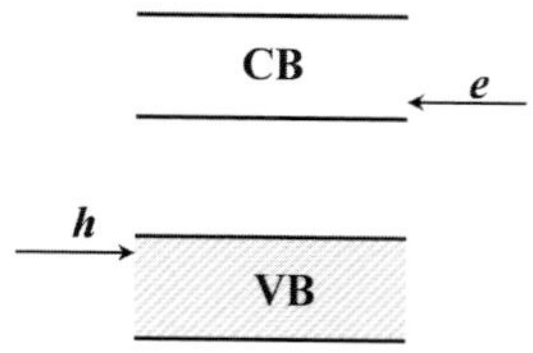

Fig. 11.2 Injection of an electron and a hole into the lowest unoccupied band (conduction band, CB, or π*-band) and highest occupied band (valence band, VB or π-band), respectively; the origin of population inversion in a semiconductor diode laser (injection laser).

that the emitted photons are at a longer wavelength than the photons that are absorbed. Because of the bond relaxation, both the excited state and the ground state energies are shifted; hence a four-level system. Thus, in a four-level system, population inversion is achieved in principle by the absorption of a single photon!

Conjugated polymers are typically four-level systems; excitation of an electron from a bonding state (in the π-band) to an antibonding state (in the π*-band) is followed by local structural relaxation. The resulting red shift in the emission wavelength is known as the "Stokes shift". There are other sources of Stokes shift in addition to bond-length relaxation; for example, since there is disorder in thin films of semiconducting polymers cast from solution, the conjugation length is not the same throughout the film, and there are associated variations in the π-π* band gap energy. Consequently, energy can be absorbed in one region of the semiconducting polymer and then transferred to another region where the band gap is smaller.

Figure 11.3 shows the absorption and emission spectra of MEH-PPV, the spectra are typical of conjugated polymers. Since the absorption spectrum is both broad and intense, $\alpha_{max} \approx (2\text{–}5)\ 10^5\ \text{cm}^{-1}$, thin films (for example with ≈ 100 nm thickness) can absorb more than 90% of the photons from a pump source. At the same time, however, photons emitted at the peak of the photoluminescence spectrum can travel large distances (several mm) without being reabsorbed.

In contrast, for wavelengths at the SE peak, self-absorption in most inorganic semiconductors is usually quite large, and transparency is only obtained at relatively high pump intensities (structural relaxation following absorption is not important in such systems because of their rigid three-dimensional band structure). Note, however, that in Fig. 11.3, there is an absorption tail that extends into the spectral region of photoluminescence. This residual self-absorption must be overcome by pumping the system to a higher density of excited states in order to have gain.

In summary, a good laser material must have high luminescence efficiency and a high chromophore density. Fortunately, in addition to being natural four-level systems, many semiconducting polymers excel in this regard. If the polymer side-chains are designed in such a way that aggregates do not form between neighboring chains, photoluminescence efficiencies higher than 60% can be obtained even in neat (not diluted) films.

The Davidov splitting, well known in molecular spectroscopy, results from exciton–exciton interaction between a pair of molecules. As a result of this interaction, the excited state is spread over the two molecules, and the two otherwise degenerate energy levels are split

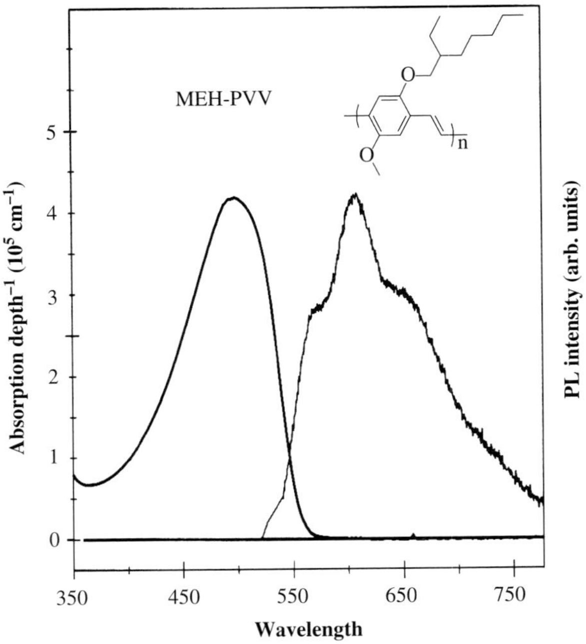

Fig. 11.3 The absorption spectrum (absorption edge at approx. 550 nm) and emission spectrum of MEH-PPV. The Stokes shift results from a combination of bond relaxation and disorder. [Reproduced from B. J. Schwartz *et al.*, *Phil. Trans. R. Soc. Lond. A*, **355**, 775, (1997) with permission from The Royal Society.]

in energy. In such a Davidov pair, the lower energy state is typically dark (nonemissive) because of the parity selection rule. Consequently, at high concentrations of molecular chromophores in solution or in the solid state, the photoluminescence is typically quenched. This is known as "concentration quenching". In macromolecular conjugated chains, however, the lowest energy excited states (excitons) are spread over many repeat units. As a result the exciton–exciton interaction and the resulting Davidov splitting are relatively small. Moreover, the disorder in such systems mixes the two Davidov states. Therefore, concentration quenching is not of major importance in semiconducting polymers; luminescence efficiencies of 60–80% are available in this class of materials.

An additional criterion for laser materials is that the excited state absorption spectrum (photoinduced absorption) should not overlap with the stimulated emission spectrum. The two transitions in the excited state that tend to be problematic for many laser materials are absorption from the radiative state into higher lying excited states (see Fig. 11.4) and triplet–triplet absorption. Cross-over to the triplet manifold and triplet–triplet absorption is particularly problematic for laser dyes because of the long lifetime of the triplet state. (The triplet problem in dye lasers is avoided by flowing the dye solution or by using a triplet quencher.) Fortunately, for most semiconducting polymers, the lowest triplet absorption is lower in energy

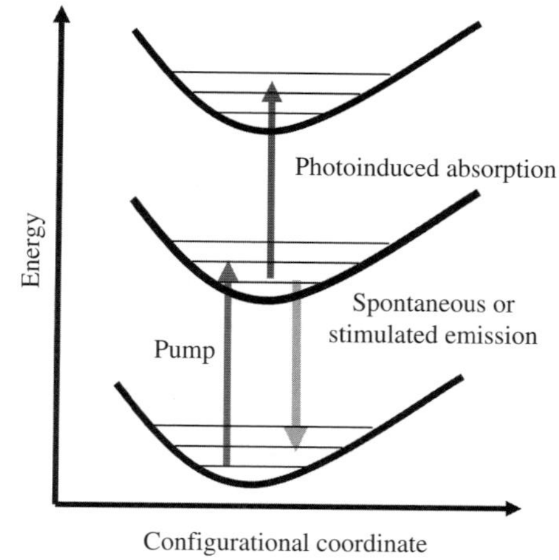

Fig. 11.4 Schematic diagram of the excited state energies as a function of the amplitude of the bond relaxation coordinate (configurational coordinate). The ground state, first and second excited states are shown. For each of these states, there are vibrational splittings (horizontal lines). Optical transitions are "vertical"; i.e. they take place without change in the configurational coordinate.

than the singlet emission. On the other hand, excited state absorption into higher lying energy levels can be a serious problem.

In summary, semiconducting polymers offer promise as laser materials.

(a) Studies of the photophysics demonstrate that semiconducting polymers are intrinsically four-level systems (see Chapter 7); as a result, population inversion can be realized with relative ease.
(b) For a laser material to have a large gain coefficient, it must have a large cross-section for stimulated emission. Semiconducting polymers typically have large cross-sections ($\sigma \approx 10^{-15}$ cm^2) because the direct (vertical in *k*-space) π–π* transition is allowed, and the quasi-one-dimensional electronic wavefunctions have a high density-of-states at the band edge (see Chapter 1). Thus, high gain from stimulated emission is expected.
(c) Structural relaxation in the excited state leads to a relatively large Stokes shift. Since the spectral overlap of the emission and absorption is small, self-absorption is significantly lower than in inorganic semiconductor.
(d) Concentration quenching is not of major importance.

11.2 Photoinduced absorption as a loss mechanism

The primary technique for measuring excited state absorption (photoinduced absorption, PA, (see Fig. 11.4) is time-resolved pump–probe spectroscopy. An ultrafast pulse of intense laser light is used to pump the sample into its lowest energy excited state and then a weaker pulse of light (at varying time delays and with varying wavelengths) is used to probe the sample.

At wavelengths where stimulated emission (SE) is dominant, the probe beam is amplified; at wavelengths where excited state absorption is dominant, the probe beam is attenuated.

For an "ideal" four-level system

(a) the SE spectrum is identical to the photoluminescence (PL) spectrum;
(b) Lifetime for SE = Lifetime for PL = Lifetime of excited state = Lifetime for PA.

However, excited state absorption is commonly observed in the spectroscopy of semiconducting polymers. Often, but not always, the excited state absorption spectrum overlaps the luminescence spectrum. When this occurs, the gain is reduced. This can be understood more clearly by comparison of contrasting examples.

As shown in Fig. 11.5, for DP6-PPV, the PL and SE spectra are nearly identical. Moreover, in this luminescent polymer (PL efficiency greater than 60%), the decay of the SE and the PL are identical as would be expected for a simple system in which the gain and the PL are directly

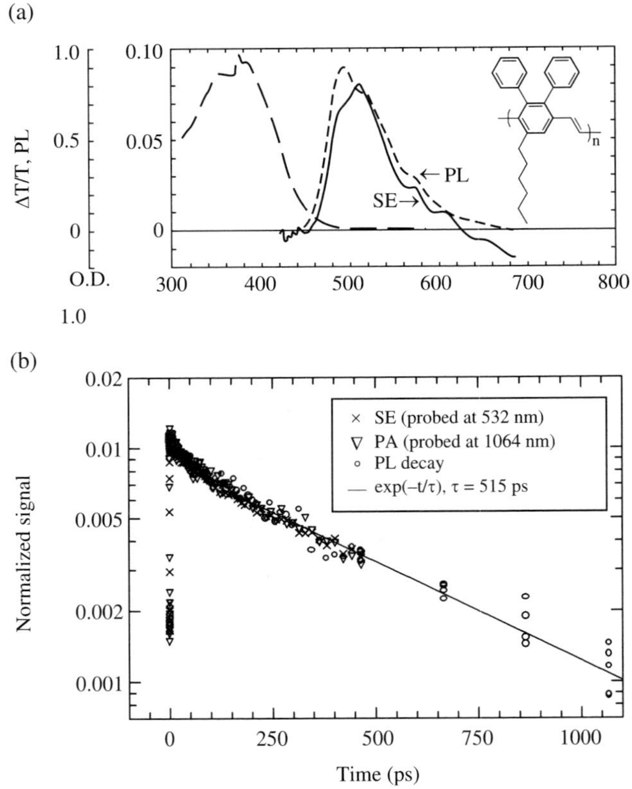

Fig. 11.5 (a) Absorption, photoluminescence (PL), and stimulated emission (SE) spectra of DP6-PPV. The inset shows the molecular structure of poly(2,3-diphenyl-5-hexyl-*p*-phenylene vinylene) (DP6-PPV). (b) Time decay of the PL, SE and photoinduced absorption (PA) in DP6-PPV. [Reproduced from A. Dogariu *et al.*, *Phys. Rev. B*, **61**, 16183 (2000) with permission from The American Physical Society.]

related, and there is no loss from PA. In such a system, the time constant for the decay of the PL and SE are identical and determined by the lifetime of the excited state.

As implied by Fig. 11.4, the photoinduced absorption from the excited state to a higher energy state should (and does) decay with the same time constant.

Similarly, in BuEH-PPV, the excited state absorption has little effect on the gain. As shown in Fig. 11.6, the PA spectrum lies farther into the infrared than the PL spectrum. Since there is little overlap of the two, BuEH-PPV is again a good example of a four-level system.

For MEH-PPV, however, the photoinduced absorption spectrum strongly overlaps the PL spectrum. Thus, although the PL extends to wavelengths beyond 750 nm (see Fig. 11.3), the PA reduces the net gain (stimulated emission greater than photoinduced absorption) over a relatively large fraction of the PL spectrum.

There have been many studies of MEH-PPV (and other luminescent polymers), and a wide variety of seemingly contradictory results have been reported. Nguyen *et al.*[3] reviewed the literature on the photophysical properties of PPV derivatives and concluded that chain morphology and interchain aggregation were the causes of the differing results. An important lesson to be learned from these studies is that the properties of conjugated polymer films are quite sensitive to chain packing morphology, which can vary dramatically depending on preparation conditions. Unfortunately, because the morphology of polymer films is difficult

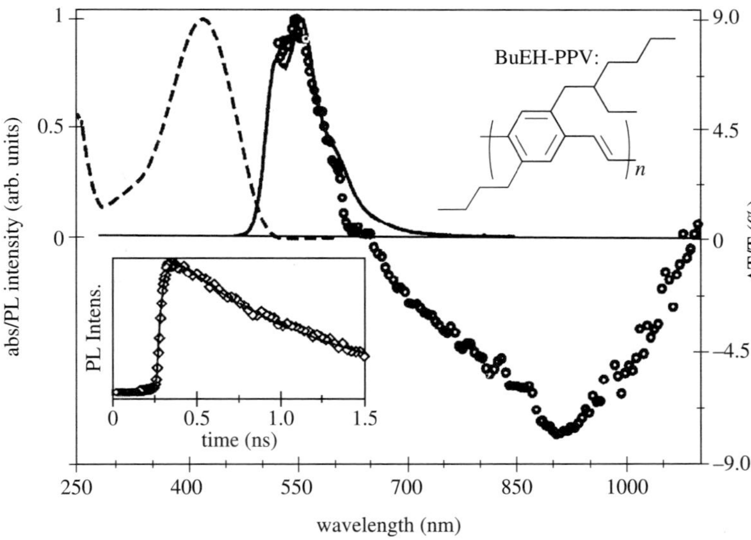

Fig. 11.6 Steady state absorption (dotted line), PL (solid line) and PA spectra of BuEH-PPV. In this case the PA cuts into the gain spectrum at long wavelengths (beyond approx 600 nm). Lower inset: spectrally integrated PL decay of BuEH-PPV excited at 440 nm. [Reproduced from B. J. Schwartz *et al.*, *Phil. Trans. R. Soc. Lond. A*, **355**, 775, (1997) with permission from The Royal Society.]

[3] T. Nguyen *et al.*, *J. Phys. Chem. B*, **104**, 23 (2000).

to quantify, most reports on the photophysical properties of conjugated polymers do not address this issue.

In summary, polymers like DP6-PPV are well suited for laser applications: they can be accurately described as having the electronic structure of a four-level system, they are highly fluorescent, they have a large stimulated emission cross-section, and they are not hindered by excited state absorption.

11.3 Amplified stimulated emission (ASE) from films of conjugated, semiconducting polymers

Although picosecond pump–probe spectroscopy provides detailed information on the nature of the excited states, it is certainly not the easiest method for characterizing a semiconducting polymer as a laser material. Photopumping a thin-film waveguide and measuring the gain or loss (by probing the intensity and linewidth of the emission from the end of the waveguide) is straightforward and directly relevant to lasing. The thin-film waveguide can be made by casting a film on a low-refractive-index substrate (see Fig. 11.7). When the pump intensity is sufficiently high that the polymer has net gain, the spontaneously emitted photons are exponentially amplified by stimulated emission as they travel through the waveguide (amplified stimulated emission, ASE).

The ASE output intensity from one end of the waveguide is given by the following expression:

$$I(\lambda) = \frac{A(\lambda)I_p}{g(\lambda)}(e^{g(\lambda)l} - 1), \tag{11.2}$$

where A is a constant related to the cross-section for spontaneous emission, Ip is the pump intensity, $g(\lambda)$ is the net gain coefficient, and l is the length of the pumped region.

Since the gain, $g(\lambda)$, is maximum near the peak of the spontaneous emission spectrum, the spectrum exhibits "gain narrowing"; i.e. because there is more amplification where gain is largest, the full width at half maximum (FWHM) of the ASE beam decreases as gl increases. Consequently, an intense, spectrally narrow beam is emitted from the end of the waveguide.

An example of ASE is shown in Fig. 11.8; the emission spectrum of BuEH-PPV narrows dramatically as the pump power is increased. The data in Fig. 11.8 were obtained by pumping

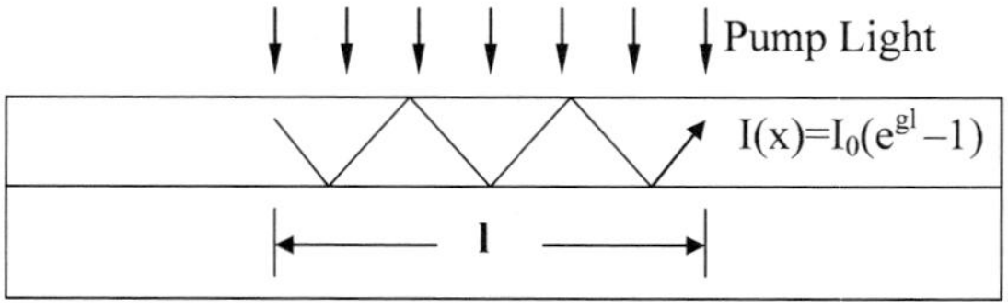

Fig. 11.7 Configuration for observing ASE; a thin-film waveguide of a luminescent polymer on glass is pumped from above at a wavelength in the absorption band (the pump beam extends over a length l). If $gl > 1$, the spontaneous emission is amplified causing an intense, spectrally narrow beam to be emitted from the ends of the waveguide.

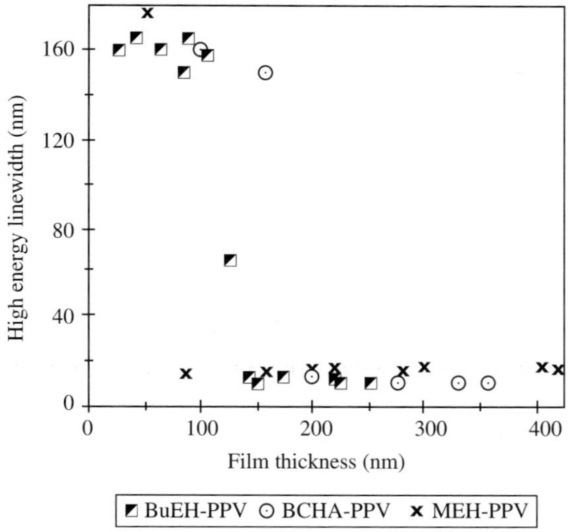

Fig. 11.8 Amplified spontaneous emission (ASE) from a thin film of BuEH-PPV on a glass substrate. The pump pulse width was 1 ns; stripe length = 2 mm. [Reproduced from F. Hide *et. al.*, *Science*, **273**, 1833 (1996) with permission from AAA.]

over a thin stripe with length of 2 mm using the instrumentation shown in greater detail Fig. 11.10 below.

Structures that exhibit ASE are sometimes referred to as "mirrorless lasers", because they have many of the properties of a laser. Although the light has been amplified by stimulated emission, the amplified light is not coherent as in a true laser. As shown later in this chapter, it is straightforward, but nontrivial, to convert a mirrorless laser into an actual laser that emits narrow band coherent radiation by incorporating resonant feedback. For this reason, most polymer characterization for laser development is done with mirrorless laser waveguide structures with the understanding that the polymers can be incorporated into a resonant structure once they have been optimized.

The proof that the mechanism sketched in Fig. 11.7 is operative was demonstrated through ASE measurements as a function of the film thickness. For asymmetric planar waveguides, glass/polymer/air, light will not propagate in the thin-film waveguide for thicknesses less than $t_{cut-off} = [\lambda/2\pi(n_f^2 - n_s^2)^{1/2}] \tan^{-1}[(n_s^2 - n_c^2)^{1/2}/(n_f^2 - n_s^2)^{1/2}]$ where n_f, n_s, and n_c are the indices of refraction of the film (f), substrate (s) and cover layer (c). Thus, there is a minimum thickness required for wave guiding and a minimum thickness required for gain narrowing.

Consequently, the threshold for ASE should increase for films with thickness such that for wavelengths near the peak in the PL, the light is beyond cut-off. This is indeed true as shown in Fig. 11.9. The cut-off thicknesses, $t_{\text{cut-off}}$ (for λ at maximum PL), inferred from the ASE data are in good agreement with the values calculated by using the measured indices of refraction, consistent with the mechanism sketched in Fig. 11.7.

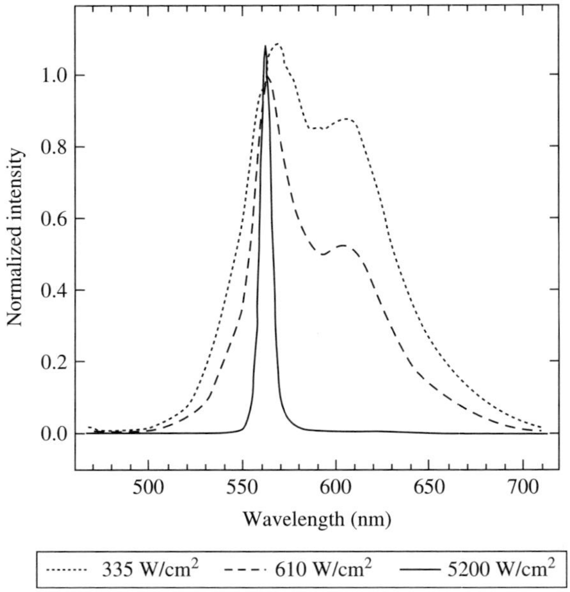

Fig. 11.9 The linewidth of the gain-narrowed emission as a function of the film thickness for BuEH-PPV, BCHA-PPV and MEH-PPV. The incomplete narrowing for films with thickness below 100 nm implies that the maximum gain is severely reduced for these thin films because the wavelength for maximum PL is beyond cut-off. [Reproduced from F. Hide *et. al.*, *Science*, **273**, 1833 (1996) with permission from AAA.]

11.4 Direct measurements of gain, loss and ASE spectra in planar waveguides of semiconducting polymers

In this section, we focus on three distinct characteristic of ASE that arises from light amplification in waveguide structure:

(a) spectral narrowing with pump intensity;
(b) threshold in both in the output power and linewidth; and
(c) light emission will be characteristic of, and influenced by, the gain medium.

As an example, we focus on data obtained with thin-film waveguides of BuEH-PPV on silica.

An adjustable slit and a cylindrical lens were used to shape the pump beam into a stripe with a width of approximately 350 μm and a length that could be varied. When ASE occurs in a long, narrow stripe, most of the light is emitted from the ends of the stripe. For this reason, the pump stripe was positioned right at the edge of the BuEH-PPV film, and the edge emission was collected. In order to observe a clean emission beam, the original edge of the sample was snapped off to remove the edge bead formed by spin-casting (see Fig. 11.10).

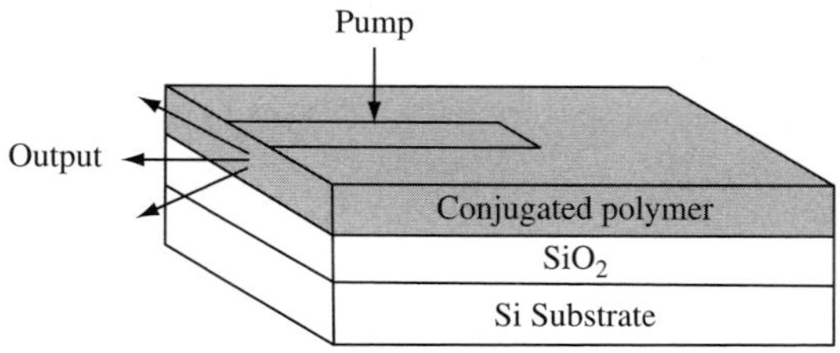

Fig. 11.10 Details of the thin-film waveguide and the strip illuminated by the pump laser.

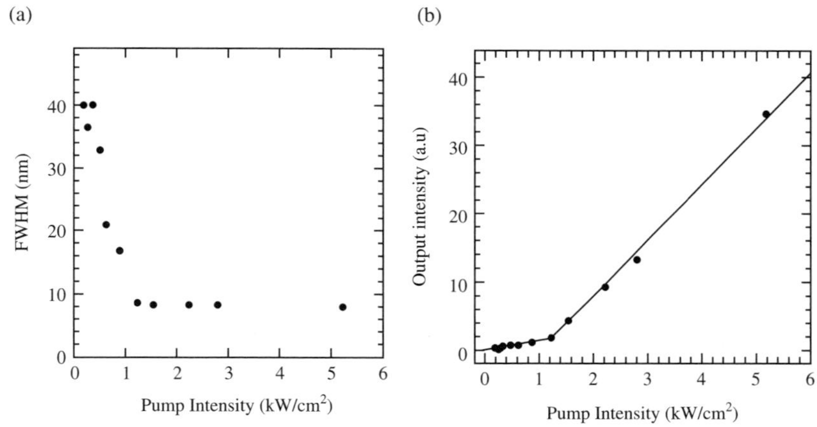

Fig. 11.11 (a) FWHM as a function of the pump intensity; the data were acquired using the instrumentation discussed in Fig. 11.9. (b) Light intensity emitted from the edge, integrated over all wavelengths, as a function of the pump intensity. [Reproduced from M. D. McGehee *et al.*, *Adv. Mater.*, **12**, 1666 (2000) with permission from Wiley-VCH Verlag.]

(a) **Spectral narrowing**

Figure 11.9 shows the emission spectra that were collected at various pump intensities with a constant pump stripe length of 2 mm. When the pump intensity was less than 0.6 kW/ cm^2, the spectrum was broad with overlapping vibronic peaks; i.e. characteristic of spontaneous emission, PL, from conjugated polymers. At these low intensities the shape of the spectrum did not depend on the pump intensity. When the pump intensity was increased above 0.6 kW/ cm^2, ASE became evident, and the intensity at the peak of the spectrum grew much more than the intensity at other wavelengths. As shown in Fig. 11.11(a), the full width at half maximum (FWHM) of the spectrum dropped to 8 nm. For BuEH-PPV, ASE occurs close to the wavelength where the spontaneous emission spectrum is strongest. For some other polymers, the ASE wavelength is shifted to larger wavelengths (relative to the peak spontaneous emission wavelength) because absorption near the band edge reduces the net gain. In general, however, for the best lasing polymers, losses do not affect the gain spectrum too seriously and ASE occurs close to the wavelength where spontaneous emission is strongest because the gain is highest there.

(b) **Threshold in both in the output power and linewidth**

Figure 11.11(b) shows the dependence of the light intensity emitted from the edge, integrated over all wavelengths, on the pump intensity. The measured relative efficiency, as inferred from the change in slope at 1 kW/ cm^2, increases by a factor of 6 at the threshold for spectral narrowing. The increase arises because only edge emission was being detected. Therefore, below the ASE threshold a significant fraction of the PL is emitted in all directions, whereas above threshold, the excitations are stimulated to emit into the waveguide mode, so a larger fraction of the light comes out the edge. These data highlight one of the advantages of lasers and even mirrorless lasers: ASE light emission is directional. Moreover, the ASE is highly polarized parallel to the substrate because, as noted above, the anisotropic polymer waveguides support only the one TE mode.

The threshold for ASE occurred at approximately 1 kW/cm^2, which corresponds to an excitation density of only 10^{17} excitations/cm^2. Thus, significant gain is achieved at relatively low excitation densities. The data are in general agreement with Eqn 11.2. In contrast, for example, if superfluorescence or biexcitonic emission were the mechanism of spectral narrowing, the emission spectrum would not depend on the size of the excited region and the output would increase only linearly with the length of the excited region (or sublinearly if waveguide losses are substantial).

(c) **Light emission will be characteristic of, and influenced by the gain**

When the pump stripe length was varied on the BuEH-PPV waveguides, the spectra did indeed narrow as the length was increased, and, as shown in Fig. 11.12, the output grew

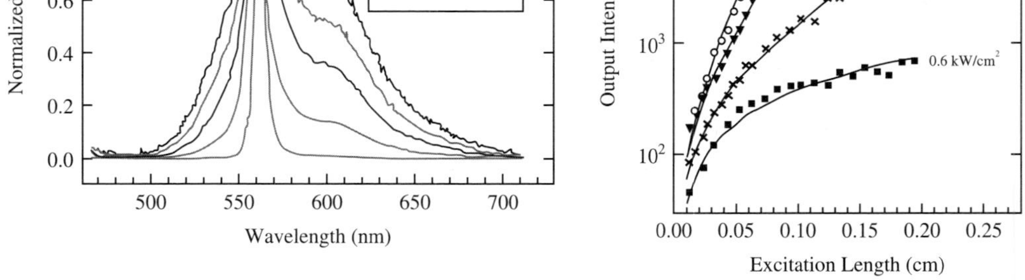

Fig. 11.12 ASE spectra (left) of BuEH-PPV at fixed pump intensity and dependence of the emission intensity on the excitation length at the indicated pump intensities (right). The solid lines are fits to the data using Eqn 11.2 with the parameters listed in Table 11.1. The net gain of the waveguide can be determined from the fits, for various stripe lengths. [Reproduced from M. D. McGehee *et al.*, *Adv. Mater.*, **12**, 1666 (2000) with permission from Wiley-VCH Verlag.]

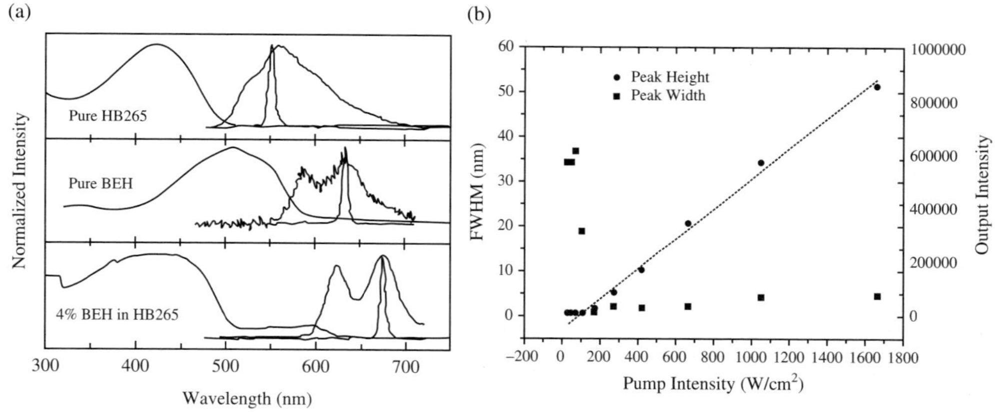

Fig. 11.13 (a) Ground state absorption (left), photoluminescence (broad curve), and ASE spectra (narrow curve on right) for pure HB265, pure BEH-PPV and 4% wt% BEH-PPV in HB265. FWHM (left axis) and output intensity (right axis) of the ASE from a blend of 4 wt.% BEH-PPV in HB265. [Reproduced from M. D. McGhee *et al.*, *Adv. Mater.*, **12**, 1666 (2000) with permission from Wiley-VCH Verlag.] (b) A soluble PPV derivative obtained from Covion (HB265). The ASE threshold occurs at approximately 100 W/cm^2. [Reproduced from R. Gupta *et al.*, *Appl. Phys. Lett.*, **73**, 3492 (1998) with permission from American Physical Society.]

exponentially in accordance with Eqn 11.2. These data provided strong evidence that spontaneous emission is being amplified by stimulated emission.

Additional evidence for ASE was obtained in separate studies by measuring the time dynamics of the spectrally narrow emission. As predicted by ASE models, the emission lifetime steadily decreases with increasing excitation density due to the increased stimulated emission rate.

Figure 11.13 shows the output intensity at the peak of the emission spectrum ($\lambda = 562$ nm) as a function of pump stripe length at four different pump intensities. Each curve fits nicely to Eqn 11.2 (at least in the low output intensity regime); from the fits the net gain of the waveguides was determined as a function of pump intensity (Table 11.1). With a gain of 62, amplification is evident after propagation through only approximately 150 μm.

At the higher pump intensities, the output intensity increased exponentially at excitation lengths of less than 1 mm, as expected, but then leveled off at longer excitation lengths. This behavior is the result of gain saturation, which occurs when the light traveling in the waveguide becomes so intense that it depletes a substantial fraction of the excitations and thereby reduces the gain coefficient. When gain saturation occurs, spectral narrowing stops. Consequently, the smallest spectral width that can be obtained with ASE is determined by gain saturation.

11.5 Lowering the lasing threshold with energy transfer (to increase the Stokes shift)

The fact that for most luminescent polymers the absorption "tail" extends into the PL spectral region implies that they do not have the absorption/emission properties of a perfect four-level

Table 11.1. Gain versus pump intensity as obtained from Fig. 11.12.

Pump intensity (kW/cm^2)	g_{net} (cm^{-1})
0.6	0.1
1.2	18
2.2	41
4.1	62

system. The self-absorption problem can be alleviated by shifting the emission to longer wavelengths where the residual absorption is smaller. This can be done by blending a small amount of a guest molecule or polymer into the host. In such blends, energy transfer by dipole–dipole coupling (Frster transfer) occurs if the emission spectrum of the host overlaps with the absorption spectrum of the guest and if the dipoles are sufficiently close to each other. Additional bonuses of using a two-component (host/guest) system as the laser active medium are that scattering losses are reduced by the redshift of the emission (proportional to $1/\lambda^4$) and residual concentration quenching is reduced (although not of major importance in conjugated polymers).

The rate of Förster transfer is given by:

$$k_{HG} = \beta_H (R_0/R_{HG})^6 \qquad (11.3)$$

where $\beta_H = 1/\tau_H$ is the inverse fluorescence lifetime of the host, R_0 is the long range resonance transfer constant (proportional to the overlap of the PL emission form the host and the absorption of the guest), and R_{HG} is the host–guest spacing. Thus, for efficient Förster transfer, there are two requirements: good overlap between the emission of the host and the absorption of the guest and small host-guest spacing, R_{HG}. The Förster range in such blends, $R_0 \approx 3$ nm as determined through time-resolved measurements of the energy transfer, is in good agreement with the value calculated from Förster's theory.

Red-emitting PPV derivatives (as guests) have been blended into green-emitting PPV derivatives (as hosts) such that there is overlap between the PL spectrum of the host and the absorption spectrum of the guest; i.e. the conditions for efficient Frster transfer are satisfied. Loss measurements were carried out using the instrumentation shown in Fig. 11.10, but moving the end of the stripe away from the end of the waveguide (with fixed strip length). By shifting the emission wavelength farther from the absorption edge, the measured loss coefficient was reduced by more than an order of magnitude to 3 cm^{-1}. Because of the reduced loss, the threshold for ASE was also reduced by an order of magnitude to only 100 W/cm^2 (compare Fig. 11.13 with Fig. 11.11b). The 100 W/cm^2 threshold for ASE corresponds to an excitation density of only 10^{16} excitations/cm^2. This remarkably low threshold provides some optimism that diode lasers will be realized with semiconducting polymers.

Time-resolved measurements of the Frster energy transfer process (with ps pump/probe spectroscopy) demonstrated that the spectral signatures of the host were replaced by those of the guest at a rate that agreed quantitatively with theoretical predictions based on Frster's model. For a 2% blend, complete energy transfer occurred within 10–20 ps, which is much less than the photoluminescence decay time. This high energy transfer rate explains why almost all of the emission comes from the guest. Consequently, by using efficient Frster transfer, ASE can be observed at low pump intensities even if the host polymer is not an efficient emitter! This fast energy transfer mechanism could be useful in electrically driven devices since the polymers that tend to transport current most effectively are frequently not efficient emitters.

11.6 Laser resonant structures and true laser emission

In order to achieve coherent emission and true lasing, it is necessary to incorporate a resonator into the gain medium. The most common approach to making a resonator is to place mirrors at the ends of the gain region (the classic Fabry–Perot structure). The processing flexibility of polymers has enabled a wide range of miniature resonators to be used, including distributed feedback (c) resonators. DFB resonators are particularly widely used and consist of a diffraction grating integrated into the device (see Fig. 11.14). They can be simply made by lithography.

With such resonator structure, specific modes are formed in the cavity at wavelengths that depend on the distance between the two mirrors. There is feedback when the emission wavelength is resonant with a cavity mode, and lasing occurs when the amplification of photons traveling in one of the modes is sufficient to overcome the losses in the gain medium and through the mirrors. The lasing mode gives the distinct properties of laser light listed below:

1. narrow linewidth emission;
2. the light output consists of a beam;
3. a clear threshold both in the output power and linewidth;
4. the light emission is influenced by the gain medium and resonator.

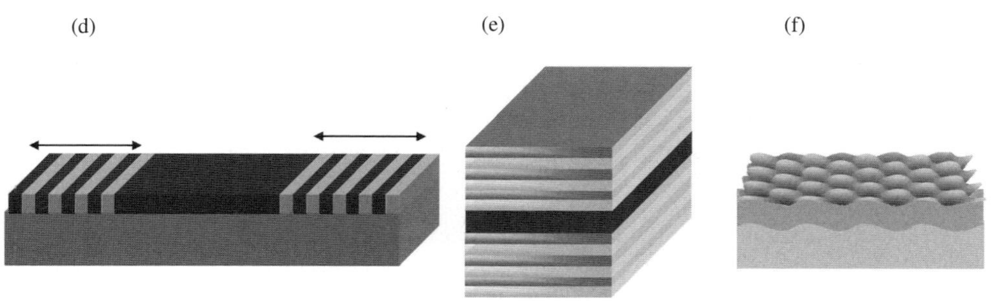

Fig. 11.14 The planar microcavity consists of a thin layer of gain material sandwiched between two mirrors (a, b, e). In the distributed feedback resonator, the gain medium is corrugated, leading to light being diffracted back on itself to achieve feedback (c, d, f).

In the following sections we focus on polymer lasers fabricated using microcavity and DBR resonator structures.

(a) Microcavity polymer laser

Tessler *et al.*[4] made the first vertical microcavity laser based on conjugated polymers as gain materials. The approach was to sandwich a film of PPV between a commercially available dielectric mirror and a thermally evaporated silver mirror to create a microcavity (Fig. 11.14b). Since even the spontaneous emission from microcavities is spectrally narrow due to quantum electrodynamical effects, it was not trivial to determine whether or not lasing occurred in their devices. Fortunately, microcavities have several spontaneous emission modes. The intensities of these modes grow linearly as the pump energy is increased until threshold. As shown in Fig. 11.15, above threshold, the mode closest to the peak of the gain spectrum grows at an even faster rate, but the other modes stop growing. This behavior is a signature of lasing; above the lasing threshold nonlasing modes stop growing because all of the additional excitations are quickly stimulated to emit into the lasing mode, which becomes the dominant mode.

In such vertical microcavity lasers the emission is highly directional with maximum intensity perpendicular to the DBR mirrors. Angular dependence of the emission from a single-mode microcavity is displayed in Fig. 11.16. The emission intensity is significantly reduced even at only a few degrees from the normal. This results in a laser emission that is highly directional.

Vertical microcavity lasers have several attractive advantages: they are easy to make; their architecture is similar to that of polymer LEDs; and they emit light that is directed perpendicular to the substrate plane, which is useful for many applications. However, the distance

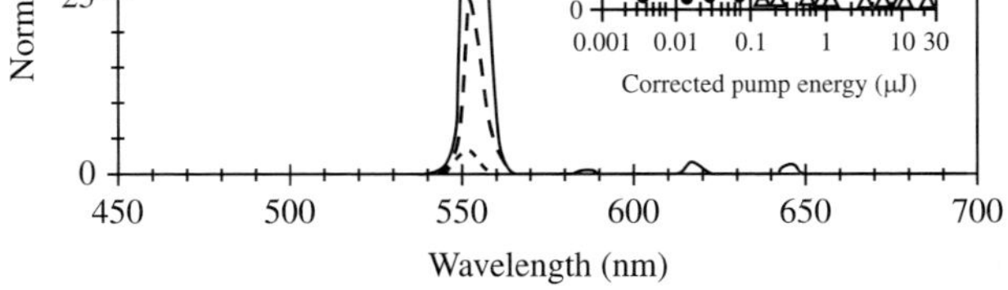

Fig. 11.15 Output intensity from a microcavity laser fabricated with BuEH-PPV. The various modes are evident. The inset compares the output of the lasing mode with the output of the cavity modes that are not resonant with the PL of BuEH-PPV. [Reproduced from M.A. Dıaz-Garcıa *et al.*, *Appl. Phys. Lett.*, **70**, 3191 (1997) with permission from The American Physical Society.]

[4] N. Tessler *et al.*, *Nature*, **382**, 695 (1996).

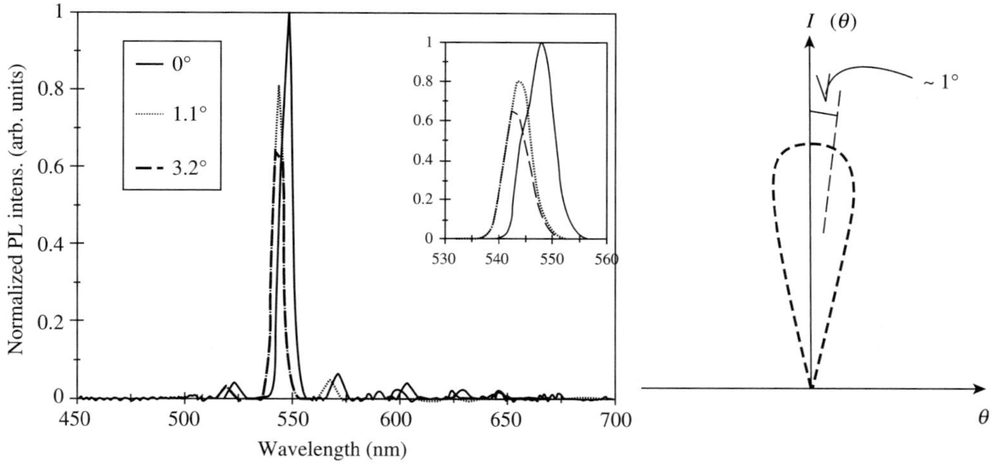

Fig. 11.16 Angular dependence of the emission from a microcavity laser (see Fig. 11.15); the data are shown on the left, whereas the figure on the right summarizes the data in a polar plot. [Reproduced from M.A. Dıaz-Garcıa *et al.*, *Appl. Phys. Lett.*, **70**, 3191 (1997) with permission from The American Physical Society.]

traveled by light during each pass through the gain region is small (<200 nm). Consequently, the lasing threshold is high unless the mirrors are extremely reflective (>99.9%).

When inorganic crystalline semiconductors are used to make waveguide lasers, mirrors are sometimes made simply by cleaving the crystal. The resulting facets at the ends of the waveguide reflect light because of the refractive index mismatch between the crystal and air. Unfortunately, polymer lasers cannot be made so simply. Since the semiconductor film is a polymer, breaking the film does not result in a high-quality facet. Furthermore, even if good facets were formed by techniques such as reactive ion etching, laser ablation, or freeze fracture, the reflectivity at the polymer–air interfaces would be less than 10% because the refractive index of luminescent polymers is relatively low (1.6–2.0).

(b) DFB polymer laser

An alternative way to reflect light in a polymer waveguide is to incorporate a periodic modulation of the refractive index so that light can be reflected by Bragg scattering. Lasers of this type are known as distributed feedback (DFB) lasers; see Fig. 11.14(c). In DFB lasers, the mirror is actually incorporated into the gain region. The DFB structure can be thought of as a one-dimensional (1D) "photonic crystal" that has a photonic band gap centered at the Bragg wavelength, $\lambda_{Bragg} = 2n_{eff}\Lambda$ (n_{eff} is the effective refractive index of the waveguide and Λ is the period of the grating). Light with energy in the 1D photonic band gap cannot propagate in the direction perpendicular to the refractive index grating. Hence, lasing does not occur at λ_{Bragg}, but at wavelengths corresponding to the edges of the photonic band gap. At these wavelengths light can propagate, but it is still reflected by the grating.

DFB lasers were fabricated by spin-casting films of BuEH-PPV onto gratings in silica; a schematic picture of the device architecture is presented in Fig. 11.14(c). The substrate (SiO_2

on Si) must be patterned into a grating with period comparable to the wavelength of visible light in order to fulfill the Bragg condition:

$$\lambda_{\text{Bragg}} = 2\Lambda n_{\text{eff}} \tag{11.4}$$

The grating is patterned as follows:

(a) pattern photoresist (SiO_2) with holographic lithography using spacing lengths in the range 170–185 nm;
(b) transfer the grating into SiO_2 with reactive ion etching;
(c) spin-cast the semiconducting polymer thin film (for example BuEH-PPV) onto the patterned substrate with thickness $d \approx 150$–300 nm.

A one-micron thick layer of polymethylglutarimide (PMG) was spin-cast over the semiconducting thin film to serve as protective cladding layer. The physical origin of the effective index of refraction, n_{eff}, is the spread of the waveguide mode into the air or PMG above and into the SiO_2 below; see Fig. 11.17.

A sketch of the DFB structure and a photograph of the laser are shown in Fig. 11.18. The light emitted from the edge of the laser is fully polarized parallel to the substrate indicating the laser operation in a TE waveguide mode.

The output intensity of the DFB laser as a function of the pump intensity is shown in Fig. 11.19(a). There is a clear change in slope at approximately 2 kW/cm^2. The spectrum of the emitted radiation is shown in Fig. 11.19(b).

Polymer lasers have also been made by incorporating polymers into photonic crystals, which are materials that have a periodic perturbation in their refractive index with a periodicity comparable to the wavelength of light. The microcavity and DFB lasers described above can be thought of as one-dimensional photonic crystals. Two-dimensional photonic crystal lasers have been made by depositing films over two-dimensional arrays of holes. Three-dimensional lasers have been made by infiltrating polymers into porous synthetic silica opals.

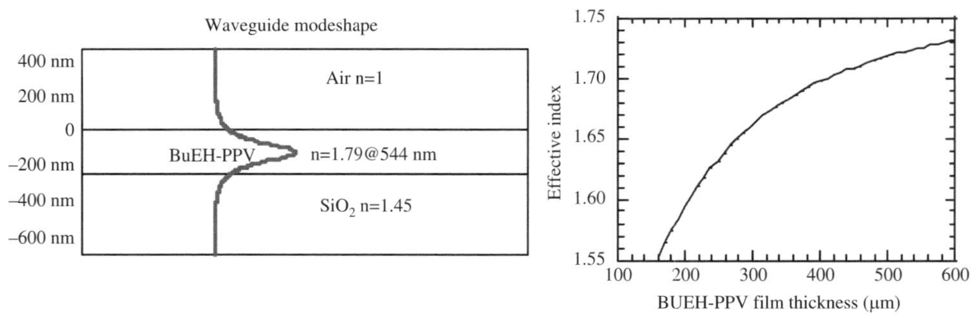

Fig. 11.17 (Left) Mode shape; (Right) n_{eff} versus thickness of polymer layer.

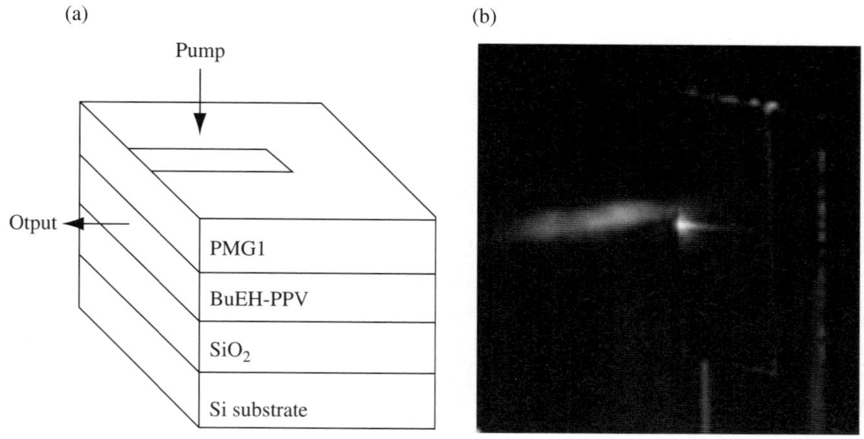

Fig. 11.18 (a) Diagram of the DFB structure; (b) photograph of the laser. [Reproduced from M. D. McGehee *et al.*, *Appl. Phys. Lett.*, **72**, 1536 (1998) with permission of American Institute of Physics.]

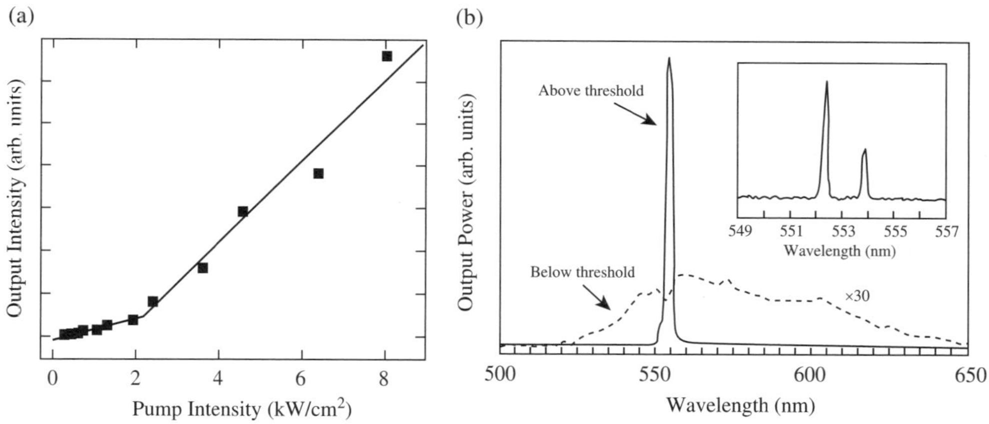

Fig. 11.19 (a) Output intensity versus pump intensity for the DFB laser (see Fig. 11.19). (b) Spectrum of the radiation emitted from the DFB laser; the inset shows the same data with better resolution. [Reproduced from M. D. McGehee *et al.*, *Appl. Phys. Lett.*, **72**, 1536 (1998) with permission of American Institute of Physics.]

11.7 Progress towards electrically pump lasing

Direct electrical pumping of polymers is a very challenging problem for several reasons. First it is difficult to pass very high current densities through most organic semiconductor devices (diodes) because of their low mobility (and in many cases substantial energy barriers to charge injection). Second, there will be substantial optical losses associated with the contacts, and

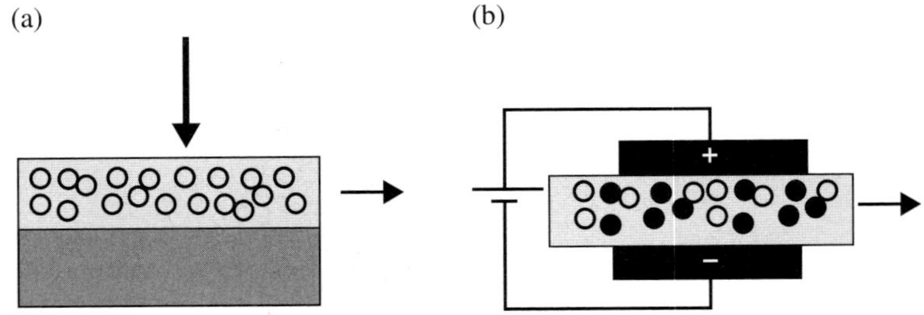

Fig. 11.20 (a) schematic of optical pumping. Pump light leads to formation of singlet excitons (open circles) in the gain medium. (b) Schematic of electrical pumping: charges (black spheres) injected from the contacts meet up and form singlet excitons (open circles). In fluorescent materials, only the singlet excitons emit light, whilst the contacts, injected charges and triplet excitons can all absorb light, making lasing difficult to achieve.

whilst configurations which mitigate this loss have been suggested, it is difficult to move the contacts far away, again because of the low mobility. Third, there is substantial absorption due to the injected charge carriers, increasing losses compared with an optically pumped laser. Finally, when opposite charges recombine in an organic semiconductor they frequently (approximately 75%) form a triplet exciton which is not emissive in fluorescent materials and may also contribute to loss via its absorption. The last three points are illustrated in Fig. 11.20, and mean that the threshold for an electrically pumped laser will be very much higher than for an optically pumped laser.

In order to reduce the optical loss in the diode structure, one strategy is to increase the thickness of the organic layers to reduce the losses from the metal electrode. However, because of the low carrier mobility of the carriers in organic semiconductors, the thickness of the diode structure cannot be increased arbitrarily.

The use of a light-emitting field effect transistor (LEFET) structure (see Fig. 11.21) to achieve the organic injection laser offers important advantages. The electrode losses are reduced by the FET structure with gate-induced injection through the spatially removed source and drain. Since the transport of injected charge carriers (polarons) is perpendicular to the waveguiding structure in which the neutral excitons of the excited polymer are confined, there are relatively few charge carriers in the light propagation pathway. Furthermore, very high current densities in the accumulation zone can be achieved in the FET structure. Because of these features, the FET structure offers specific advantages compared to the conventional diode structure toward the fabrication of the injection laser. In this structure, the only electrode interfering with the guided wave is the gate electrode.

In the following sections we demonstrate the implementation of the DFB resonator structure in FET structure using nanoimprint lithography techniques and show that the optical losses are minimal on conducting ITO gate electrode.

Figure 11.22 (left) shows schematic fabrication steps for the DFB structure. A rigid nanostructure template, a "Master", is pressed at onto the light-emitting polymer film and heated to glass transition temperature. Upon heating, the polymer becomes fluid and fills the empty space of the Master, forming a nanopattern which is a negative of the master. Finally, the

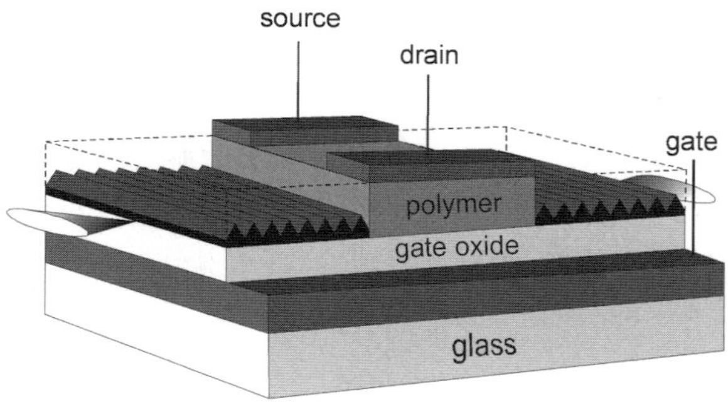

Fig. 11.21 Light-emitting field effect transistor (LEFET) structure with a Bragg reflector on either end.

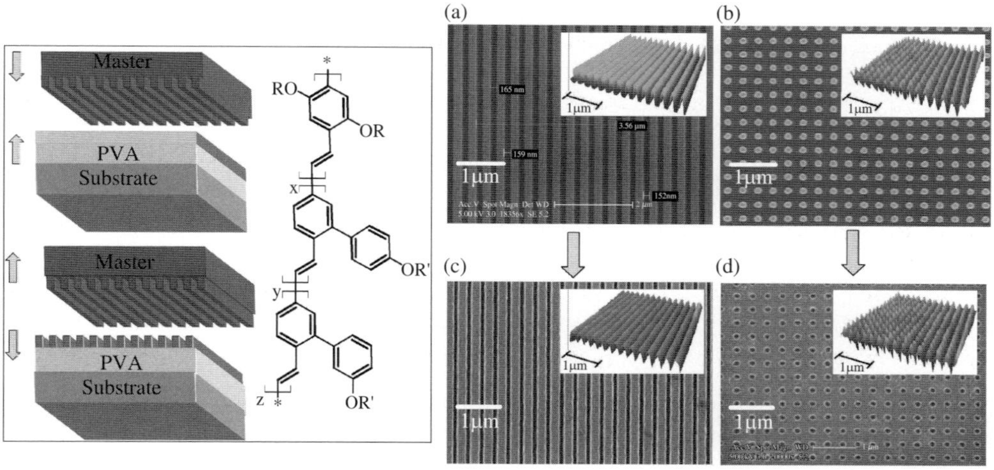

Fig. 11.22 Schematic fabrication steps for DFB structure and the molecular structure of superyellow (left); SEM images of (a) the 1D Master, (b) 2D Master, (c) nanoimprinted 1D DFB structure on SY, and (d) nanoimprinted 2D DFB structure on SY (inset shows corresponding AFM images). [Reproduced from E. B. Namdas *et al.*, *Adv. Mater.*, **21**, 799 (2009) with permission from Wiley InterScience.]

sample is carefully separated from the Master. Figure 11.22 (right) shows that both the 1D and the 2D master are faithfully replicated on the samples using this technique.

Figure 11.23 shows the emission output of the 1D and 2D DFB laser on the SY/PVA/Quartz, and SY/PVA/ITO samples as a function of excitation pump energy. The data indicate that the lasing threshold for the 2D resonator on the ITO substrate drops significantly. In summary, 1D or 2D DFB gratings can be directly incorporated into the FET device structure by nanoimprint lithography. Both the 1D and the 2D gratings have merits. Although the 2D grating yielded a lower threshold in these initial experiments, the 1D grating can provide a high

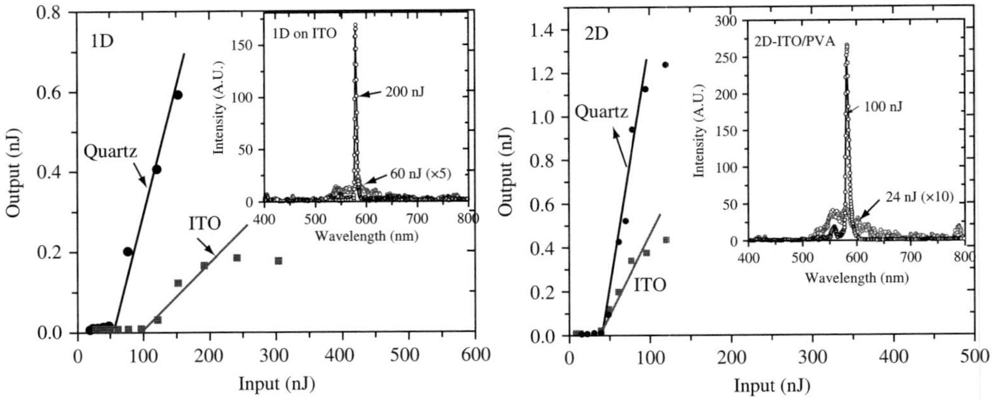

Fig. 11.23 Output energy as a function of pump energy for 1D and 2D DFB on PVA/quartz and PVA/ITO substrates. The inset shows laser spectra below and above the lasing threshold. [Reproduced from E. B. Namdas *et al.*, *Adv. Mater.*, **21**, 799 (2009) with permission from Wiley InterScience.]

degree of chain orientation for the organic semiconducting polymers in the source–drain channel.

Two-dimensional DFB structures exhibited lasing thresholds of approximately 35 nJ / pulse. This corresponds to a photon density of 7.2×10^{12} /cm^2 (or 3.6 μJ/cm^2) in the pump laser. Assuming that there are no other optical losses, e.g. from carrier-induced absorption, then the lasing threshold for the injection current can be estimated by the following equation,

$$J_{th} \approx \frac{No \times q}{\Phi_{\text{int}-\text{el}} \times \tau_{\text{rad}}} \tag{11.5}$$

where No is density of photons at the lasing threshold (1×10^{12} /cm^2), q is electrical charge, $\Phi_{\text{int-el}}$ is the internal quantum efficiency for the electroluminescence, and τ_{rad} is the radiative lifetime of the excitons. With $\Phi_{\text{int-el}} \sim 2.5\%$, and τ_{rad} for SY$\sim$ 5 ns, we estimate the lasing threshold for the injection current to be $\sim$1.2 kA/cm^2.

This current density is difficult to achieve in the OLED structure but controllable in an appropriate FET configuration.

11.8 Laser lookalikes and cases of "mistaken identity"

The criteria for the laser appear straightforward, but some of the features can arise in other ways, confusing the picture – for example interference effects, which are characteristic of coherence can be observed perfectly well using a lamp and a pair of slits. One phenomenon sometimes confused with lasing is amplified spontaneous emission (ASE). In this process, spontaneously emitted photons are amplified by stimulated emission in a gain medium. Accordingly they acquire to some extent many of the properties of laser light (narrower linewidth, polarization, directionality). The difference from laser operation is that there is no resonator, and so these properties are less fully developed – for example the typical linewidth of ASE in an organic semiconductor is 10 nm.

Waveguiding is another phenomenon that can lead to emission with some of the properties of laser light, and this is a particular problem as a waveguide is a commonly used laser geometry. For example, waveguides can be excellent spatial filters – waveguided light even if spontaneously emitted can emerge from the waveguide in a near diffraction limited spot. Light may resonantly leak into the substrate, and these "leaky waves" can propagate next to the waveguide giving a narrow emission spectrum. Waveguiding is of course a linear optical process, so one powerful way of distinguishing between waveguided spontaneous emission and lasing is to look for a threshold.

A further structure that can exhibit similar properties to laser light is a luminescent microcavity. The spontaneous emission from a molecule can be strongly affected by its local environment. Wavelength-scale microstructures and microcavities can alter the spatial, spectral and temporal properties of light emission through interference effects. This can lead to both narrow linewidth and directional emission. The microcavity reduces the total number of allowed photonic modes for spontaneous emission, and thus channels more light into a few favored modes. For an electrically driven microcavity or wavelength scale microstructure devices one powerful way of distinguishing between spontaneous emission and lasing is to compare the threshold for optical pumping. The threshold for electrical pumping will be higher than for optical pumping as a result of additional losses due to contacts and injected charges. There is a recent false claim of electrically pumped lasing in an organic semiconductor sandwiched between dielectric mirrors.[5] The threshold is this device structure is five orders of magnitude smaller than equivalent values for closely related structures.

A quite different way of obtaining a very narrow emission spectrum is inadvertently exciting a transition in a contact material. For example there is an atomic line of indium at 410 nm, which is the probable cause of the emission at this wavelength observed in a device containing green and red emitting organic materials.[6]

11.9 Concluding comments

Thin-film waveguides of semiconducting polymers and their blends exhibit gain when photo-pumped to excitation densities as low as 10^{17} cm^{-3}. The key photophysical properties that make these polymers good laser materials are the shift between the absorption and emission spectrum that leads to low self-absorption, the high photoluminescence efficiency combined with high chromophore density, the high stimulated emission cross-section, and the absence of excited state absorption in the luminescent spectral region. Gain can be obtained at even lower excitation densities (10^{16} cm^{-3}) when energy is transferred from a host polymer to a guest polymer in order to move the emission wavelength farther from the absorption edge. ASE and lasing have been demonstrated at this low excitation density with a steady state pump intensity of only 100 W/cm^2.

[5] X. Y. Liu *et al.*, *Optics Lett.*, **34**, 503 (2009).
[6] L. El-Nadi *et al.*, *Chem. Phys. Lett.*, **286**, 9 (1998).

Efforts to make polymer diode lasers have been unsuccessful because of charge induced absorption and electrode induced absorption. Structures that minimize electrode induced absorption have been demonstrated, so the primary remaining challenge is to eliminate charge induced absorption and to fabricate a light-emitting FET with hole injecting source and electron injecting drain (or vice versa). Incorporating the 1D or 2D grating structure into the FET design can be easily achieved by nanoimprint lithography. These imply that by further optimizing the electroluminescence efficiency (for example, by using polymer blends and the Frster energy transfer mechanism) using DFB structure inside a suitable FET, and pulsing to higher voltages, organic injection lasing might well be achievable.

An alternative approach to the all-polymer electronically pumped laser is to fabricate a two-component diode laser; i.e. an optically pumped polymer laser that is optically pumped by an electrically driven compact InGaN light source, the latter being capable of emitting at the necessary intensity levels to pump the polymer laser above threshold. This approach was proposed a decade ago,[7] and it has only recently been demonstrated by Yang *et al.*[8] They show that by separating the charge transporting and lasing regions in a compact package combining an indium gallium nitride light-emitting diode LED with a semiconducting polymer distributed feedback DFB laser, an electrically pumped hybrid polymer laser can be made.

Further reading

V. M. Agronovich, G. F. Bassani (eds.), *Electronic Excitations in Organic Nanostructures*, Elsevier, Amsterdam, 2003.

U. Lemmer, A. Haugeneder, C. Kallinger and J. Feldmann, in: *Semiconducting Polymers*, edited by G. Hadziioannou and P. F. van Hutten, Wiley-VCH, Weinheim, 2005, Chapter 10.

M. D. McGehee and A. J. Heeger, *Adv. Mater.*, **12**, 1655 (2000).

R. C. Polson, A. Chipouline and Z. V. Vardeny, *Adv. Mater.*, **13**, 760 (2001).

I. D. W. Samuel and G. A. Turnbull, *Chem. Rev.*, **107**, 1272 (2007).

N. Tessler, *Adv. Mater.*, **11**, 363 (1999).

[7] F. Hide *et. al.*, *Appl. Phys. Lett.*, **70** 2664 (1997).
[8] Y. Yang, *et al.*, *Appl. Phys. Lett.* **92** (16) (2008).

12

Photoinduced charge transfer from semiconducting polymers to acceptors

12.1 Introduction

Figure 12.1 shows schematically the photoinduced electron transfer reaction from a soluble PPV derivative (as donor) to C_{60} (as acceptor). The incident photon with energy $\hbar\omega$ promotes an electron from the π-band of the semiconducting polymer to the π^*-band, leaving a hole in the π^*-band. If the nearby C_{60} is a sufficiently good acceptor (i.e. if the electron affinity is sufficiently large), the excited electron will transfer into the lowest energy unoccupied molecular orbital (LUMO) of the acceptor. The photophysics of the photoinduced electron transfer reaction in composites of semiconducting polymers with suitable acceptors is the subject of this chapter.

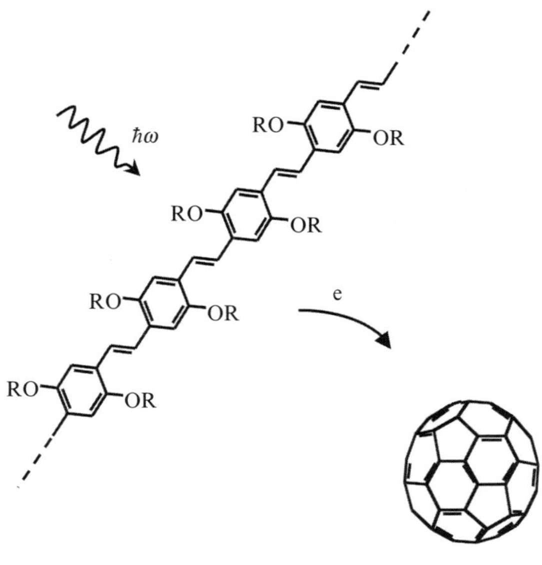

Fig. 12.1 Schematic diagram of the ultrafast photoinduced electron transfer reaction from a PPV derivative to C_{60}.

12.2 Photophysics of conjugated polymer/acceptor composites

Photoinduced electron transfer will occur only if the electron transfer reaction is energetically favorable. Specifically, electron transfer will take place only when the total free energy is lowered by the electron transfer:

$$\begin{aligned} \Delta F &\leq 0 \\ I_p^{donor} - E_A^{acceptor} - U_{Coulomb} &\leq 0 \end{aligned} \tag{12.1}$$

where I_p^{donor} is the ionization potential of the excited donor (i.e. the energy required to remove an electron from the bottom of the π^*-band), $E_A^{acceptor}$ is the electron affinity of the acceptor, and $U_{Coulomb}$ is the Coulomb interaction between the hole on the donor and the electron on the acceptor. Analogous criteria can be written to describe the energetics of photoinduced hole transfer.

The requirement for photoinduced electron transfer is that the LUMO energy level of the acceptor must be below the LUMO energy level of the donor: $E_{CT} = E_{LUMO}{}^{acc} - E_{LUMO}{}^{don} > 0$; see Fig. 12.2. If the excited electron and hole in the semiconducting polymer are bound into a neutral exciton, E_{CT} must be sufficiently large to overcome the exciton binding energy, E_b; i.e. $E_{CT} > E_b$. Thus, strong exciton binding inhibits photoinduced electron (or hole) transfer.

Note, however, that if the LUMO energy level of the acceptor is below that of the HOMO of the donor, spontaneous electron transfer would occur in the ground state (i.e. without the need for photoexcitation). The result would be chemical doping of the polymer as discussed in detail in Chapter 2.

In the photoinduced electron transfer reaction, the donor/acceptor system passes through several intermediate states before full charge separation occurs (a parallel set of intermediate states can be described for the analogous photoinduced hole transfer):

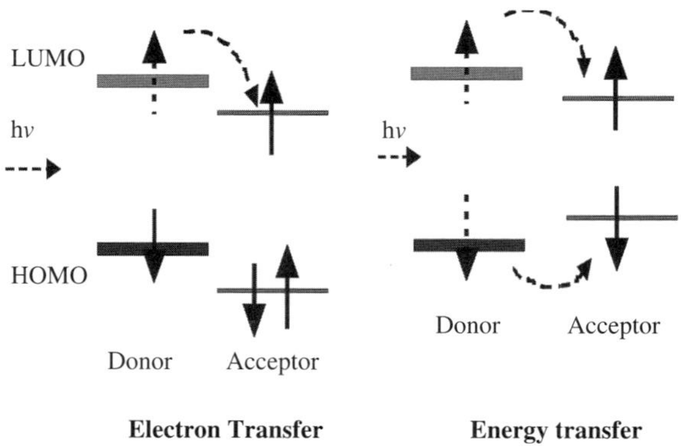

Fig. 12.2 The HOMO and LUMO energy levels of the donor and acceptor; photoinduced electron transfer will occur when the LUMO of the acceptor is lower in energy than the LUMO of the donor. Energy transfer will occur if the excitation energy of the donor is greater than that of the acceptor.

Intermediate steps in the photoinduced electron transfer reaction

1.	$D + A + h\upsilon \rightarrow {}^{1,3}D^* + A$	π-π^* excitation of the donor
2.	${}^{1,3}D^* + A \rightarrow {}^{1,3}(D \ldots A)^*$	Excitation delocalized on D-A complex, (exciplex formation)
3.	${}^{1,3}(D \ldots A)^* \rightarrow {}^{1,3}(D^{\delta+\bullet} \ldots A^{\delta-\bullet})^*$	Formation of charge transfer complex
4.	${}^{1,3}(D^{\delta+} \ldots A^{\delta-})^* \rightarrow {}^{1,3}(D^{+\bullet} \ldots A^{-\bullet})^*$	Radical pair formation
5.	${}^{1,3}(D^{+\bullet} \ldots A^{-\bullet})^* \rightarrow D^{+\bullet} + A^{-\bullet}$	Complete charge separation

Here, D and A denote donor and acceptor, respectively. The excitation pathway is written in terms of donor singlet and triplet excitations (in principle both are possible) denoted as the superscripts1,3, respectively. The partial charges are denoted as $\delta+$ / $\delta-$ and the • denotes a radical ion. In each of these intermediate steps, the excitation can relax to the ground state, either radiatively or nonradiatively.

Photoluminescence is in direct competition with photoinduced electron transfer. If photoinduced electron transfer takes place on a time scale faster than the luminescence decay time, separation of the electron (onto the acceptor) and the hole (on the donor) will efficiently quench the photoluminescence. Both channels, photoluminescence and photoinduced electron (or hole) transfer, are interesting and both are potentially useful (Fig. 12.3).

The importance of the luminescence channel in Fig. 12.3 is quite clear; luminescence and electroluminescence represent major opportunities for a variety of applications, including displays, lighting and lasers (see Chapters 9, 10 and 11). By charge separation through ultrafast photoinduced charge transfer, one enables the possibility of collecting the separated charges at electrodes, resulting in photodetectors and photovoltaic cells (see Chapter 13 for details). The

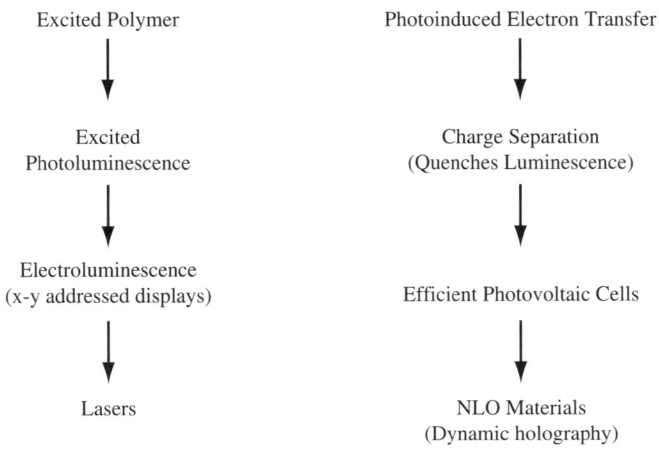

Fig. 12.3 Photoluminescence and photoinduced electron transfer are competing processes; both are interesting and both are potentially useful.

large local electric fields created by the separated charges can be useful for nonlinear optical phenomena; e.g. in photorefractive materials.[1] Furthermore, the efficient quenching of the photoluminescence can used to create biosensors with high sensitivity.[2]

12.3 Photoinduced electron transfer from semiconducting polymers to C_{60}

Phase separated films comprising acceptors such as fullerenes in one phase mixed with semiconducting polymers in a second phase exhibit ultrafast photoinduced electron transfer from the semiconducting polymer (as donor) onto the fullerene (as acceptor). The forward transfer is ultrafast; the electron transfer takes place at times less than 100 femtoseconds (rate $> 10^{13}\ s^{-1}$). As a result, the photoluminescence (PL) in such composites is quenched because the PL decay time is typically several hundred ps (Fig 12.4).

Remarkably, however, this charge transfer reaction is highly asymmetric; the "back reaction" occurs in the microsecond time scale at room temperature (and exponentially slower at low temperatures). As a result of this asymmetry, there is sufficient time to collect the

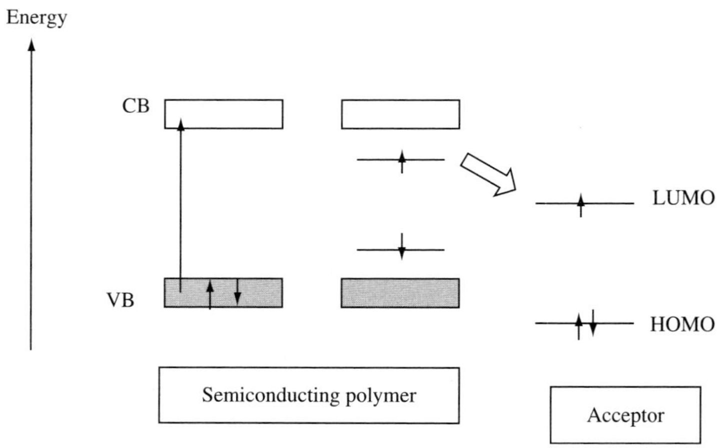

Fig. 12.4 Energy diagram of the photoinduced electron transfer reaction from a semiconducting polymer to an acceptor such as C_{60}. Step 1: After photoexcitation, the bond lengths of the polymer relax to form polarons and/or excitons. Step 2: The excited electron transfers to the previously unoccupied LUMO of the acceptor. For semiconducting polymers as donors, Step 2 is sufficiently fast that the transfer competes (in time) with the bond relaxation. In the transferred state, bond relaxation leads to polaron formation on both the polymer (positive polaron) and the acceptor (distorted C_{60}^{-}).

[1] See, for example, Y. Zhang *et al.*, *Phys. Rev. B*, **46**, 9900 (1992); W. E. Moerner *et al.*, *Ann. Rev. Mater. Sci*, **27**, 585 (1997); *Optical Limiting Materials*, US Patent 5741442, 1995.

[2] See US patents 6743640 (2001) and 7122383 (2002).

separated charges at electrodes, thereby opening a new route to efficient "plastic" solar cells and high detectivity photodetectors.

12.4 Light induced electron spin resonance (LESR)

After complete charge separation (Step 5 in the "intermediate steps" described above), two LESR signals should be observed: one from the radical cation on the electron donor (an unpaired electron on the semiconducting polymer), one from the hole on the semiconducting polymer, ($CP^{+\bullet}$) and a second from the radical anion on the electron acceptor ($C_{60}^{-\bullet}$). The independent precession of the spins on the $CP^{+\bullet}$ and the $C_{60}^{-\bullet}$, respectively, provides unambiguous proof of complete charge transfer and charge separation.

In organic molecules, the *g*-factors of the radical ions are very close to the free electron value, 2.0023; because of the low atomic numbers of C and H, the spin–orbit interaction is weak. As a result, it is typically difficult to resolve the separate ESR signals from radical cations and anions. Fortunately, C_{60}^{-} has an unusually low *g*-factor, slightly below 2.000. Therefore, the ESR signal from the $CP^{+\bullet}$ (positive polaron) ESR resonance at $g \approx 2.002$ is well separated from that of $C_{60}^{-\bullet}$ ($g<2$) even in X-band ESR experiments with magnetic fields near 0.3 T. As shown in Fig. 12.5, the $C_{60}^{-\bullet}$resonance signal appears at a higher magnetic field than that from the $CP^{+\bullet}$; the two are well resolved.

As additional proof of the creation of two independent spin systems ($CP^{+\bullet}$ and $C_{60}^{-\bullet}$), the two ESR signals exhibit very different relaxation behavior. At microwave powers below that

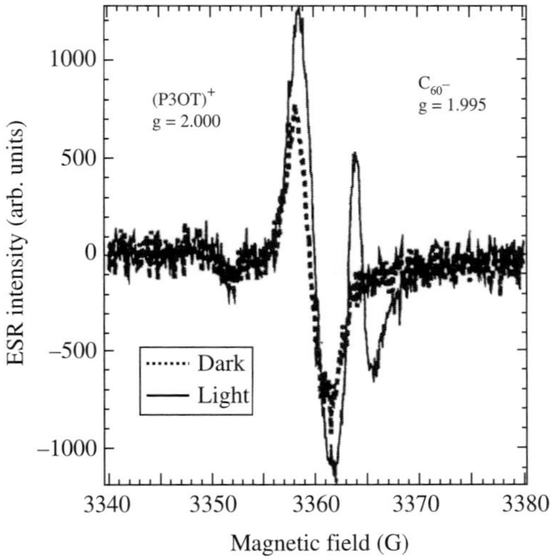

Fig. 12.5 Light-induced ESR (LESR) spectrum obtained from MEH-PPV/C_{60} composite (room temperature). Because of the different *g*-values, the ESR lines from the $CP^{+\bullet}$ and the $C_{60}^{-\bullet}$ are well resolved. The signal observed from the semiconducting polymer in the dark results from weak residual doping by impurities. The integrated areas of the two LESR signal are approximately equal. [Reproduced from L. Smilowitz, *et al.*, *Phys. Rev. B*, **47**, (20) 13835 (1993) with permission of The American Physical Society.]

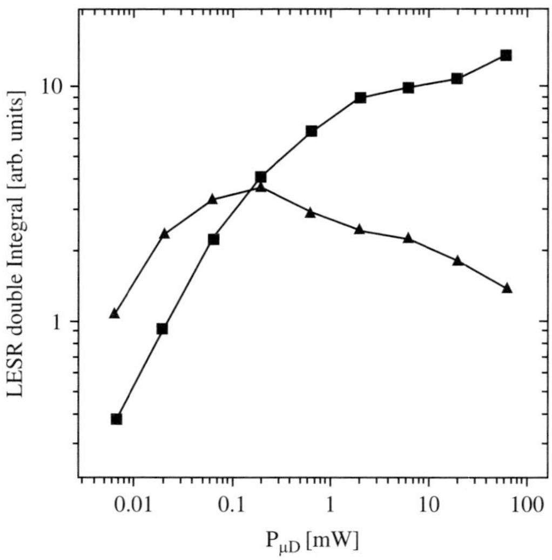

Fig. 12.6 Saturation behaviour of the LESR signals in polymer/fullerene composite films. [Reproduced from V. Dyakonov *et al.*, *Phys. Rev. B*, **59**, 8019 (1999) with permission of The American Physical Society.]

required for saturation, the maximum of the ESR signal is proportional to $(T_1T_2)^{1/2}$, where T_1 is the longitudinal (spin–lattice) relaxation time and T_2 is the transverse (spin–spin) relaxation time. However, when $(WT_1) > >1$ (W is the transition probability, proportional to the microwave power), the populations in the upper and lower energy states of the spin-1/2 system become equal and there is no net absorption because the number of "up" transitions is equal to the number of stimulated "down" transitions. As shown in Fig. 12.6, the two LESR signals have different relaxation times. While the $CP^{+\bullet}$ line is easily saturated, the $C_{60}{}^{-\bullet}$ line cannot be saturated at similar microwave power levels. The shorter T_1 for $C_{60}{}^{-\bullet}$ arises from the relatively rapid relaxation of the $C_{60}{}^{-\bullet}$ spins through coupling to the Jahn–Teller distorted structure and splitting of the LUMO levels of C_{60}.

12.5 Ultrafast photoinduced charge transfer

It is important to demonstrate that the photoinduced charge transfer occurs at ultrafast time scales and that the characteristic transfer time, τ_{tr}, is orders of magnitude less than τ_{PL}, where τ_{PL} is the natural decay time of the photoluminescence in the pristine polymer. In fact, this is clear from two aspects of the earliest measurements related to the photoinduced charge transfer in the semiconducting polymer/C_{60} system:

(1) quenching of the PL; and
(2) changes in the photoinduced absorption spectrum.

As shown in Fig. 12.7, the photoluminescence is quenched by the addition of C_{60}. The effect is dramatic; the photoluminescence quantum efficiency is reduced by more than four orders of magnitude. Figure 12.8 demonstrates that a reduction in the PL decay time and the PL yield can be observed even at concentrations of only 0.1% C_{60}. The decay of the PL arises from two parallel channels; the natural PL decay and the photoinduced electron transfer which separates the electrons and holes and thereby depletes the photoexcited state. Since the rates add,

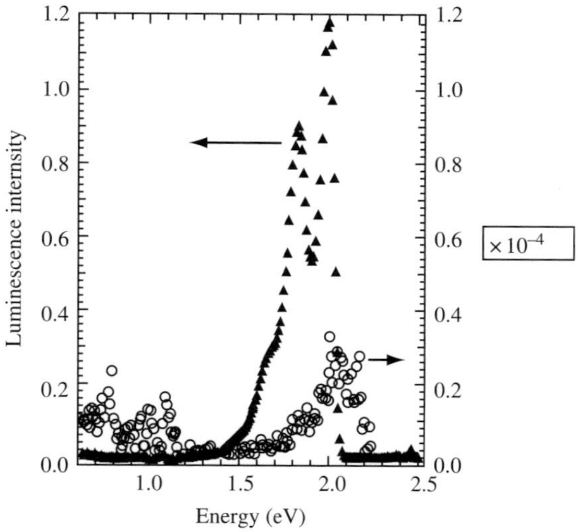

Fig. 12.7 Comparison of the PL intensity in pristine MEH-PPV (left axis) with that after mixing with C_{60} (50%). [Reproduced from N.S. Sariciftci *et al.*, *Science*, **258**, 1474 (1992) with permission of the AAAS.]

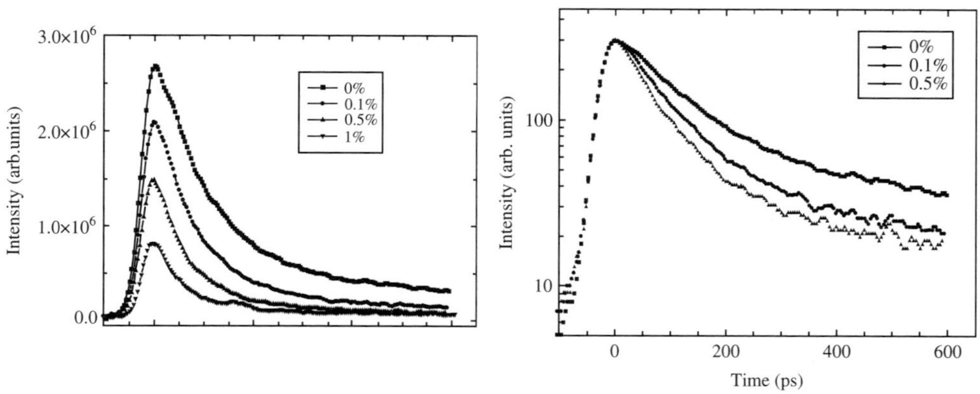

Fig. 12.8 Photoluminesence decay for composites of MEH-PPV with various concentrations of C_{60}. The quenching and associated reduction in the PL decay time are observed even at concentrations of C_{60} as low as 0.1%. On the right, the data are normalized to the peak value to demonstrate clearly the faster decay with the addition of C_{60}.

$$1/\tau_{PL} = 1/\tau_o + 1/\tau_{tr}. \tag{12.2}$$

When $1/\tau_{tr} >> 1/\tau_o$, the photoluminescence decay is dominated by the charge transfer;

$$1/\tau_{PL} \approx 1/\tau_{tr}, \tag{12.3}$$

and the photoluminescence intensity is correspondingly reduced:

$$(S_{PL})/(S_{PL})_0 \tilde{} \tau_{tr}/\tau_0. \tag{12.4}$$

Thus, one can obtain an approximate determination of τ_{tr}/τ_{PL} from the ratio of the photoluminescence signal in the pure polymer $(S_{PL})_0$ to that of the photoluminescence signal in the phase separated composite (S_{PL}). At concentrations of 10–20% C_{60} in composites with MEH-PPV, $(S_{PL}) / (S_{PL})_0 \approx 10^{-3}$. Thus, since $\tau_{PL} \approx 300$ ps for MEH-PPV, $\tau_{tr} << 1$ ps.

Although initial ultrafast pump–probe measurements did not have sufficient temporal resolution to time-resolve the electron transfer, the data shown in Fig. 12.9 demonstrate that the electron transfer time is significantly less than 1 ps, consistent with the quenched photoluminescence.

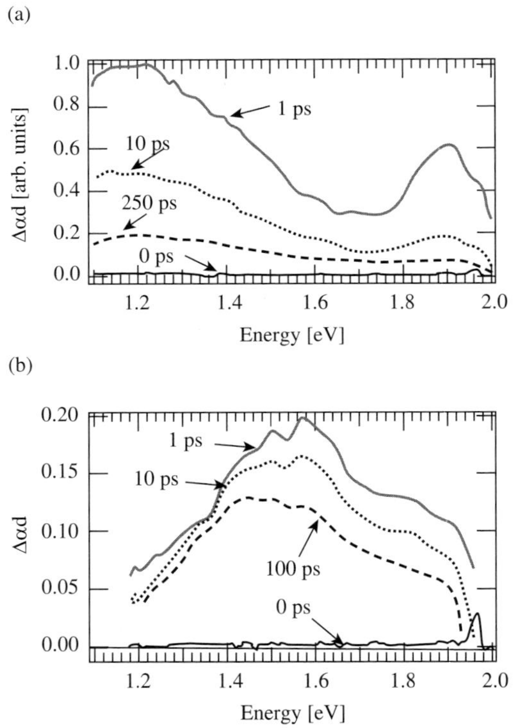

Fig. 12.9 Top: photoinduced absorption spectrum for pristine poly(3-octylthiophene). Bottom: photoinduced absorption spectrum for a composite of poly(3-octylthiophene) mixed with 1% C_{60} composite. [Reproduced from B. Kraabel *et al.*, *Phys. Rev. B*, **50**, 18543 (1994), with permission of The American Physical Society.]

Figure 12.9(top) is the photoinduced absorption spectrum of pristine poly(3-octylthiophene) obtained at a series of time delays following the initial excitation. The double peaked photoinduced absorption is a clear "signature"; this characteristic spectrum is observed at the earliest times and subsequently decays without any change in spectral shape over times of order 100 ps. Figure 12.9(bottom) is the photoinduced absorption spectrum of poly(3-octylthiophene) mixed with 1% C_{60} at different time delays. Note that even at the earliest times that could be resolved with this instrumentation, the spectrum of the polythiophene/1% C_{60} phase separated composite is clearly different from that of pristine poly(3-octylthiophene). The implication is that at 1 ps time delay, charge transfer has already occurred in polythiophene/1% C_{60}.

Subsequently, Brabec *et al.* were able to time-resolve the electron transfer process. By pumping with ultrashort pulses (pulse width of a few fs), all macromolecules in the absorption range of the thin film sample are simultaneously promoted to the excited state; the electron is promoted from the bonding π-band to the antibonding π^*-band. As a result, the bond lengths in all the macromolecules in the thin film begin to relax simultaneously and coherently. Since there is relatively little dissipation during this relaxation, the bond-length changes "overshoot" and a slowly decaying bond-length oscillation occurs. The phase of the bond-length oscillation is the same on all macromolecules in the sample, because all were excited with the same ultrashort pump pulse.

The ringing is analogous to that which occurs following the plucking of a violin string. The violin string oscillates with amplitude in many harmonics. In the case of the bond-length oscillations, the "harmonics" are the amplitude normal modes of oscillation; i.e. the totally symmetric Ag vibrations of the backbone (the Raman modes) of the polymer chain. The bond-length oscillations modulate the excited state absorption (PIA) and the stimulated emission (SE) through associated local changes of the bond length on going from the ground state to the excited state. Consequently, as shown in Fig. 12.10, the coherent bond-length oscillations at the Raman mode frequencies are clearly observed in the ultrafast SE signal (at 610 nm, $\Delta T/T > 0$ implying modulated SE rather than modulated PIA). The inset to Fig. 12.10 compares the Fourier transform of these oscillations to the resonant and nonresonant Raman spectra, with excellent agreement.

Upon mixing the fullerenes in high concentration ratios (dotted curve in Fig. 12.10) the photoinduced luminescence (SE) is quenched and photoinduced absorption is observed ($\Delta T/T < 0$); the PIA originates from the electronic transitions of the photoexcited polaron. The electron transfer from MDMO-PPV to a C_{60} acceptor starts at approx. 45 fs and occurs at random times. As a result, although each individual macromolecule in the sample is undergoing lattice relaxation, the phase is random and the coherent modulation seen clearly in the pristine MDMO-PPV is not observed in the phase separated composite with C_{60}.

A number of interesting questions arise when considering the photoinduced electron transfer reaction from a semiconducting polymer to an acceptor. The total energy must be conserved in such a reaction; i.e. the energy gain from the electron transfer, E_{CT}, must be compensated by an equal increase in the energy elsewhere in the system How can this be done efficiently within 100 fs? The forward electron transfer time is 7–8 orders of magnitude faster than the back electron transfer and relaxation to the ground state. This unusually large asymmetry is especially interesting for such a relatively simple two-component system.

Because of the quasi-continuous density of states in the band structure of the semiconducting polymer component, energy conservation does not cause a "bottleneck" in the electron

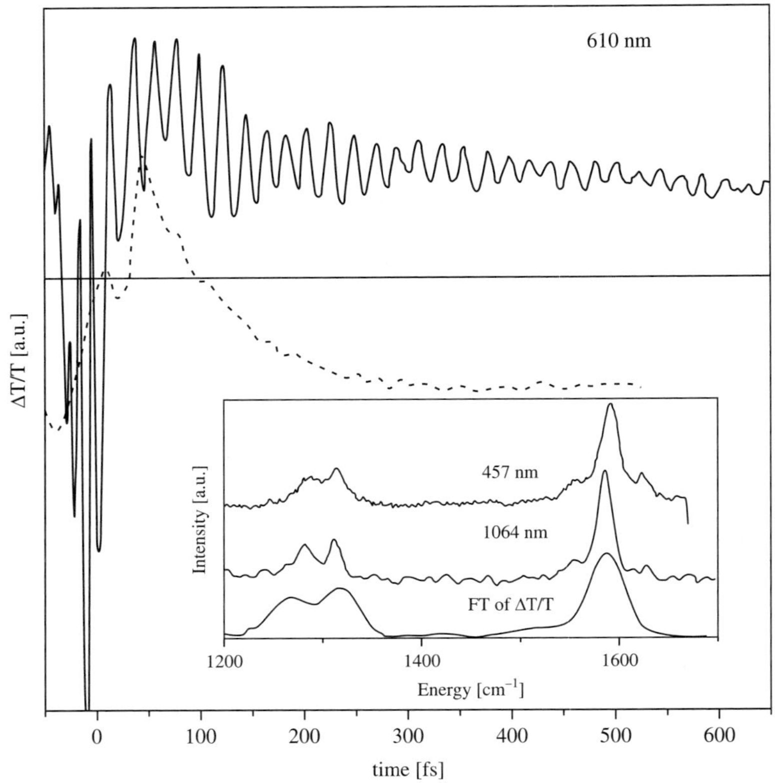

Fig. 12.10 Time evolution of the pump–probe signal in MDMO-PPV (solid curve with oscillations) and MDMO-PPV /C_{60} (dotted curve). For pristine MDMO-PPV, the coherent oscillations are clearly observed. The inset compares the Fourier transform of these oscillations to the resonant and nonresonant Raman spectra, with excellent agreement. For MDMO-PPV /C_{60}, the oscillations are not observed. The electron transfer process starts at approx. 45 fs; electrons transfer to the C_{60} at random times. As a result, although each individual macromolecule in the sample is undergoing lattice relaxation, the phase is random and coherent modulation is not observed. [Reproduced from J. Christoph *et al.*, *Chem. Phys. Lett.*, **340**, 232 (2001) with permission of the Elsevier Publishing.]

transfer process; the hole on the polymer can take up energy by promotion to excited states deep in the π-band. The importance of the electron–hole Coulomb interaction in the energy conservation process was first pointed out by Michael Rice.[3]

The asymmetry of the forward and back electron transfer rates remains an interesting problem. The ultrafast forward electron transfer implies a finite overlap of the excited state wavefunctions between donor (semiconducting polymer) and acceptor (fullerene). After the transfer, however, structural relaxation occurs on both the $CP^{+\bullet}$ and the $C_{60}^{-\bullet}$ with the associated formation of polarons. A "belt polaron" on $C_{60}^{-\bullet}$ (caused by the Jahn–Teller

[3] M. J. Rice *et al.*, *Phys. Rev. B*, **53** 10764 (1996).

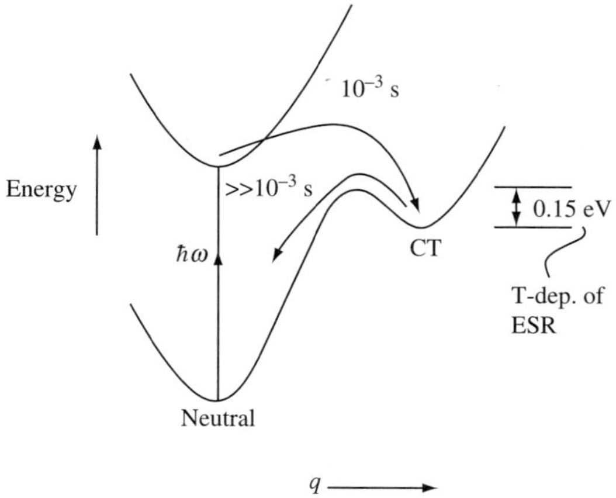

Fig. 12.11 Diagram of the energy of the semiconducting polymer/C_{60} composite as a function of the generalized coordinate that describes the charge transfer. Polaron distortions create a local minimum in the total energy and thereby provide a mechanism for inhibiting the back charge transfer (return to the ground state). The magnitude of this energy barrier can be estimated from the temperature dependence of the lifetimes of the LESR signals[5].

distortion on $C_{60}^{-\bullet}$) was suggested by Kirova *et al.*[4] As sketched in Fig. 12.11, the polaron distortions create a local minimum in the total energy and thereby provide a mechanism for inhibiting the back charge transfer to the ground state. The magnitude of this energy barrier can be estimated from the temperature dependence of the lifetimes of the LESR signals.

12.6 Sensitization of photoconductivity

The photoinduced electron transfer reaction creates positive polarons on the semiconducting polymer and $C60^{-\bullet}$ ions. Since the polarons are mobile, the photoinduced electron transfer serves to sensitize the photoconductivity. As noted above, the electron transfer is ultrafast, orders of magnitude faster than any competing process. Thus, the quantum efficiency for charge separation and the formation of polarons on the conducting polymer approaches unity! The photoinduced electron transfer is equivalent to transient "photodoping" of the semiconducting polymer. In addition, the electrons on the $C_{60}^{-\bullet}$ can hop along the C_{60} pathways.

Because of the asymmetry of the forward and back charge transfer, the lifetime of the photogenerated carriers is much longer than the lifetime in the pristine polymer. Both effects, the increased quantum yield and the extended lifetime, increase the magnitude of the photoconductivity. Thus, the addition of C_{60} into the conjugated polymer matrix serves to sensitize the photoconductivity. The effect is so strong that upon mixing only a few percent of fullerene

[4] N. Kirova *et al.*, *Synthetic Metals*, **86**, 2385 (1997).

[5] The LESR signal decays with an Arrhenius energy barrier of 100–120 meV in such composites; see L. Smilowitz *et al.*, *Phys. Rev. B*, **47**, 13835 (1993).

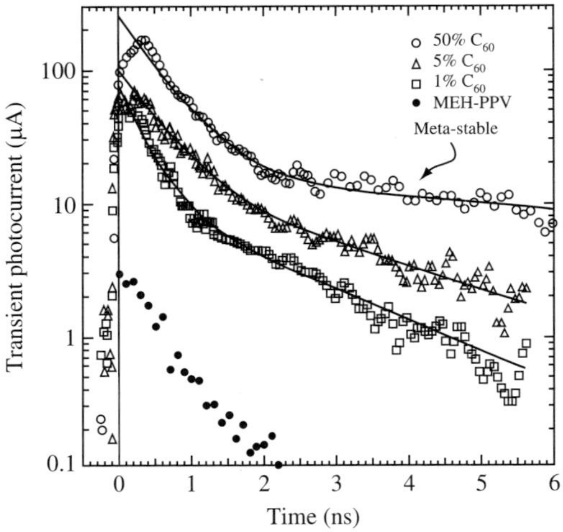

Fig. 12.12 Time resolved (transient) photoconductivity of MEH-PPV for different concentrations of C_{60}. The data were obtained using the Auston switch experimental method. [Reproduced from C. H. Lee *et al.*, *Phys. Rev. B*, **48**, 15425 (1993), with permission of The American Physical Society.]

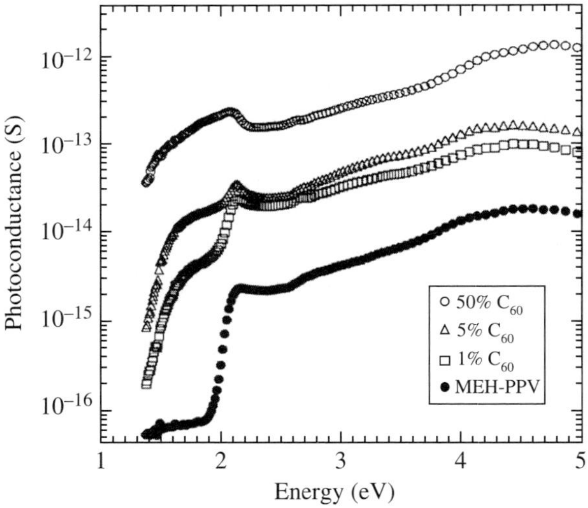

Fig. 12.13 Steady state photoconductivity of MEH-PPV for different concentrations of C_{60}. [Reproduced from C. H. Lee, *et al.*, *Phys. Rev. B*, **48**, 15425 (1993), with permission of The American Physical Society.]

acceptors into the semiconducting polymer matrix, the peak photoconductivity increases by more than an order of magnitude, and the lifetime of the photoexcited charge carriers increases by several orders of magnitude. The data are shown in Fig. 12.12. Therefore, as shown in Fig. 12.13, with the addition of C_{60} into the conjugated polymer matrix, the steady state photoconductivity increases by several orders of magnitude.

Further reading

C. Brabec, V. Dyakonov, J. Parisi and N. S. Sariciftci (eds.), *Organic Photovoltaics*, Springer, Berlin, 2003.

G. Hadziioannou and P. F. Hutten (eds.), *Semiconducting Polymers*, Wiley-VCH, Weinheim, 2000.

G. Hadziioannou and G. G. Mallarias (eds.), *Semiconducting Polymers*, Vols. 1–2, Wiley-VCH, Weinheim, 2007.

H. Haken and H. C. Wolf, *Molecular Physics and Elements of Quantum Chemistry*, Springer, Berlin, 2004.

M. Klessinger and J. Michl, *Excited States and Photochemistry of Organic Molecules*, VCH Publishers, Weinheim, 1995.

V. May and O. Kühn, *Charge and Energy Transfer Dynamics in Molecular Systems*, Wiley-VCH, Weinheim, 2004.

H. S. Nalwa (ed.), *Handbook of Organic and Conductive Molecules and Polymers*, John Wiley, Chichester, 1997.

L. Rothberg (ed.), *Photoinduced Charge Transfer*, World Scientific, Singapore, 2000.

J. Shinar, Z.V. Vardeny and Z. Kafafi (eds.), *Optical and Electronic Properties of Fullerenes and Fullerene Based Materials*, Marcel Dekker, New York, 2000.

T. A. Skotheim and J. R. Reynolds (eds.), *Handbook of Conducting Polymers*, 3rd edition, CRC Press, Taylor and Francis, Boca Raton, 2007.

S.S. Sun and N. S. Sariciftci (eds.), *Organic Photovoltaics: Mechanisms, Materials and Devices*, Taylor and Francis, Boca Raton, 2005.

A. Moliton, *Optoelectronics of Molecules and Polymers*, Springer, Berlin, 2006.

S. S. Sun and L. Dalton (eds.), *Introduction to Organic Electronic and Optoelectronic Materials and Devices*, Taylor and Francis, Boca Raton, 2008.

13
Photovoltaic cells and photodetectors

13.1 Dual function polymer diodes

An LED emits photons when a voltage is applied across a thin film of a semiconducting luminescent polymer; a polymer LED creates light from electrical power (Fig. 13.1a). A photovoltaic cell converts light energy into electrical energy; electrons and holes are photogenerated and subsequently collected at the electrodes where the electricity can be delivered to an external load (Fig. 13.1b). The operation of a photodetector is similar to that of a photovoltaic cell. Typically, however, a photodetector is operated under short circuit conditions or under reverse bias to improve the collection efficiency of the photogenerated electrons and holes; for a photodetector, the goal is to detect the incident photons rather than extract useful energy into an external circuit. Throughout this chapter we will typically use the terms "electrons and holes" to mean "electron polarons and hole polarons". Only when the details of the polaron energy or the associated chain distortion are specifically important will these details be emphasized.

Three steps are required for generation of electrical power via the absorption of incident light:

(i) absorption of photons by the photoactive material;
(ii) photoinduced charge separation and the generation of mobile carriers;
(iii) collection of electrons at one electrode and holes at the opposite electrode.

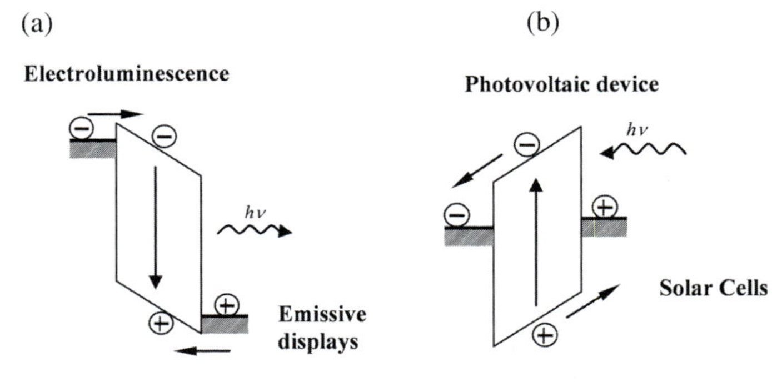

Fig. 13.1 (a) MSM diode in forward bias; electrons and holes are injected at the cathode and anode, respectively, and radiatively recombine in the polymer. Light is emitted.
(b) MSM diode in reverse bias; photogenerated electrons and holes are collected at the cathode and anode, respectively.

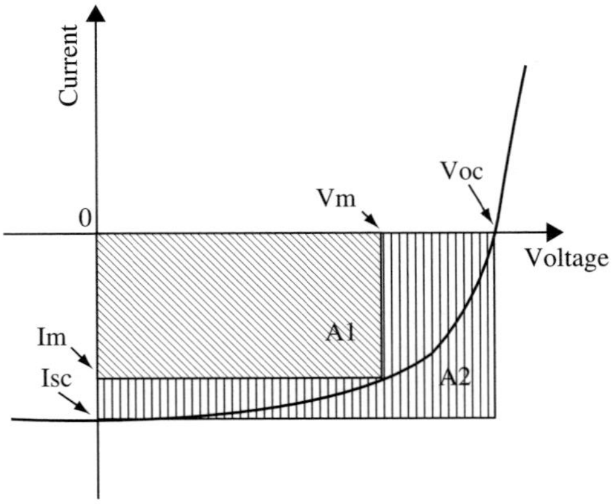

Fig. 13.2 Current density versus voltage (j versus V) of a solar cell. The maximum power point is determined by V_M and I_M. The fill factor is defined by FF =A1/A2.

For a direct band gap semiconductor, photons with energy $\hbar\omega \geq E_g$ create electron–hole pairs by promoting electrons across the energy gap, E_g, from the π-(valence)-band to the π*-(conduction)-band. Thus, the onset of charge separation as observed through photodetectors and photovoltaic cells fabricated from a *single* material is an established and accepted method for determining the energy gap of a semiconductor.

A typical current density versus voltage (j versus V) curve for a solar cell is shown in Fig. 13.2.

Figure 13.2 is not specific to bulk heterojunction solar cells; it describes any solar cell. The three parameters that determine the power conversion efficiency are the short circuit current (I_{sc}); the open circuit voltage (V_{oc}); and the fill factor (FF).

(a) **The short circuit current** (I_{sc})
I_{sc} is the current that flows when there is no external field applied; the charges are drifting because of the internal field. I_{sc} is determined by the quantum efficiency for charge separation, the transport of the charge carriers through the material and the loss of carriers through recombination during the transport to the electrodes.
(b) **The open circuit voltage** (V_{oc})
V_{oc} is the maximum possible voltage delivered by a solar cell. At this voltage the current is zero. V_{oc} is determined by the electronic structure of the component materials.
(c) **Fill factor**
The fill factor (FF) is the ratio between areas A1 and A2 in Fig. 13.2. For maximum fill factor, these two areas should be similar. The fill factor (FF) is defined by the following equation (see Fig. 13.2)

$$\mathrm{FF} = V_m I_m / V_{oc} I_{sc} \tag{13.1}$$

The power conversion efficiency of a solar cell is given as

$$\eta = \frac{P_{\text{out}}}{P_{\text{in}}.} = \frac{I_{SC}.V_{OC}.F.F}{P_{\text{in}}} \tag{13.2}$$

where P_{in} is the incident light power and P_{out} is the electrical power generated by the cell. The incident light power is usually standardized to be equivalent to the AM 1.5 solar spectrum.

The operation of a solar cell can be modeled with the equivalent circuit shown in Fig. 13.3.

The current versus voltage relationship obtained from this equivalent circuit is given by the following expression:

$$I = Io.\left\{\exp\left(\frac{e}{nkT}(U - IRs)\right) - 1\right\} + \frac{U - IRs}{R_{SH}} - I_{PH} \tag{13.3}$$

where I_o is the dark current, e is the electron charge, U is the applied voltage, n is the diode ideality factor, R_s is the series resistance and R_{sh} is the shunt resistance. The series resistance originates from a sum of contributions, including the resistivity of the active materials (see Section 13.14 for a microscopic understanding of FF in terms of the material parameters of the BHJ layer), the contacts between semiconductor and the metal, and the ohmic resistance of the electrodes. The series resistance needs to be minimized for maximum solar cell efficiency. The series resistance decreases with decreasing thickness. The shunt resistance quantifies the leakage current in the device. R_{sh} has to be maximized to reach maximum efficiency. The shunt resistance decreases with decreasing the film thickness (often caused by microshorts).

In a photovoltaic cell or a photodetector, the short circuit current is proportional to the intensity of the absorbed light (the number of photons absorbed per unit area per second). Note, however, that the useful thickness cannot be increased beyond the exciton diffusion length or the charge carrier recombination length (the carrier diffusion length beyond which photogenerated carriers typically recombine). Because of the relatively low mobility of both the excitons and charge carriers in semiconducting polymers, the maximum useful thickness (approx. 100–200 nm) is comparable to the absorption depth (ca. 100 nm).

Charge separation is necessary, but not sufficient, for the conversion of light to electricity. In the third and final step, the separated charges must be transported to and selectively collected by the electrodes.

As described in Chapters 8 and 9, a polymer diode can be fabricated as a metal–semiconductor–metal (MSM) "sandwich" device, the classic structure of the polymer LED. Figure 9.6 shows the electronic structure of a pristine polymer MSM device under different operating conditions. Figure 9.6(a) shows forward bias injection leading to operation as a light-emitting diode. Figure 9.6(b) shows the electronic structure with a small reverse bias applied. At zero

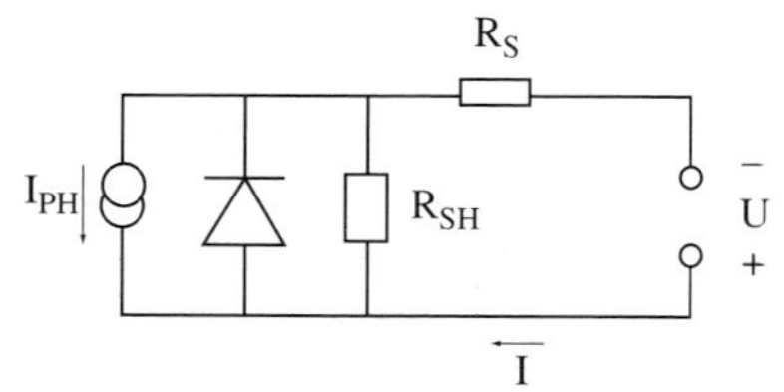

Fig. 13.3 Equivalent circuit of a solar cell.

bias (Fig. 9.6b), the device operates as a solar cell. Under reverse bias (Fig. 9.6c), the collection efficiency increases leading to improved performance as a photodetector.

Based upon the low density of electrically active impurities as measured by the capacitance versus voltage (see Chapter 8), the depletion width (W) is typically larger than the film thickness (d) for thin films of pure semiconducting polymers. When $W > d$, the difference in work functions creates a built-in potential and a uniform electric field across the semiconductor layer (see Figs 8.4 and 8.6). Although this is typically a valid assumption with the device in the dark, the electric field is not necessarily uniform under high intensity illumination. For relatively thick films under high intensity illumination, the MSM device must be modeled as two Schottky diodes in series.

Neutral excitons are not affected by the relatively small built-in electric field from the difference in the work functions of the two electrodes; this built-in potential is generally not sufficiently large to field ionize the exciton to create a separated electron–hole pair.

When a negative bias is applied (positive contact to the Al and negative contact to the ITO), the diode functions as photodetector. Because the external potential adds to the built-in potential, the external field improves the collection efficiency (although the reverse bias might also increase the noise). Photodetectors do not need mobile carriers with both signs of charge. If only one sign of photoinduced charge carrier is mobile, a photocurrent that is sufficiently large for useful applications can be generated.

Thus, the simplest semiconducting polymer device in the MSM configuration has "dual function"; such a device is an LED when biased in the forward direction (carriers injected at the electrodes) and a photodetector when biased in the reverse direction (photogenerated carriers collected at the electrodes).

13.2 Donor–acceptor bilayer heterojunctions

Charge separation is an efficient process at a donor–acceptor interface where a photoexcited electron on in the π^* band of the donor can lower its energy by transferring to the π^* band (or LUMO) of the acceptor. Thus, a bilayer consisting of a donor thin film deposited onto an electron acceptor thin film in sandwich geometry (D-A heterojunction) with metal electrodes will function as a solar cell (see Fig. 13.4). As demonstrated by Tang *et al.*,[1] using phtalocyanine/perylene bilayers, such solar cells show photovoltaic action and a relatively small dark current under reverse bias in the photodetector mode.

In devices fabricated from inorganic semiconductors, charge separation takes place only in regions where there is a built-in potential, for example in a p-n junction or in a Schottky junction. In organic materials, the analogue of the p-n junction is the donor–acceptor bilayer interface. There is, however, an important difference. In the p-n junction there are charge carriers in both the p-type and the n-type doped regions. The evolution from p-type to n-type doping takes place over a relatively long distance; a distance much larger than the interatomic spacing. On the contrary, in organic donor–acceptor (D-A) bilayers there are no charge carriers in either the donor or in the acceptor because the donor and the acceptor materials are not purposely doped. Therefore, the width of the D-A heterojunction is limited to at most a few intermolecular spacings. All excitations

[1] C. W. Tang *et al.*, *Appl. Phys. Lett.*,**48**, 183 (1986).

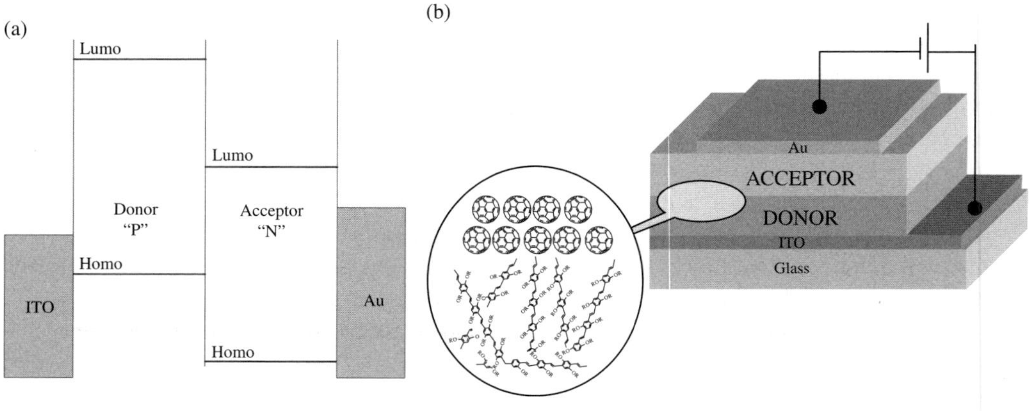

Fig. 13.4 Bilayer photovoltaic device consisting of a fullerene thin film deposited onto a semiconducting polymer thin film in sandwich geometry (i.e. a D-A heterojunction) with metal electrodes (Tang cell).

created at distances far away from the relatively sharp D-A interface (distances greater than the exciton diffusion length and/or the carrier recombination length) will recombine before reaching the D-A interface where charge separation takes place.

Bilayer devices exhibit selective transport of electrons and holes in the two materials (holes in the donor and electrons in the acceptor) to the corresponding electrodes resulting in relatively high rectification ratios. For efficient operation, however, the thicknesses of the organic layers in bilayer solar cells must not exceed the exciton diffusion lengths in either of the two materials.

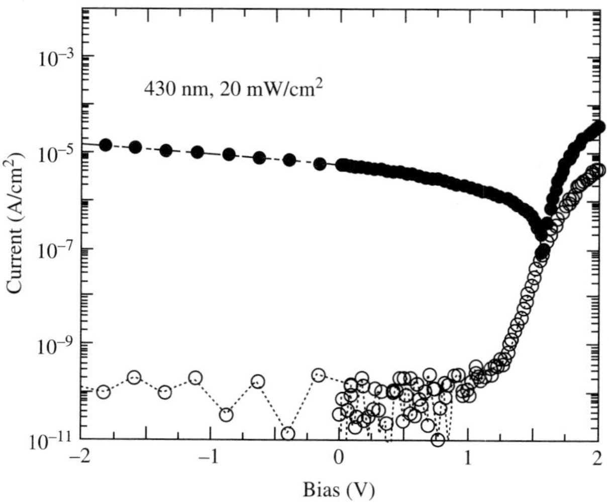

Fig. 13.5 Photoresponse (I versus V) from a bilayer solar cell fabricated from MEH-PPV as the donor layer and C_{60} as the acceptor layer; (o) I-V in the dark; (•) I-V under 20 mW/cm^2 of light with 430 nm wavelength. ITO was used as the anode (in contact with the MEH-PPV layer) and Al was used as the cathode (in contact with the C_{60} layer). [Unpublished data.]

Because the bilayer heterojunction device utilizes semiconducting films with thicknesses comparable to the absorption depth of the incident light and a reflecting metallic cathode, interference will create intensity maxima and minima within the device structure. Thus, one must optimize the thicknesses of the layers. As noted by B. V. Andersson *et al.*,[2] to ensure maximum charge separation efficiency, the intensity maxima must spatially coincide with the donor–acceptor interface.

The short circuit current in the bilayer cell is far too small for useful photovoltaic conversion (see Fig. 13.5). If the film is sufficiently thin that all excitons are created within an exciton diffusion length from the D-A interface, the resulting film has such a low optical density that most of the photons pass through without being absorbed. For thicker films, excitons created far from the DA interface recombine prior to reaching the interface and do not contribute to the photocurrent.

13.3 Self-assembly of BHJ nanomaterials by spontaneous phase separation

To overcome the limitations of bilayer devices that have only a planar interface between the donor and the acceptor thin films, the concept of the "bulk heterojunction" material was introduced. A bulk heterojunction (BHJ) material is a solid state mixture (see Fig 13.6) of two components (donor and acceptor) with nanostructured morphology formed by spontaneous phase separation: the donor and acceptor components self-assemble to form bicontinuous interpenetrating networks. Because the exciton diffusion lengths and the carrier recombination lengths are typically less than 10 nm, the length scale for this self-assembly must be in the 10–20 nanometer regime. The formation of interpenetrating networks requires that the component materials phase separate, that the interfacial energy favors high surface area and that each of the two components is fully percolated with connecting pathways to the electrodes.

The photophysics of BHJ materials comprising semiconducting polymers (the donor component) and fullerenes (the acceptor component) is described in Chapter 12. In a BHJ material, there is a donor–acceptor interface within a few nm of any point within the volume. Because the ultrafast electron transfer occurs at donor–acceptor interfaces throughout the volume (bulk), the nanoscale interpenetrating network ensures that nearly every photon absorbed creates a pair of separated charge carriers and that the photogenerated carriers have pathways to the charge collecting electrode.

13.4 Nanomorphology of BHJ solar cells

Spontaneous phase separation occurs in BHJ materials because "like likes like" and because there is no entropy of mixing for high molecular weight polymers (the entropy of mixing goes to zero as the polymerization index approaches infinity). Thus, even though the donor and acceptor components are each soluble in the solvent used during processing, the BHJ networks spontaneously phase separate as the solvent evaporates. When necessary, the interfacial energy can be controlled using "surfactants"; i.e. additives such as amphiphilic molecules that are

[2] B. V. Andersson *et al.*, *Appl. Phys. Lett.*, **94**, 043302 (2009).

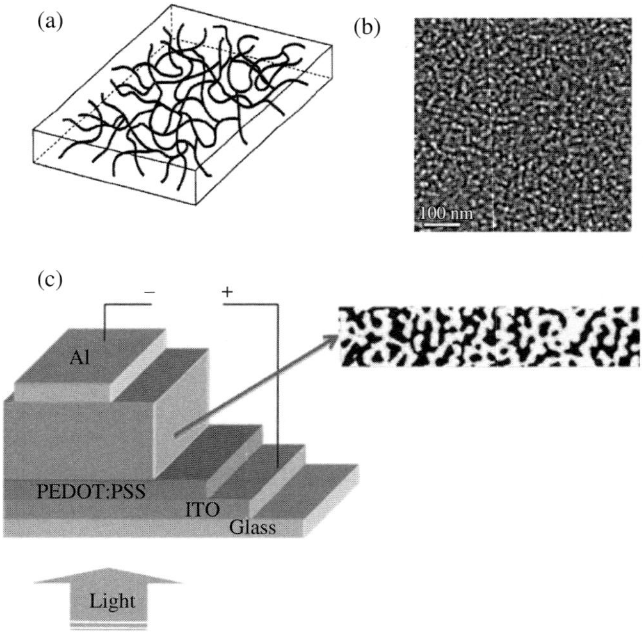

Fig. 13.6 Bulk heterojunction material. (a) Conceptual drawing of the bicontinuous interpenetrating networks (the black and white regions correspond to the fullerene and semiconducting polymer regions, respectively). (b) Phase contrast TEM micrograph that shows the phase separated BHJ material with interpenetrating bicontinuous networks (note the 100 nm scale bar). (c) Structure of the BHJ solar cell.

designed so that one side "likes" the donor and the opposite side "likes" the acceptor (see Chapter 10, Fig. 10.9). Optimization of the nanomorphology and the interpenetrating networks is required for efficient charge collection; the length scale must be approximately 20 nm, and "dead-ends" within the interpenetrating networks must be avoided.

The morphology of the BHJ blend can be fine-tuned by controlling the liquid–liquid phase separation during the film formation or by subsequent treatment of the solid BHJ film. The following parameters and approaches have been identified as significant for their influence on the nanoscale morphology of the BHJ blends:

(a) solvent from which the BHJ composite is cast;
(b) ratio between the polymer and fullerene in the BHJ film;
(c) chemical additives (with differential solubility)
(d) concentration of the solution (% solids);
(e) thermal and/or solvent annealing;
(f) molecular structure and intermolecular ordering of the materials.

The molecular structures of the polymer and fullerene determine their solubility in organic solvents and their miscibility in solution. The solvent influences the drying time during film formation. Thermal and/or solvent annealing enable the crystallization and diffusion of one or both components in the BHJ blend leading to de-mixing and coarsening of the phase separation.

13.5 Optimizing the nanomorphology by thermal annealing

Post-fabrication thermal annealing of P3HT:PCBM BHJ solar cells has been used to improve the nanomorphology. The effect of thermal annealing on j_{sc}, V_{oc} and FF obtained from P3HT: PCBM photovoltaic cells is demonstrated in Fig. 13.7(top). These current versus voltage curves can be modeled with the equivalent circuit shown in Fig. 13.2. The increase in FF following thermal annealing corresponds to a decrease in R_s from 113 ohms to 7.9 ohms (see Section 13.1).

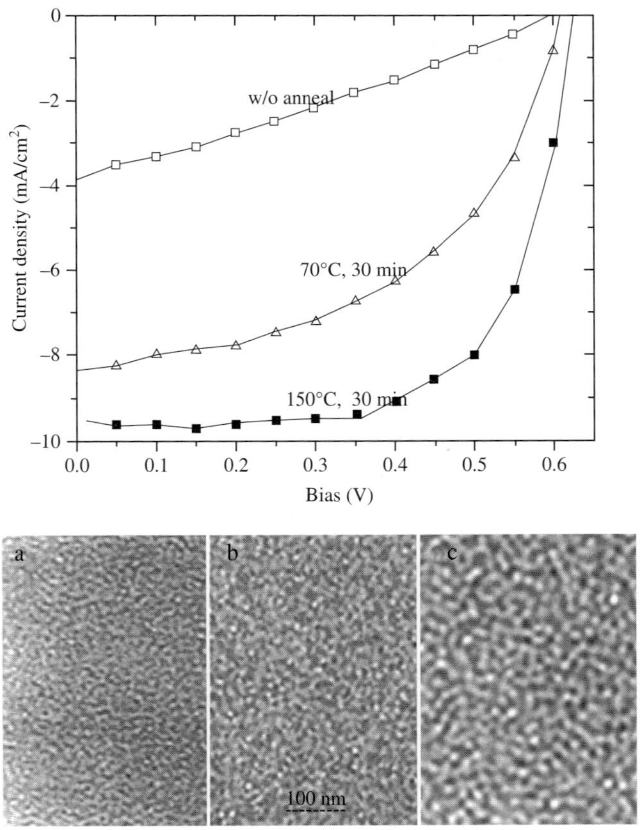

Fig. 13.7 Top: current voltage curves for P3HT:PCBM obtained before and after thermal annealing; bottom: phase contrast TEM images of P3HT:PCBM film. The scale bar on the center bottom (100 nm) is the same for all three images; left: prior to annealing; center: annealed at 150 °C for 10 min; right: annealed at 150 °C for 2 hours. [Reproduced from W.L. Ma *et al.*, *Adv. Funct. Mater.*, **15**, 1617 (2005) with permission from Wiley-VCH Verlag GMBH.]

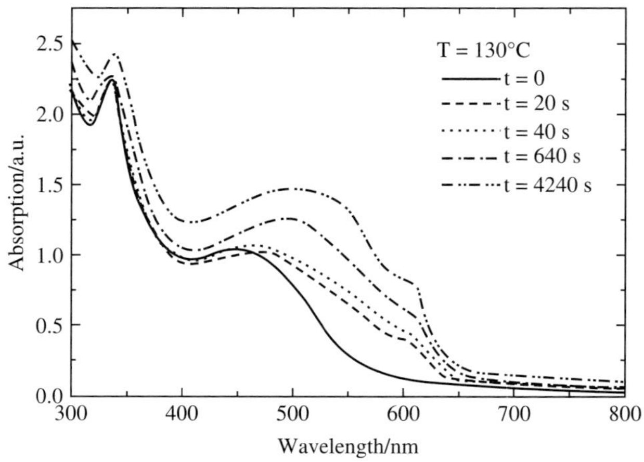

Fig. 13.8 Absorption spectra of a P3HT:PCBM composite film as cast (solid curve) and after four successive thermal annealing steps. The PCBM concentration is 67%. [Reproduced from H. Hoppe and N. Sariciftci, *J. Mater. Chem.*, **16**, 45 (2006) with permission from The Royal Society of Chemistry.]

Phase contrast TEM images of P3HT:PCBM film before and after thermal annealing are shown in Fig. 13.7(bottom). Prior to annealing at elevated temperatures, the interpenetrating networks are not well developed, and the donor/acceptor (D/A) domains are somewhat difficult to distinguish. After thermal annealing for 30 minutes and two hours, respectively, the morphology of the interpenetrating D/A networks becomes clear and is easily visible with a typical feature size of ~10–20 nm. The nanomorphology of BHJ materials composed of different donor polymers and fullerenes is different; e.g. the morphology of the P3HT:PCBM BHJ blends is smaller in length scale and quite different from the spherical nanoclusters observed in PPV.

Annealing at high temperatures improves the crystallinity within the phase separated networks, facilitates charge transport to the electrodes and increases the efficiency. Stable solar cell performance is observed for P3HT:PCBM even after annealing for hours at 150 °C. This remarkable thermal stability of the nanoscale interpenetrating donor–acceptor networks suggests that long lifetimes might be possible with BHJ solar cells.

The changes in the morphology are accompanied by changes in the optical absorption (Fig. 13.8) suggesting improvements in the local structural order. The changes induced by thermal annealing are permanent; they remain after cooling the films to room temperature.

13.6 Characterization of the BHJ nanomorphology

The nanomorphology of BHJ materials has been studied using a number of experimental approaches, including scanning probe microscopy and both scanning and transmission electron microscopy (SEM and TEM). We focus here on phase contrast TEM because it offers the possibility of seeing through the bulk of the BHJ film rather than just probing the surface.

Phase contrast TEM provides a method for spatially imaging the polymer and acceptor regions of the BHJ nanomorphology. In TEM, the probability of direct collision of the beam electrons with the nuclei of the material being studied is small. The electrons in the beam undergo multiple small angle scattering from the nuclei (and the corresponding electron density) of the materials. The net result is a change in the local electron velocity that is analogous to the effect of the index of refraction for optical beams. As a result, the TEM beam exhibits negligible absorption losses. The dominant effect is a local change in the phase of the electron wave that is sensitive to the density of nuclei within the materials. Thus, for BHJ materials, phase contrast TEM enables contrast between the semiconducting polymer and the fullerene; the fullerene has a higher density of carbon nuclei. As sketched in Fig 13.9(b), to obtain phase contrast TEM images, the TEM beam is defocused, passes through the sample and is then focused well below the plane of the thin film sample. Figure 13.9(d) shows a defocused cross-sectional TEM image of a 120 nm slice. The phase separation within rrP3HT: PCBM material can be seen in this image. The cross-sectional image of the part of the sample that was trimmed even thinner (Fig. 13.9d) reveals the same structure even more clearly.

Spatial Fourier transforms and calculation of the power spectral density can be used to analyze the phase contrast TEM digital images of the P3HT:PCBM bulk heterojunction material in two different ways:

1. TEM beam passing through the spin-cast film at normal incidence ("top-down");
2. TEM beam passing through cross-sections of the BHJ film cut using a focused ion beam (see Fig. 13.9).

The observation of structural features at the tens of nanometers length scale requires that the corresponding spatial frequencies be within the region of spatial resolution of the phase contrast electron microscopy. The contrast transfer function $CTF(\vec{u})$ describes the relation between the spatial frequency and the defocus value (Eqn 13.4)

$$CTF(\vec{u}) = A(\vec{u})E(\vec{u})2\sin(\pi\Delta Z\lambda u^2 + \frac{1}{2}\pi\lambda^3 C_s u^4) \qquad (13.4)$$

where $\bar{u}$ is the spatial frequency for a particular direction, $A(\vec{u})$ is the aperture function, $E(\vec{u})$ is the envelope function, λ is the electron wavelength, C_s is the spherical aberration and ΔZ is the defocus value.

When $CTF(\vec{u}) = 0$, there is no contrast in the TEM image for the frequency region $\bar{u}$; the contrast increases as $CTF(\vec{u})$ increases. When $\Delta Z \approx 0$, $CTF(\vec{u}) \approx 0$ at spatial frequencies associated with length scales of tens of nm, and there is no contrast. By increasing the defocus, $CTF(\vec{u})$ increases and good contrast can be obtained in this spatial frequency region. Negative defocus yields positive phase contrast in which regions of higher nuclear density appear dark in a bright background. Figure 13.9(b) shows the ray diagrams for in-focus and negatively defocused TEM conditions.

The power spectral density obtained from the spatial Fourier transform of the TEM images provides a detailed and quantitative description of the phase separated BHJ material. The two-dimensional power spectral density (PSD) of the Fourier transform of the TEM images is defined as:

$$PSD(f_x, f_y) = \frac{1}{L^2}\left|\int_0^L\int_0^L e^{-i2\pi x f_x} e^{-i2\pi y f_y} I(x,y)\,dx\,dy\right|^2 \qquad (13.5)$$

where L is the image size, and f_x and f_y are the spatial frequencies along the x and y axes. The relevant range lies between the inverse image size ($1/L$) and the high-frequency limit $N/2L$, and

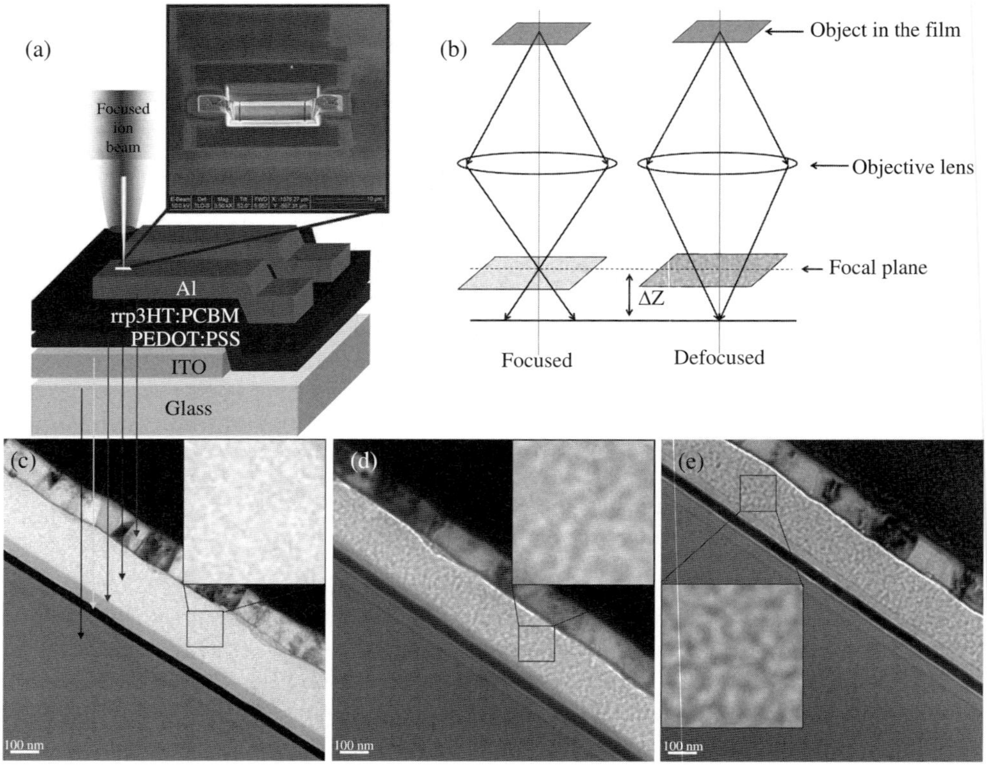

Fig. 13.9 (a) Schematic showing the cross-section of the rrP3HT:PCBM BHJ device; inset: SEM image of the cross-section of the device while still embedded in the substrate after the FIB process is complete. (b) Ray diagram showing focused and defocused configurations. (c) Focused cross-sectional TEM image; inset: the magnified image of the rrP3HT:PCBM BHJ layer. (d) Defocused cross-sectional TEM image of the 120 nm-thick sample; inset: the magnified defocused image of the rrP3HT:PCBM BHJ layer observed through the 120 nm slice. (e) Defocused cross-sectional TEM image for the thinner region (trimmed to ≈ 60 nm); inset: the magnified defocused image of the rrP3HT:PCBM BHJ layer observed through a thinner (≈ 60 nm) slice. [Reproduced from J. S. Moon *et al.*, *Nano Lett.*, **9**, 231 (2009) with permission from the American Chemical Society.]

N is the number of pixels per scan line. $I(x,y)$ is the phase contrast intensity in the TEM image. Using this equation, one obtains a 2D FT in f_x and f_y. By averaging over θ, the PSD can be represented as a function of radial frequencies f to yield the radial 1D PSD:

$$PSD(f) = \frac{1}{2\pi}\int_{0}^{2\pi} PSD(f,\theta)d\theta \tag{13.6}$$

Figure 13.10 shows the power spectral density versus spatial frequency obtained from P3HT:PBCM using phase contrast TEM with the beam passing through the spin-cast film at

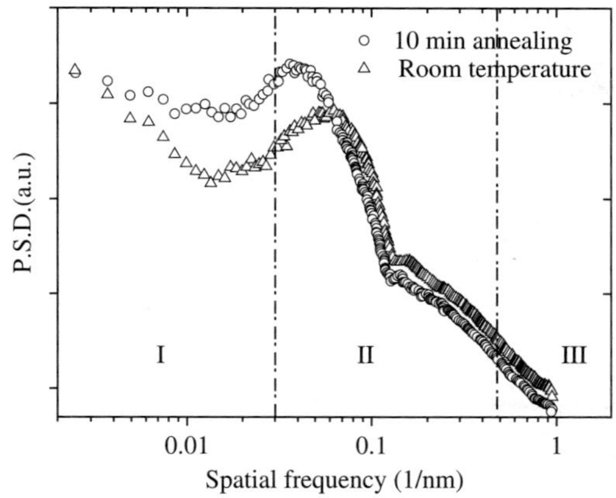

Fig. 13.10 Power spectral density versus radial spatial frequency of P3HT:PCBM bulk heterojunction material before (open triangles) and after annealing (open circles) for 10 min at 150 °C. [Reproduced from W. Ma *et al.*, *Adv. Mater.*, **19**, 1387 (2007) with permission from Wiley-VCH Verlag GMBH.]

normal incidence ("top-down"). The spectra can be divided into three main regions. The highest frequency region (III) is of little importance since it is dominated by noise and aliasing. The useful information related to the demixing scale and the details of the morphology is contained primarily in the middle frequency region (II). A maximum is observed in II, indicating a quasi-periodic structure characteristic of the phase separated morphology. Region I indicates the absence of any specific length scale beyond 100 nm; inelastic scattering also contributes to region I.

The shift in the PSD peaks before and after annealing indicates quasi-periodic phase separation with "periods" of 14 nm and 20 nm, respectively. X-ray diffraction data provide additional evidence of demixing of the BHJ network after annealing, consistent with improved transport and the reduced series resistance exhibited in Fig. 13.7.

The phase separation length scale affects the photovoltaic cell performance in two ways: Efficiency of charge separation at the BHJ interfaces and charge transport within the bicontinuous interpenetrating networks.

As emphasized in Chapter 12, high efficiency charge separation occurs near the donor/acceptor boundary where the ultrafast charge transfer occurs in less than 100 fs. This is comparable to or even faster than the relaxation of "hot" electrons and holes created by absorption of photons with energies larger than the band gap; i.e. comparable to the electron and hole relaxation to the π-band and π*-band edges. Thus, for most of the photons that are absorbed, the charge transfer from the polymer to the fullerene probably takes place prior to the formation of excitons. Some fraction of the photoexcited electron hole pairs will relax to the band edge and subsequently form exciton bound states.

In low mobility semiconducting polymers cast from solution, however, the excitation diffusion length is limited by disorder to less than 10 nm, which, therefore, defines the optimum length scale for phase separation. Morphology features with "periods" beyond this limit will decrease the quantum efficiency because of inefficient charge separation. On the contrary, if the morphology length scale is too small to form a connected interpenetrating network, the charge transport will be hampered (charge transport with minimal recombination can only occur inside a single-phase material).

Therefore the maximum distance from the P3HT component of the bulk heterojunction material to the interface of P3HT:PCBM is within the range of the excitation diffusion length. As shown in Fig. 13.7, the nanoscale phase separated network is improved after annealing and provides both efficient charge separation and enhanced charge transport (higher mobility).

Phase contrast TEM through thin sections cut from films of the P3HT:PCBM BHJ material provides information on the cross-sectional morphology. Subsequent to photoinduced charge separation, the photogenerated carriers must move toward the electrodes by traveling across the film thickness rather than parallel to the film surface. The "column-like" structures in the defocused phase contrast TEM cross-sectional images are the vertical pathways for charge transport across the film thickness. Through analysis of the cross-sectional TEM images, the power spectral density and the autocorrelation function of the vertical pathways can be calculated.

To obtain thin sections ("slices") of the BHJ film, a focused ion beam (FIB) was used (see Fig. 13.9a). The sample was cut to the shape of a thin hexahedron with length 10 μm, width 4 μm and thickness 120 nm. Over a part of this section, the slice was further trimmed to approximately half the original thickness; i.e. to a thickness of approximately 60 nm. The TEM images were obtained with the electron beam passing through the 120 nm and the $\approx$ 60 nm thick regions, respectively.

Figures 13.11(a) and (b) show cross-sectional images for the 120 nm slice and 60 nm slice respectively. The dark and bright regions are typically connected across the film, thereby forming "column-like" pathways from the top to the bottom of the BHJ layer. The binary enhanced images directly below the original images show these "column-like" pathways more clearly and can be used as a guide to the eye. To obtain the binary enhanced image, the original TEM image was subjected to Gaussian blur with a 2.5 nm radius to reduce noise.

A threshold brightness was chosen that results in the same ratio of white and black as the known volume fraction (1:0.7) of rrP3HT to PCBM within the BHJ material. Figures 13.11(a) and (b) show many well-connected vertical pathways through the BHJ layer. The shortest distance within a single pathway is typically only slightly larger than the film thickness.

Although certainly not perfect, this "column-like" structure results in efficient transport of carriers through the BHJ film as required for high efficiency BHJ solar cells. Moreover, because the optical absorption depth is greater than the film thickness, light is absorbed and separated charges are generated throughout the film thickness. Thus, the imperfect nature of the somewhat elongated and distorted columns is not very important.

Figures 13.12(a) and (b) show the averaged PSD spectrum as a function of the spatial frequency. The PSD spectra calculated with the cross-section TEM images from 120 nm-thick part and the thinner trimmed part are consistent. These spectra are also consistent with that obtained from analysis of TEM images taken "top-down" through the BHJ film. The inverse of the spatial frequency at the peak represents the "period" of this quasi-periodic structure ($\approx$ 25 nm).

The autocorrelation function (ACR) also provides important information on quasi-periodic structures. From the ACR, one can determine not only the "period" but also the domain size. The ACR is defined as:

$$ACR(x) = \frac{1}{L_x}\int_0^{L_x} I(x)I(x-x^{'})dx \tag{13.7}$$

where L_x is the lateral length of the image, and $I(x)$ is the phase contrast intensity along the x-direction for each line of pixels. The averaged ACR was calculated from the ACR from a single line of pixels by averaging over 165 separate pixel lines. In the ACR, the most probable "period" of the pathway is given by the position of the first maximum after the autocorrelation triangle. The value is again $\approx$ 25 nm, in good agreement with the PSD result (Fig. 13.12c, d). When a straight line is drawn from the self-correlation point, ACR(0), with initial slope equal to that of ACR(x) (see the red dotted lines in Fig. 13.12c and d), the line intersects the x-axis at approximately 10 nm implying that the average domain size of the PCBM regions is $\approx$ 10 nm. Therefore the widths of pathways for electrons and holes in the vertical direction are estimated to be $\approx$ 10 nm and $\approx$ 15 nm, respectively.

The fact that results from the 120 nm thick and the thinner trimmed areas are the same provides confidence that the contrast in the images does not arise from quasi-periodic variations in the thickness of the BHJ film.

13.7 The origin of the open circuit voltage, V_{oc}

The roadmap to low-cost plastic solar cells with high efficiency must include the synthesis of electronegative polymers with deeper HOMO energies to enable higher V_{oc} and smaller bandgaps for better harvesting of the solar spectrum. The energy gap required for highest efficiency is in the range 1.2–1.3 eV, i.e. close to the band gap of silicon.

As indicated by the arrows in Fig. 13.13, the open circuit voltage is determined by the energy difference between the LUMO of the acceptor and the HOMO of the donor.

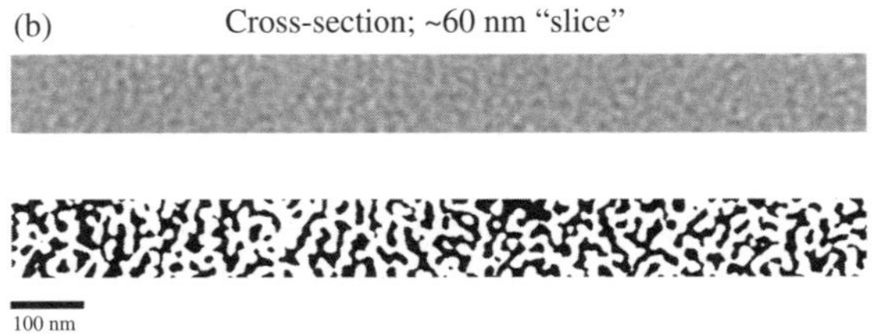

Fig. 13.11 Defocused cross-sectional images of the rrP3HT:PCBM BHJ layer (a) 120 nm-thick slice and for the thin part of the sample; (b) $\approx$ 60 nm thinned slice. [Reproduced from J. S. Moon *et al.*, *Nano Lett.*, **9**, 231 (2009) with permission from the American Chemical Society.]

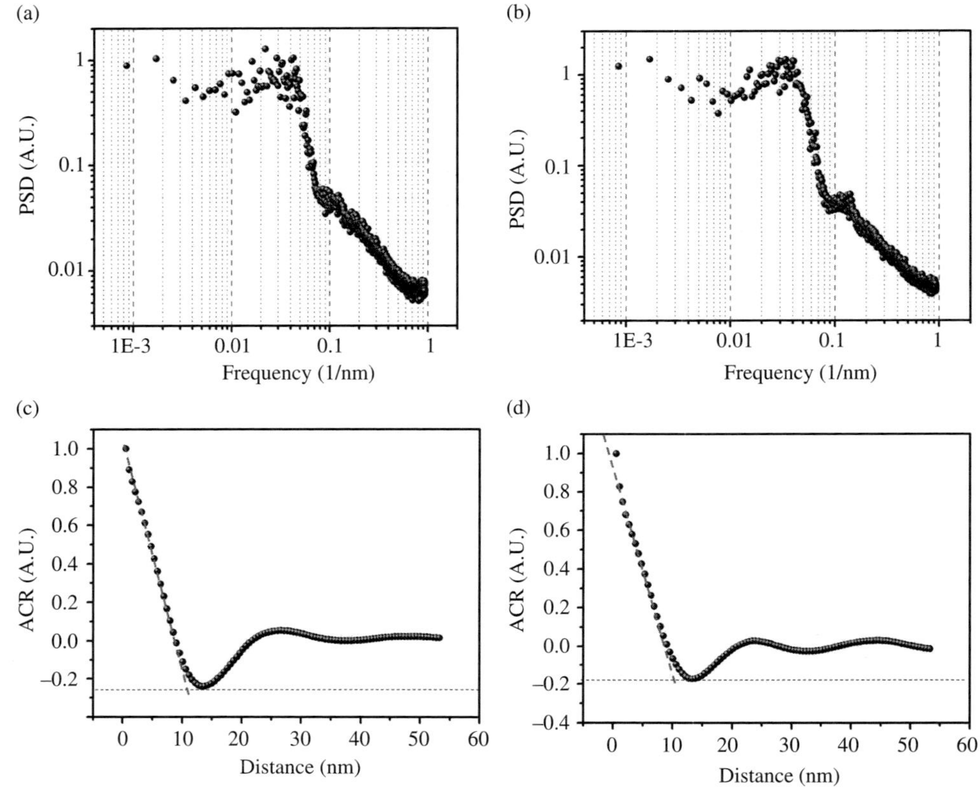

Fig. 13.12 (a) The power density spectrum and (c) the autocorrelation function obtained from the cross-sectional defocused TEM image of the 120 nm slice. (b) The power density spectrum and (d) the autocorrelation function from the thinner region of the slice (≈ 60 nm). [Reproduced from J. S. Moon *et al.*, *Nano Lett.*, **9**, 231 (2009) with permission from the American Chemical Society.]

$$V_{OC} \approx E_a(\text{LUMO}) - E_d(\text{HOMO}) \tag{13.8}$$

Under high illumination intensity (for example, under 1 sun), there is a relatively high density of holes in the π-band of the polymer and an equivalent density of electrons in the LUMO band of the acceptor. Thus under high illumination, the Fermi energy in the donor domain, μ_d, is located near the π-band edge of the polymer, and the Fermi energy in the fullerene acceptor domain, μ_a, is located near the LUMO of the fullerene. In each of the two regions, however, the band edge is not precisely sharp because of band tailing as a result of disorder. The Fermi energy in either of the two domains is, therefore, in the respective band tail. The difference between the two Fermi energies, $\mu_a - \mu_d$, is equal to the open circuit voltage, $V_{OC} \approx E_a(\text{LUMO}) - E_d$ (HOMO), with a small correction expected from occupation in the band tails (see Fig 13.13). The goal for V_{oc} should be that V_{oc} approach the magnitude of the band gap; i.e. the LUMO offset (acceptor relative to polymer) should be just large enough to enable photoinduced charge transfer.

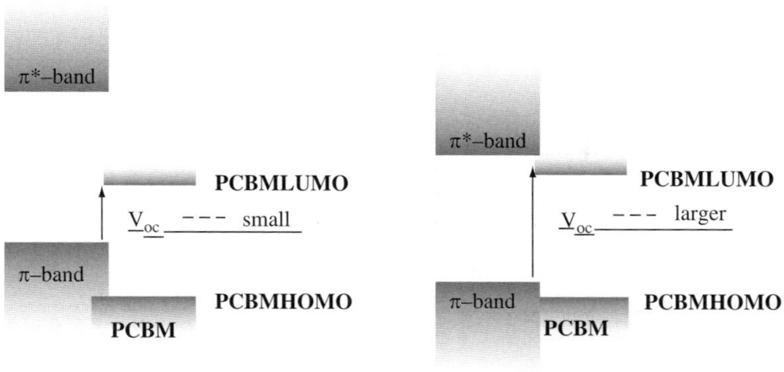

Fig. 13.13 Electronic structure of BHJ materials with two different semiconducting polymers. The diagram on the right assumes a more electronegative semiconducting polymer (deeper HOMO). The open circuit voltage is determined by the energy difference between the HOMO of the semiconducting polymer and the LUMO of the acceptor.

Figure 13.14 shows a simulated image of how the open circuit voltage is established. Following photogenerated charge separation, electrons and holes accumulate on the fullerene and polymer sides of the interface to form a "parallel plate capacitor"; the density of charges per unit length along the "capacitor" increases until the voltage difference across the interface becomes equal to V_{oc}. For this interface charge density, the difference in the Fermi levels in the donor and acceptor regions is compensated by the voltage across the interface and steady-state conditions are established. Note that a general feature of such a parallel plate capacitor is that the electric field is large in the interface region and zero outside (truly zero in a parallel plate capacitor with dimensions much larger than thickness of the interface). Therefore, there will be no electric field in the phase separated regions.

Two questions then must be answered:

1. Why do the + and – charges across the interface simply not recombine?
2. How does this microscopic picture of the origin of V_{oc} translate to the macroscopic V_{oc} measured at the electrodes of the BHJ solar cell?

The + and – charges across the interface do not readily recombine when the applied voltage is zero (short circuit condition) because of the band off-set in Fig. 13.13. The lower energy of the electron following charge transfer, the difference in the two LUMO energies shown in the heterojunction band diagrams, provides a barrier to recombination. As the applied voltage approaches V_{oc}, however, the charge collected at the electrodes goes to zero. At $V \approx V_{oc}$, the probability of recombination of the photogenerated carriers must approach unity (see, for example, Section 13.14).

This microscopic picture of the origin of V_{oc} is translated to the macroscopic V_{oc} measured at the electrodes of the BHJ solar cell by a combination of symmetry breaking by the use of two different contacts with different work functions and by using charge selective transport

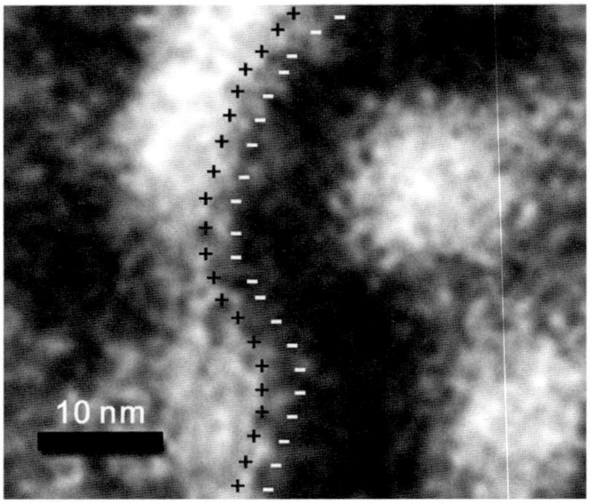

Fig. 13.14 Simulated image showing the formation of a "parallel plate capacitor" at the polymer: fullerene interface.

layers; i.e. an electron transport/hole blocking layer on the cathode side and a hole transport/electron blocking layer on the anode side. This combination automatically connects the BHJ domains in parallel and establishes the macroscopic V_{oc}. Insertion of a TiO_x layer between the BHJ film and the metal electrode is a known example; the TiO_x layer functions as an electron transport/hole blocking layer (see Section 13.10). Indeed cell performance is significantly improved with the use of the TiO_x layer. PEDOT, however, does not function as an electron blocking layer. Thus, a replacement for PEDOT that functions as a hole transport/electron blocking layer on the anode side is expected to improve performance.

13.8 Dynamics of the BHJ cell

The time dynamics of the operation of the BHJ solar cell are summarized in Fig. 13.15.

- Photoinduced charge transfer takes place in the subpicosecond time scale; the ultrafast photoinduced charge transfer is described in detail in Chapter 12.
- At longer times, the mobile carriers are swept out by the internal voltage (built-in electric field) in competition with recombination during the sweep-out transit time.

Note that not all the absorbed photons lead to mobile carriers; some of the initially photo-generated carriers fall into interfacial traps or form bound interfacial charge transfer excitons (see Fig. 13.15), and some carriers are trapped in localized states deep within the π-band tail of the donor polymer or the LUMO band tail of the acceptor.

Figure 13.15(b) shows the evolution from sweep out (at or near short circuit conditions where the internal built-in voltage is large) to recombination decay near open circuit (where the internal voltage is compensated by the external applied voltage). The data shown in Fig. 13.15(b) were

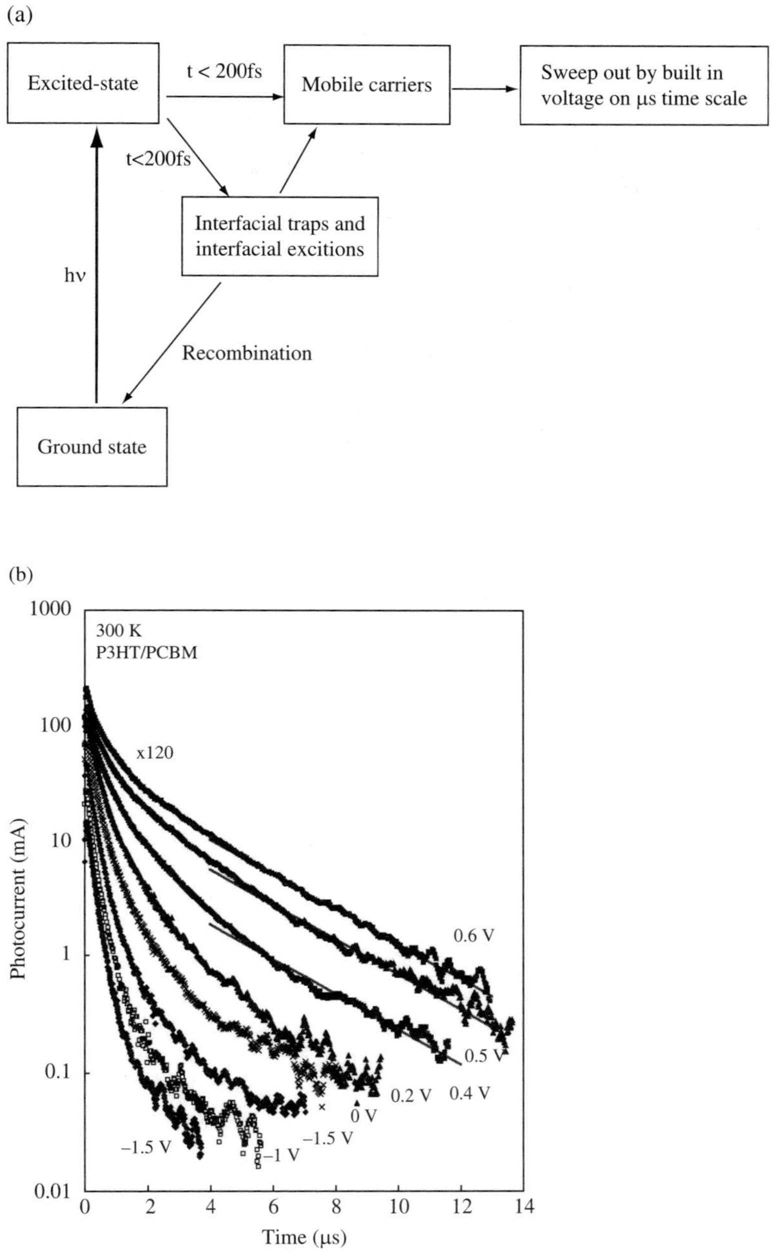

Fig. 13.15 (a) Time dynamics during the operation of the BHJ solar cell. Photoexcitation is followed by ultrafast charge transfer that creates mobile carriers in the donor and acceptor domains. These carriers are subsequently either swept out to the electrodes by the built-in electric field or recombine to the ground state. The sweep-out time is of order a μs at short circuit where the recombination probability is small. For efficient solar cells the recombination time must be long compared to the sweep out time. (b) The evolution from sweep out (at short circuit) to recombination at open circuit is demonstrated through transient photoconductivity measurements carried out at different external applied voltages for a P3HT/PCBM solar cell at room temperature. The data are offset vertically for clarity; the vertical axis corresponds to the data at −1.5 V. Solid lines through the data near open circuit (where the internal field is near zero) show an exponential decay with 2.9 μs decay time. [R. Street, S. Cowan and A.J. Heeger, Phys. Rev. Lett. (in press).]

obtained from solar cells fabricated from P3HT/PCBM.[3] The BHJ layer within the cell was about 100 nm thick. The excitation source was a nitrogen pumped dye laser emitting a pulse of <10 ns duration at ~520 nm.

The characteristic sweep-out time, τ_{sw}, is given by the thickness (t) divided by the drift velocity of the mobile carriers, μE, where μ is the average carrier mobility, $E = V/t$, and V is the internal voltage ($|V| \sim V_{oc}$ at short circuit, and $V \to 0$ as $V_{oc} \to |V|$). The factor of 2 in Eqn 13.9a arises because the carrier generation is uniform throughout the film; on average, mobile carriers must travel through only approximately half the film thickness.

$$\tau_{SW} = t^2/2\mu V \tag{13.9}$$

Some carriers are lost via recombination during the sweep-out transit time and therefore do not reach the electrodes. At open circuit, all carriers must recombine within the BHJ layer. Assuming uniform optical absorption across the device and nondispersive mobility, the transient photocurrent is given by the sum of the separate contributions from electrons and holes,

$$I_{Pi}(t, V) = N_0 e\mu_i E\, exp(-t/\tau_{Ri})(1 - \mu_i Vt/d^2);\; i = e, h \tag{13.10a}$$

The analysis can be extended to a specific absorption profile and dispersive mobility. The exponential term describes the monomolecular recombination of carriers with a single lifetime, and the term in brackets on the right describes the sweep-out of carriers to the contacts. At open circuit, the recombination decay in Fig. 13.15(b) shows evidence of nonlinear (bimolecular) decay at high carrier densities, but transitions to a single exponential at carrier densities below that corresponding to 1 sun.

The limiting cases of Eqn 13.10a are

$$\tau_{R1} >> t_T > t;\; I_p(t, V) = N_0 e\mu_1 E(1 - \mu_1 Vt/d^2) \;\; \text{sweep} - \text{out transit limit;} \tag{13.10b}$$

$$\tau_{R1} << t_T;\; I_p(t, V) = N_0 e\mu_1 E exp(-t/\tau_{R1}) \;\; \text{recombination time limit.} \tag{13.10c}$$

For efficient collection of the photogenerated carriers, the carrier recombination time (τ_r) must be significantly longer that the sweep-out time. Times longer than both the sweep-out time and the carrier recombination time are of little importance to the operation of the BHJ solar cell. Deeply trapped carriers contribute relatively little to the current and come out to the electrodes very slowly (if at all).

The various recombination processes are discussed in Section 13.14.

13.9 Optimizing the performance of new materials: Processing additives

New materials are needed to achieve the required improvements in power conversion efficiency. The energy gap of P3HT is too large; P3HT cannot harvest the large fraction of the energy incident in the infrared part of the solar spectrum. Moreover, the LUMO of P3HT is more than 500 mV higher than required for electron transfer to the fullerene. As a result, a large fraction of the photon energy is lost during charge transfer.

[3] R. Street, S. Cowan and A.J. Heeger, *Phys. Rev. Lett.* (in press, 2010).

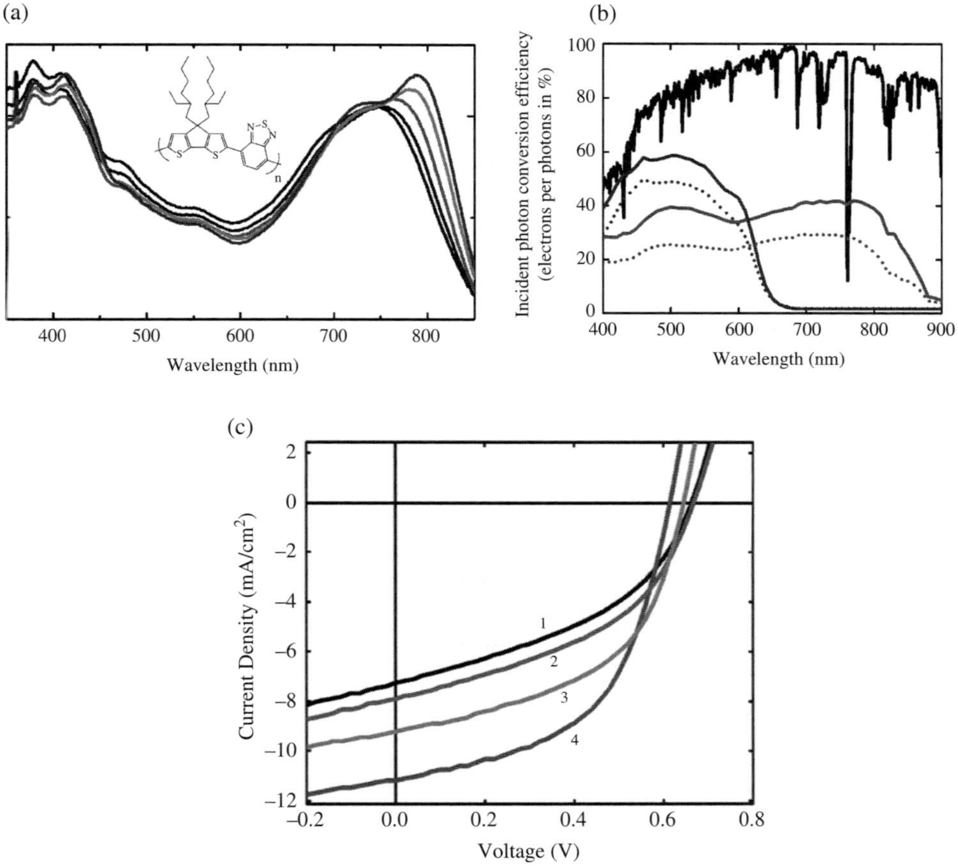

Fig. 13.16 (a) Optical absorption spectra for PCPDTBT/PC71BM blends cast from pristine chlorobenzene and cast from chlorobenzene containing 2% by volume of various alkane dithiols. The PCPDTBT structure is shown in the inset where the R represents an ethylhexyl group. (b) EQE specra of solar cells composed of P3HT/PC61BM before and after thermal annealing, and PCPDTBT/PC71BM cast from chlorobenzene and cast from chlorobenzene containing octanedithiol. The AM 1.5G solar spectrum is shown in dark black for reference. Note that the absorption edge for P3HT is at ~650 nm; the absorption edge for PCPDTBT is at ~ 900 nm. [Reproduced from J. Peet *et al.*, *Nature Materials*, **6**, 497 (2007) with permission from Nature Publishing Group.] (c) Current density vs. voltage curves (under simulated AM1.5 radiation at 80 mW/cm2) for a series of PCPDTBT/C71PCBM solar cells. The PCPDTBT/C71PCBM films were cast at 1200 RPM from pristine CB (1) and CB containing 2.5% butanedithiol (2), hexanedithiol (3), or octanedithiol (4). [Reproduced J. Peet et al., Nature Material **6** 497 (2007) with permission from NPG]

An example of a new material which has been introduced as a candidate for higher performance solar cells is PCPDTBT (structure shown in Fig. 13.16a). PCPDTBT is a "push–pull" copolymer which uses alternating electron withdrawing and electron donating components to increase the double bond character between the units and thereby stabilize the quinoid form of the polymer to decrease the band gap.

PCPDTBT illustrates the concept of "tuning" the HOMO and LUMO: The HOMO energy is largely derived from the electron donating part of the repeat unit, and the LUMO energy is largely derived from the electron withdrawing part of the repeat unit. PCPDTBT has approximately the same HOMO energy value as P3HT, and thus the same V_{OC} is observed in BHJ solar cells. Because of the lower lying LUMO, PCPDTBT harvests light and generates photocurrent between 300 and 900 nm (compared with 300–650 nm for P3HT). Thus, the polycarbazole family of which PCPDTBT is the first example has a significantly greater potential for high efficiency solar cells than does P3HT.

By incorporating small concentrations of alkanedithiols in the solution from which the bulk heterojunction films are cast, polymer/fullerene phase separation and nanomorphology can be modified. This approach offers the potential to introduce morphology control to BHJ materials during device fabrication and without subsequent thermal annealing.

Current–voltage characteristics obtained under 80 mW/cm^2 for devices fabricated from PCPDTBT:PC$_{70}$BM are shown in Fig. 13.16c. The various colors correspond to the data obtained from devices processed from chlorobenzene and from chlorobenzene with 2.5% by volume alkanedithiols with different chain lengths. Note that both I_{sc} and FF are significantly increased by processing with the 1,8-octanedithiol processing additive. When devices are fabricated using octane dithiol, the external quantum efficiency (EQE) improves across all wavelengths (Fig. 13.16b) within the absorption band and the efficiency nearly doubles.

Two criteria were identified for processing additives introduced to control the morphology of BHJ materials: (i) selective (differential) solubility of the fullerene component and (ii) higher boiling point than the host solvent. Control of the BHJ morphology by selective solubility of the fullerene component was demonstrated. The best results were obtained with 1,8-diiodooctane as the processing additive.[4]

13.10 Design rules toward bulk heterojunction solar cells with > 10% PCE

Dennler, Scharber and Brabec carried out an analysis of the theoretical maximum efficiency for a single junction BHJ solar cell assuming thermalized carriers. Excess energy from photon energy larger than the semiconductor band gap is assumed to be dissipated as heat. Two sources of energy loss are included:

(i) energy loss during photoinduced charge-transfer, ΔE_{CT}, and
(ii) energy loss due to polaron formation, ΔE_{Pol}.

ΔE_{CT} is approximately equal to the difference in the donor–acceptor LUMO energies. Figure 13.17 assumes $\Delta E_{CT} = 0.25$ eV and $\Delta E_{Pol} = 0.25$ eV (although charge transfer has been reported in BHJ materials with LUMO energy differences as small as 0.1 eV).

Figure 13.17 (left) assumes that both FF $= 1$ and EQE $= 100\%$. Figure 13.17 (right) assumes FF $= 0.70$ and EQE $= 90\%$. The highest EQE value reported for an organic solar cell is approximately 70% (with internal quantum efficiency approaching 100%), and FF $> 70\%$

[4] J. K. Lee *et al.*, *J. Am. Chem. Soc.*, **130**, 3619 (2008).

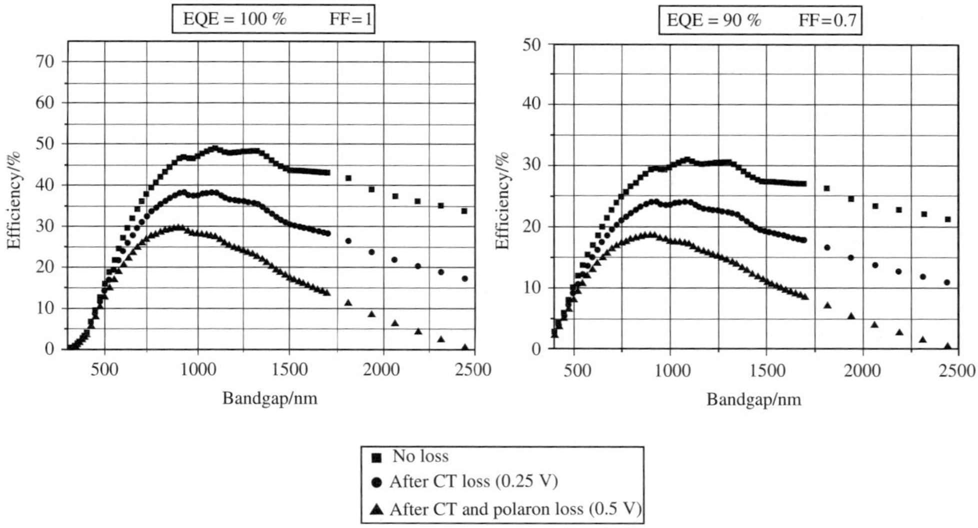

Fig. 13.17 Left: theoretical maximum efficiency versus band gap for BHJ solar cells assuming FF = 1 and EQE = 100%. Right: practical maximum efficiency versus band gap for BHJ solar cells assuming FF = 0.70 and EQE = 90%. [Unpublished data.]

have been reported. For each loss mechanism, the optimum band gap shifts slightly and the absolute efficiency numbers are reduced. Even in the most unfavorable case, including all V_{oc}, EQE, and FF losses, efficiencies approaching 20% appear to be realistic.

13.11 Titanium oxide as multifunctional components in BHJ solar cells

The introduction of a thin layer (or layers) of TiO_x into the structure by sol–gel chemistry enables a number of opportunities for improving the performance of bulk heterojunction photovoltaic cells. The TiO_x layers serve six separate functions.

1. When deposited between the charge separation layer and the aluminum cathode, the TiO_x layer functions as an optical spacer that redistributes the light intensity to optimize the efficiency.
2. By introducing a TiO_x layer between the charge separating layer and the aluminum cathode, excellent air stability has been demonstrated. The TiO_x layer improves the lifetime of unpackaged devices exposed to air by nearly two orders of magnitude.
3. The TiO_x functions as an electron transport layer. As a result of the oxygen deficiency, the TiO_x layer is n-type doped. As a result, the inclusion of the TiO_x layer between the charge separating layer and the aluminum cathode does not result in an increase in the series resistance. Moreover, since the lowest energy states at the bottom of the conduc-

tion band of TiO_x are well matched to the Fermi energy of aluminum, there is facile electron transfer from the TiO_x electron transport layer to the aluminum cathode.

4. The TiO_x layer breaks the symmetry, functions as a charge selective collection layer and, thereby, contributes to the open circuit voltage (see Section 13.7).
5. The TiO_x functions as a hole-blocking layer since the top of the valence band of TiO_x is sufficiently electronegative, 8.1 eV below the vacuum, to block holes.
6. The TiO_x layer enables the fabrication of tandem cells (see next section). The transparent TiO_x layer is used to separate and connect the front cell and the back cell. The TiO_x layer serves as an electron transport and collecting layer for the first cell and as a stable foundation that enables the fabrication of the second cell to complete the tandem cell architecture.

The reflection for the metal electrode leads to the formation of a "dead zone" as sketched in Fig. 13.18. In order to overcome this problem in BHJ solar cells, one might simply increase the thickness of the active layer to absorb more light. Because of the low mobility of the charge carriers in the polymer/C_{60} composites, however, the increased internal resistance of thicker films will inevitably lead to a reduced fill factor. An alternative approach is to change the device architecture with the goal of spatially redistributing the light intensity inside the device by introducing an optical spacer between the active layer and the Al electrode.

The layer must be a good acceptor and an electron transport material with a conduction band edge lower in energy than that of the lowest unoccupied molecular orbital (LUMO) of C_{60}, the conduction band edge must be above (or close to) the Fermi energy of the collecting metal electrode, and the material must be transparent to light with wavelengths within the solar spectrum.

A low temperature solution-based sol–gel process has been used to fabricate a titanium oxide (TiO_x) layer on top of the polymer–fullerene active layer. Because the titanium oxide layer was treated at temperatures below 100 °C, the film is amorphous as confirmed by X-ray diffraction.

Using this solution-based sol–gel process for casting TiO_x layers, the device lifetime, unpackaged and in air, is significantly enhanced. Figure 13.19 compares the lifetimes of

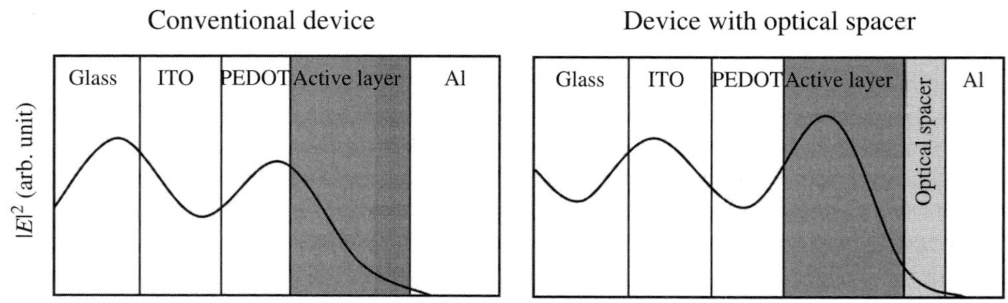

Fig. 13.18 Distribution of the squared optical electric field strength $|E|^2$ inside the devices with a structure of ITO/PEDOT/Active Layer/Al (left) and ITO/PEDOT/Active Layer/Optical Spacer/Al. The dotted region in the left figure denotes the dead-zone.

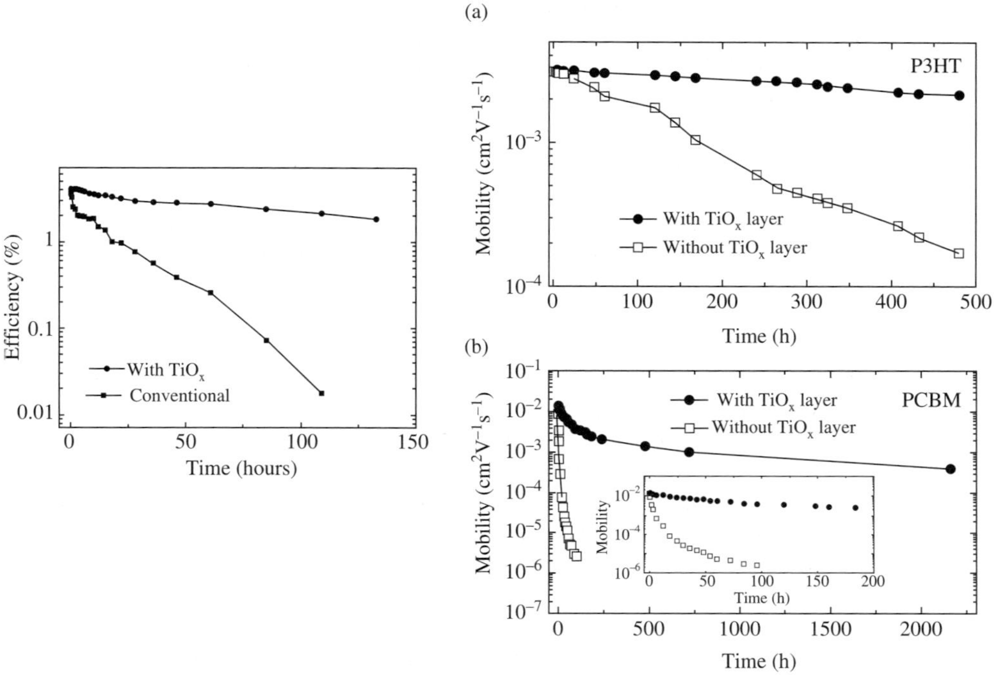

Fig. 13.19 Left: comparison of the power conversion efficiencies as a function of storage time for polymer solar cells with and without the TiO_x layer. [Reproduced from K. Lee *et al.*, *Adv. Mater.*, **19**, 2445 (2007).] Right: comparison of the mobilities as a function of storage time in air: (a) p-type P3HT FET with and without the TiO_x layer; (b) n-type PCBM FET with and without the TiO_x layer. [Reproduced from S. Cho *et al.*, *Adv. Mater.*, **21**, 1941 (2009) with permission from Wiley-VCH Verlag GMBH.]

devices fabricated with and without the TiO_x layer; the lifetime in air of the polymer solar cells is enhanced by the TiO_x layer by two orders of magnitude.

The electron mobility obtained from devices with the TiO_x layer shows a decrease of less than one order of magnitude over the same period of exposure (see Fig 13.19 right). Even after 90 days exposure to air, the electron mobility remains at 3.9×10^{-4} $cm^2/V.s$.

The achievement of "air-stable" polymer devices implies that the promise of high performance "flexible" and "printable" devices can be realized. Since the TiO_x layer is generated by a sol–gel process, all fabrication steps are carried out by processing the component materials from solution; i.e. consistent with printing and coating technologies. The TiO_x layer reduces the sensitivity to oxygen and water vapor to a point where simple barrier materials might be sufficient to enable the lifetimes required for printed, flexible, plastic electronics.

13.12 BHJ tandem solar cells

In tandem solar cells, two solar cells are deposited on the top of each other. Figure 13.20 shows the architecture of a tandem cell consisting of two distinct devices stacked on top of each other and connected in series by a thin recombination (nonabsorbing) layer.

The thin recombination layer enables the alignment of the quasi-Fermi level of the acceptor of the bottom cell with the Fermi level of the donor of top cell. This intermediate layer allows the recombination holes coming from one cell with electrons coming from another. According to Kirchhoff's law, this type of connection implies that the voltage across the tandem cell device is equal to sum of the voltages across each device. In other words, the open circuit voltage of the tandem cell is $V_{oc}(\text{tandem}) = V_{oc1} + V_{oc2}$.

By using the solution-processed TiO_x as the separator layer between the front cell and back cell, polymer tandem cells have been successfully demonstrated, with each layer processed from solution. The structure of the multilayer polymer tandem solar cell is shown in Fig. 13.21.

The current density versus voltage (*J-V*) characteristics of single solar cells and the tandem solar cell using PCPDTBT:PCBM and P3HT:PC_{70}BM composites under AM 1.5G illumination from a calibrated solar simulator with irradiation intensity of 100 mW/cm^2 are shown in Fig. 13.22. The PCPDTBT:PCBM single cell yields $J_{sc} = 9.2$ mA/cm^2, $V_{oc} = 0.66$ V, $FF = 0.50$, and $\eta_e = 3.0\%$, and the P3HT:PC_{70}BM single cell yields $J_{sc} = 10.8$ mA/cm^2, $V_{oc} = 0.63$ V, $FF = 0.69$, and $\eta_e = 4.7\%$.

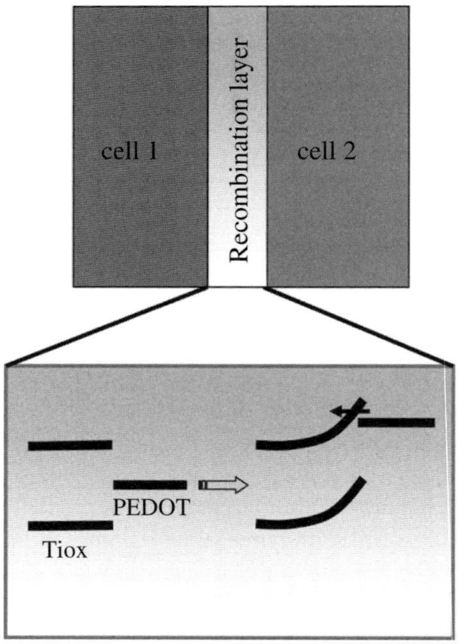

Fig. 13.20 Tandem solar cell made of two subcells connected in series via a recombination layer. Simplified band diagram of recombination layer (for example TiO_x and PEDOT) before and after contact is shown in the lower panel.

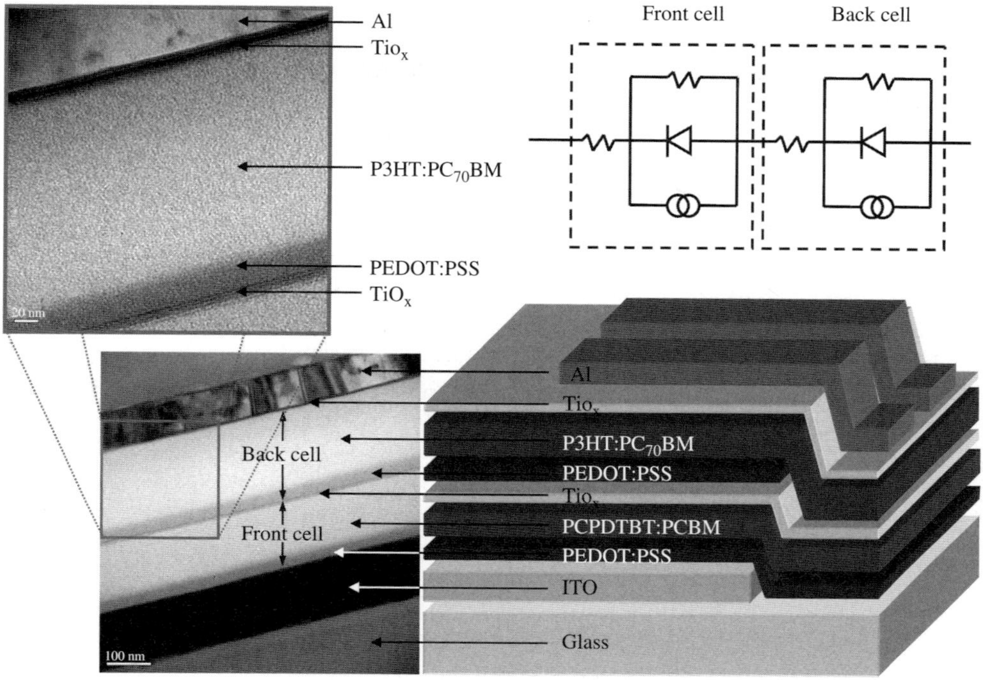

Fig. 13.21 The tandem cell structure (right) and TEM cross-section image (left). The scale bar is 100 nm in the lower image and 20 nm in the upper image. [Reproduced from J. Y. Kim *et al.*, *Science*, **317,** 222 (2007) with permission from AAA.]

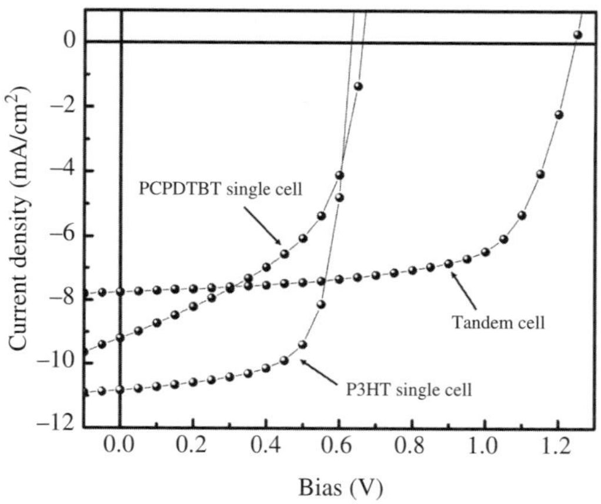

Fig. 13.22 *J-V* characteristics of single cells and tandem cell using PCPDTBT:PCBM and P3HT: $PC_{70}BM$ composites under AM 1.5G illumination from a calibrated solar simulator with irradiation intensity of 100 mW/cm^2. [Reproduced from J. Y. Kim *et al.*, *Science*, **317**, 222 (2007) with permission from AAA.]

With two subcells stacked in series, the current that is extracted from the tandem cell is determined by the current generated in either the front or back cell, whichever is smaller. Matching the currents from the two cells is the challenge. Optimization led to the inverted tandem cell architecture depicted in Fig. 13.21. Using this inverted structure, typical performance parameters were as follows; $J_{sc} = 7.8$ mA/cm^2, $V_{oc} = 1.24$ V, $FF = 0.67$ and $\eta_e = 6.5\%$.

Ideally, the TiO_x/PEDOT:PSSA bilayer would form a Schottky diode and thereby shift the Fermi energy of the PEDOT:PSS up equal that of the bottom of the conduction band of TiO_x. That shift in the Fermi energy is critical; without the shift in the Fermi energy the two cells will not perform in tandem (the voltages will not add).

Although tandem cells with architecture similar to that shown in Fig. 13.21 have been reproduced in the literature,[5] the present status is not satisfactory; much more work is needed. The acidity of the PEDOT:PSSA causes serious problems that show up as a need for lengthy "light soaking". Progress with BHJ tandem cells will require replacing the PEDOT:PSS with a stable, nonreactive transparent conductor.

13.13 BHJ solar cells with internal quantum efficiency approaching 100%

Alternating copolymers based on poly(2,7-carbazole) derivatives,[6] with a suite of electron deficient moieties to choose from, are particularly promising. These electron deficient moieties can be used to tune the electronic energy gap of the semiconducting polymer while the deep HOMO of the carbazole leads to higher values for V_{OC}. The implied flexibility in the synthesis is expected to enable synthesis of smaller bandgap polymers that can harvest a larger fraction of the solar radiation spectrum (Section 13.9), and a deeper HOMO energy that increases the open circuit voltage (Section 13.7).

Using BHJ composites with PCDTBT and $PC_{70}BM$, solar cells with 6% power conversion efficiency were demonstrated with $V_{oc} = 0.88$ and internal quantum efficiency (IQE) close to 100%, implying that essentially every absorbed photon results in a separated pair of charge carriers and that all photogenerated carriers are collected at the electrodes.[7]

Figure 13.23 shows the structure of the PCDTBT:PC70BM BHJ solar cell together with the molecular structures and an energy level diagram of the component materials. From the fundamental physics of the open circuit voltage associated with the donor–acceptor heterojunction (see Section 13.7), the relatively deep HOMO energy of PCDTBT, approximately 5.5 eV, is expected to yield a higher open circuit voltage. Moreover, since the spherical symmetry of the fullerene has been lifted in $PC_{70}BM$ (compared to PCBM), the PCDTBT/$PC_{70}BM$ BHJ material has higher absorption and consequently, enhanced photocurrent.

The titanium suboxide (TiO_x) layer was introduced between the BHJ layer and the top metal electrode (Section 13.10). In the PCDTBT:$PC_{70}PM$ solar cells, the thickness of the active layer is about 80 nm, and the TiO_x layer thickness is about 10 nm. Note that the energy of the bottom of the conduction band of TiO_x is well-matched to that of the LUMO of PCBM.

[5] S. Sista *et al.*, *Adv. Mater.*, **21**, 1 (2009); J. Gilot *et al.*, *Appl. Phys. Lett.*, **90**, 143512 (2007).
[6] N. Blouin *et al.*, *J. Am. Chem. Soc.*,**130**, 732 (2008); N. Blouin *et al.*, *Adv. Mater.*, **19**, 2295 (2007).
[7] S. H. Park *et al.*, *Nature Photonics*, **3**, 297 (2009).

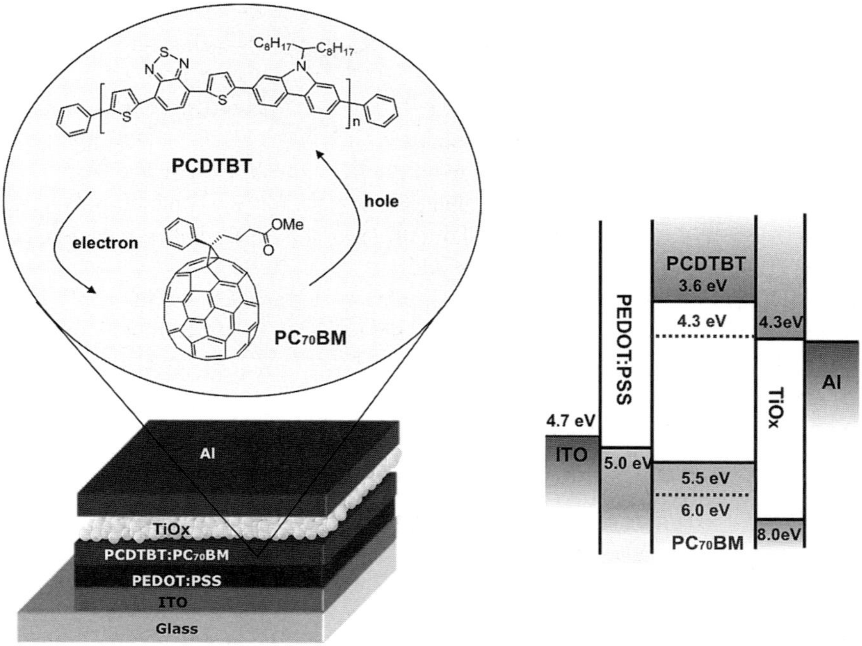

Fig. 13.23 The device structure is sketched together with energy level diagram. The titanium oxide (TiO_x) composite layer is introduced as an optical spacer on top of the BHJ layer. Note the deep HOMO energy of PCDTBT.

Finally, the relatively high electron mobility of $PC_{70}BM$ and the hole-blocking feature of TiO_x enable efficient electron collection without significant increase in the series resistance.

Figure 13.24(a) shows the absorption spectra of PCDTBT solar cells with and without the TiO_x layer. The total absorption by the active layer (including the doubled path length in the BHJ layer as a result of reflection from the Al electrode) was measured in reflection geometry as sketched in the inset of Fig. 13.24(a). Comparing two devices with the same active layer thickness (80 nm), a substantial enhancement in absorption is observed in the device with the TiO_x layer. Consequently, the device with the TiO_x layer demonstrates higher IPCE (incident photon-to-current efficiency; equivalent to the external quantum efficiency, EQE) throughout the visible range compared with the device without the TiO_x layer. Since the integration of the product of the IPCE with the AM 1.5 solar spectrum is equal to the short circuit current, the higher short circuit current (see Fig. 13.24c) obtained with the TiO_x layer is consistent with the higher IPCE values.

Measurements of the total absorption spectrum and the IPCE of the PCDTBT/$PC_{70}BM$ solar cells enable the calculation of the internal quantum efficiency (IQE) spectrum shown in Fig. 13.25(a), which indicates the total absorption of the 1:4 device and the IPCE spectrum. The PCDTBT/$PC_{70}BM$ solar cells have remarkably high IQE values, i.e. high efficiency of collected carriers per incident absorbed photon. The IQE of the PCDTBT/$PC_{70}BM$ solar cells approaches 100% around 450 nm and stays near or even above 90% throughout the entire absorption spectrum (400–650 nm). Such a high IQE is truly remarkable and indicates that

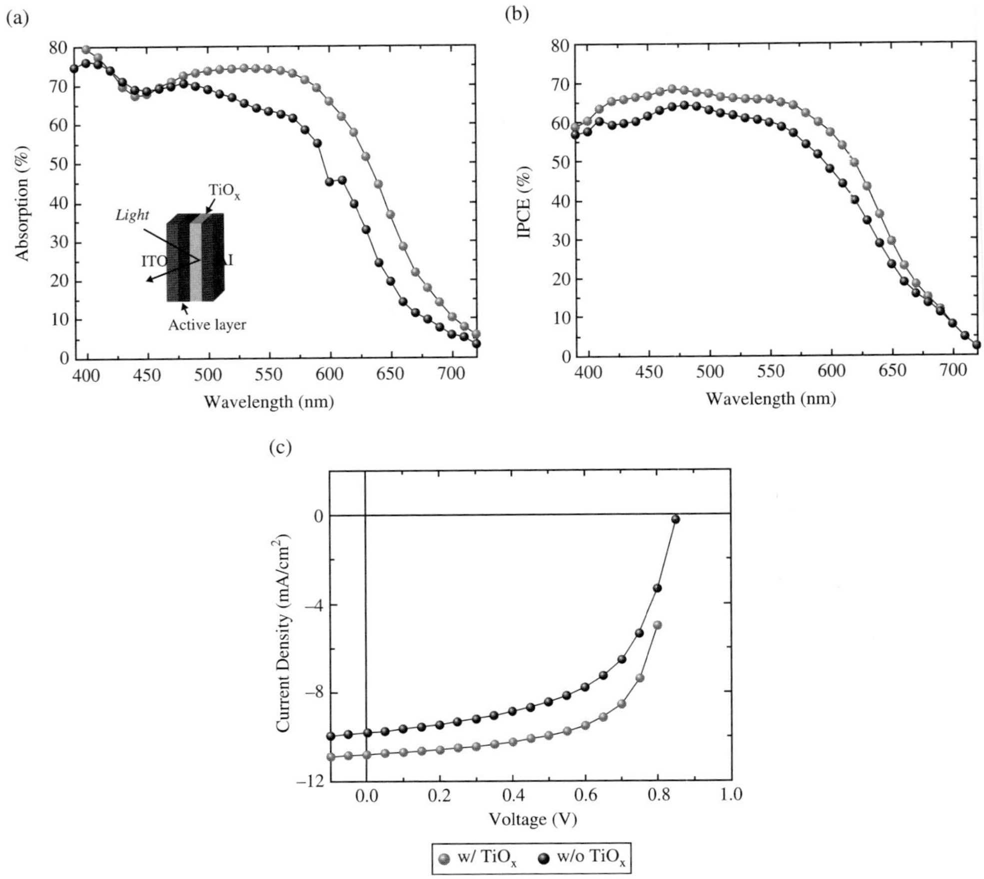

Fig. 13.24 (a) Total absorption in the active layer; (b) IPCE spectra; (c) current density versus voltage with and without TiO_x layer. [Reproduced from S. H. Park *et al.*, *Nature Photonics*, **3**, 297 (2009) with permission from Nature Publishing Group.]

nearly every absorbed photon leads to a separated pair of charge carriers and that *all* photo-generated carriers are collected at the electrodes.

The high efficiency of the PCDTBT/PC_{70}BM solar cells is further demonstrated when illuminated with monochromatic light within the absorption band (see Fig. 13.24b). Figure 13.25 (b) shows the *J-V* characteristics of a PCDTBT/PC_{70}BM BHJ device under monochromatic illumination (532 nm) with an intensity of 19.67 mW/cm^2. The power conversion efficiency is $\eta_e = 17\%$. Note that the measured 17% efficiency is consistent with IQE $\approx$ 95% at 532 nm when calculated directly from the *J-V* curve and the relatively low optical density at 532 nm (at 532 nm, approximately 30% of the incident light is reflected out through the front surface).

The observation of $\eta_e = 17\%$ when illuminated by monochromatic light within the absorption spectrum demonstrates without ambiguity that efficiencies in excess of 10–15%

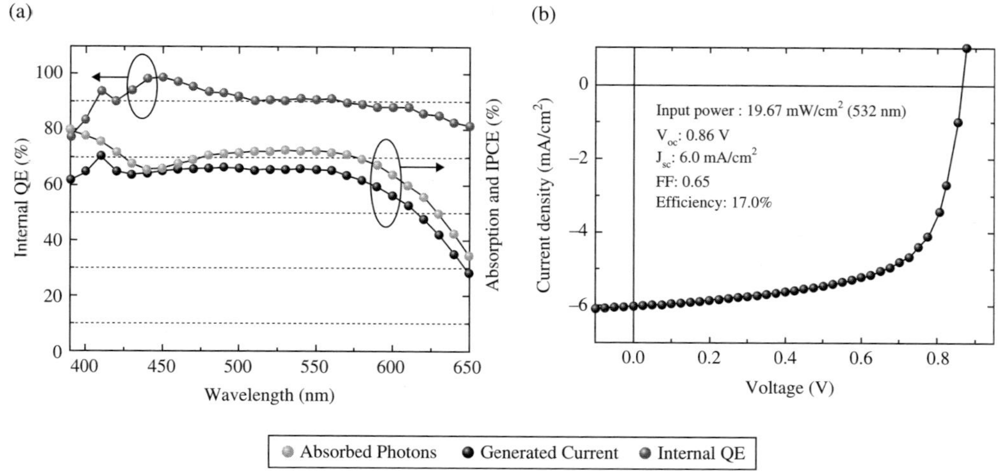

Fig. 13.25 (a): IQE values of 1:4 device; film cast from DCB. (b): *J-V* characteristics of the device under monochromatic 532 nm light. [Reproduced from S. H. Park *et al.*, *Nature Photonics*, **3**, 297 (2009) with permission from Nature Publishing Group.]

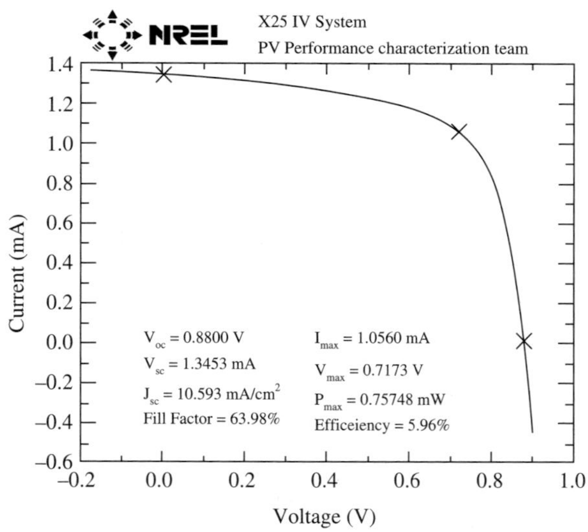

Fig. 13.26 NREL certified *J-V* characteristics of a PCDTBT:$PC_{70}BM$ solar cell; V_{oc}= 0.88 V and PCE = 6%. [Reproduced from S. H. Park *et al.*, *Nature Photonics*, **3**, 297 (2009) with permission from Nature Publishing Group.]

under AM 1.5 solar radiation should be achievable with BHJ cells fabricated using semiconducting poly-carbazole derivatives with energy gaps in the 1.2–1.4 range (see Fig. 13.17).

PCDTBT/PC_{70}BM solar cells exhibit high power conversion efficiency. Figure 13.26 shows the *J-V* characteristics of a PCDTBT/PC_{70}BM solar cell under air mass 1.5 global (AM 1.5G) illumination as certified by the National Renewable Energy Laboratory. The power conversion efficiency is $\eta_e = 6\%$. Higher molecular weight PCDTBT has increased the PCE to $\eta_e = 6.5\%$. Using a different new polymer, scientists at Solarmer Inc. have demonstrated power conversion efficiencies of approximately 8%.[8]

13.14 Recombination in PCDTBT: An example

The solar cell current, J_G, under illumination, G (in watts/cm^2), and at voltage, V, can be expressed as,

$$J_G(\mathrm{V}) = J_{Dark}(\mathrm{V}) - |\mathrm{e}|P_C(V)G_{PH} \tag{13.11}$$

where $|e|$ is the electron charge, G_{PH} is the flux of photogenerated carriers (proportional to absorbed illumination intensity) and J_{Dark} is the dark diode current. As defined in Eqn 13.11, $|e|P_C(V)G_{PH}$ is the photocurrent, so that $P_C(V)$ is the probability of charge collection as a function of bias voltage. An ideal cell has $P_C(V) = 1$ at short circuit ($\tau_r >> \tau_{sw}$). In real devices, however, $P_C(V)$ is typically reduced due to carrier recombination. As shown in Fig. 13.25, IQE > 90% over the range of photon energies that is absorbed by PCDTBT:PC_{70}BM, implying that $P_C \approx 1$ at short circuit. Values even closer to unity can be obtained under reverse bias of a few tenths of a volt.

The intensity dependence of the current versus voltage curves is shown in Fig. 13.27. According to Eqn 13.11, one can obtain $P_C(V)$ by subtracting off the dark current and then

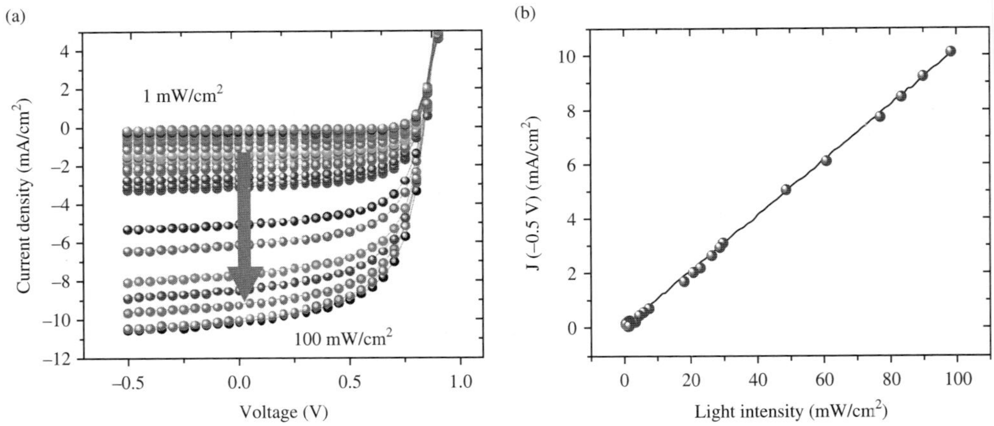

Fig. 13.27 (a) *j* versus *V* curves for a series of light intensities increasing from 1 mW/cm^2 to 100 mW/cm^2; (b) *j* (–0.5 V) versus light intensity from 1 mW/cm^2 to 100 mW/cm^2. [S. Cowan, unpublished data.]

[8] Y. Liang *et al., Adv. Mater.*, **22**, 1 (2010).

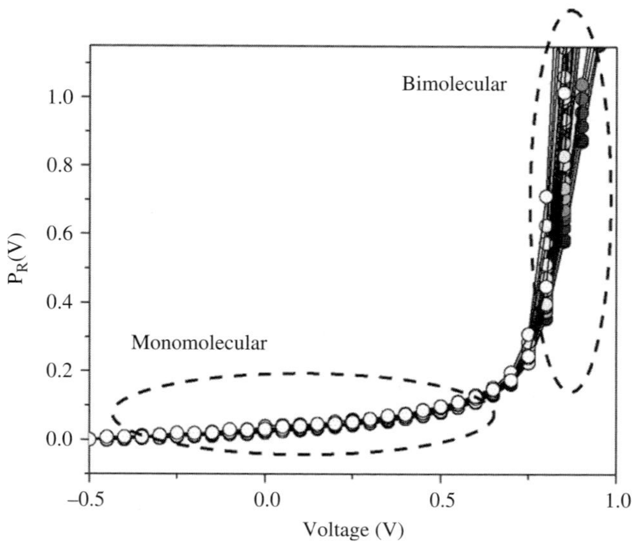

Fig. 13.28 Recombination probability, $P_R(V) = 1 - P_C(V)$, versus V. All the data from Fig. 13.27(a) collapse onto a single curve. [S. Cowan, unpublished data.]

dividing each of the j versus V curves by the light intensity used for that j versus V curve. The resulting recombination data, $P_R(V) = 1 - P_C(V)$, are shown in Fig. 13.28. The values of $P_R(V)$ run from 0 (IQE $\approx$ 1 at $V = 0$) to 1 (photocurrent is zero $V = V_{oc}$). Note that since the photoluminescence has been heavily quenched, nonradiative recombination processes must dominate. Figure 13.28 demonstrates that all the data from Fig. 13.27(a) collapse onto a single curve for PCDTBT:PC_{70}BM, although bimolecular recombination appears to become important as V approaches V_{oc} (some intensity dependence is observed very near V_{oc}). Thus, the recombination probability, $P_R(V)$, is independent of the light intensity except very near V_{oc}.

The voltage dependence of the recombination probability determines FF and is therefore critically important to the goal of achieving high power conversion efficiency. The most important parameter is the product $\mu\tau_r$; increasing $\mu\tau_r$ decreases the series resistance in the equivalent circuit model (Fig. 13.3) and enables efficient sweep-out.

In order to obtain an understanding of the voltage dependence of $P_R(V)$, Street *et al.*[9] considered the various recombination mechanisms shown in Fig. 13.29. The Auger process was ruled out because it would have to take place prior to charge transfer (times < 100 fs), and it is not strongly voltage dependent. Nongeminate is inconsistent with Fig. 13.28 (except very near V_{oc}); the probability of nongeminate recombination is proportional to the product np, where n and p are the densities of photoexcited electrons in the PC_{70}BM and photoexcited holes in the PCDTBT, respectively. Thus, nongeminate recombination would be "bimolecular" recombination and strongly dependent on the intensity. Recombination via geminate interfacial exciplexes or via interfacial trap states are both consistent with the data in Fig. 13.29.

[9] R. A. Street *et al.*, *Phys. Rev. B*, **17**, 3984 (1978).

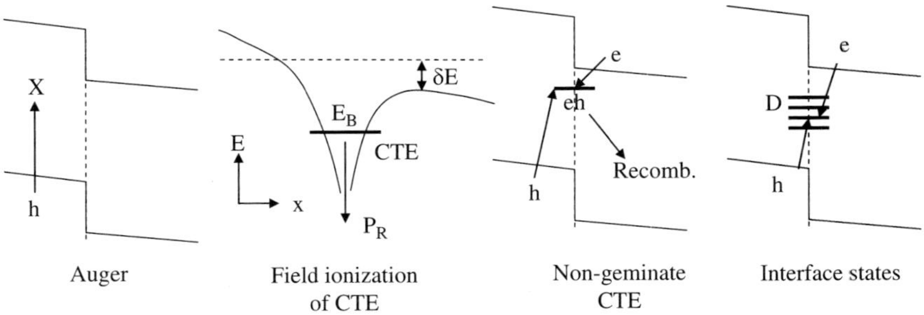

Fig. 13.29 Illustration of four possible bias-dependent recombination mechanisms.

Street *et al.*[10] argued, however, that the temperature dependence is inconsistent with the geminate interfacial exciton mechanism and conclude that for PCDTBT: $PC_{70}BM$, interfacial traps dominate the recombination. As noted in Section 13.7, the interfaces are far from perfect, implying that a relatively high density of interfacial traps is to be expected.

This kind of analysis of the recombination mechanisms is important and the answer might well be different for different BHJ materials. If indeed interfacial traps are important, then one might hope to be able to reduce the trap density by improving the interfaces within the phase separated nanomorphology.

13.15 Photodetectors

Diodes fabricated from bulk heterojunction materials are promising as photodetectors. The very high resistivity of such thin film devices in the dark is a major advantage for the detectivity (low dark current). In photodetector applications, the collection efficiency of the photogenerated carriers can be significantly higher than in photovoltaic cells because of the applied voltage bias. As photodetectors, BHJ diodes show performance that is competitive with that of commercial silicon detectors within the spectral regime in which they operate (Fig. 13.30).

To demonstrate the potential of the bulk heterojunction as a photodetector, a linear photodiode array was fabricated for use as a full color scanner. The scanner consists of a series of 120 photodiodes, linearly placed adjacent to each other and fabricated using a simple etching process to structure the transparent conducting electrode. All the other layers are continuous. Since the diode array can be considered as a series of capacitors with thickness of approximately 100 nm and with an interpixel spacing of several microns, the electric field is confined to the area defined by the patterned ITO electrode with minimal spreading of the electric field from pixel to pixel.

To make the color-sensitive recording of an image, red, green and blue color filters were used. After consecutive recording of the same image with different color filters, a full-color image was reconstructed by superimposing the color selected photodetector signals. The image is reproduced in Fig. 13.31.

[10] R. A. Street *et al.*, *Phys. Rev. B* (in communication 2010).

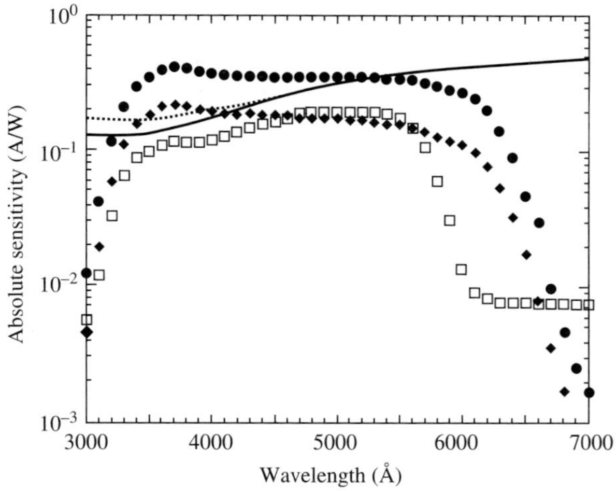

Fig. 13.30 The absolute sensitivity of bulk heterojunction diodes as photodetectors is compared to that obtained from UV-enhanced Si. (♦): Ca/MEH-PPV:C_{60}/ITO at –10 V; (●): Au/P3OT/ITO at –15 V; (□): Au/P3OT/ITO at –10 V. [Reproduced from G. Yu *et al.*, *Appl. Phys. Lett.*, **64**, 3422 (1994) with permission from The American Physical Society.]

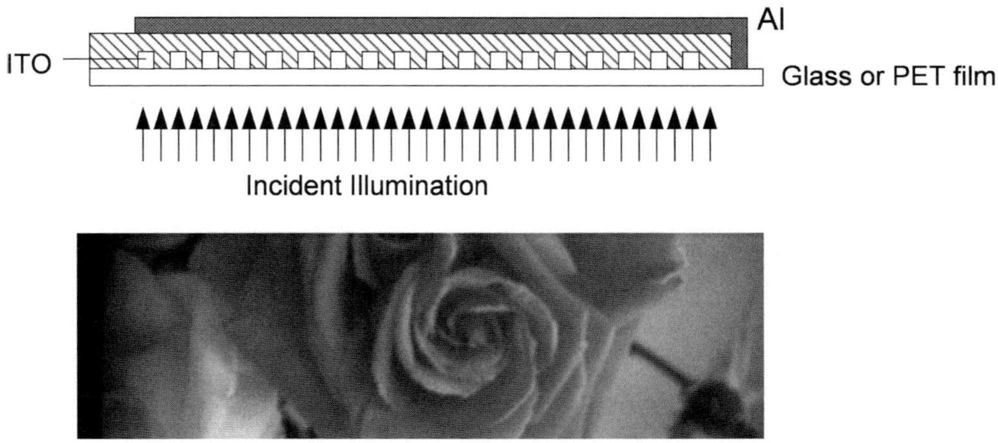

Fig. 13.31 Top: a linear array of photodiodes for use as a scanner. The BHJ material was P3HT:PCBM. Bottom: color image recorded by the scanner described above. (This figure is reproduced in color in the color plate section.)

Polymer photodetectors, processed from solution and operating at room temperature, have been demonstrated to exhibit high detectivity over the remarkably wide spectral range extending from the UV (300 nm) well into the infrared out to 1450 nm (see Fig. 13.32). Over this entire spectral range, the detectivities are greater than 10^{12} cm $Hz^{1/2}$/W.

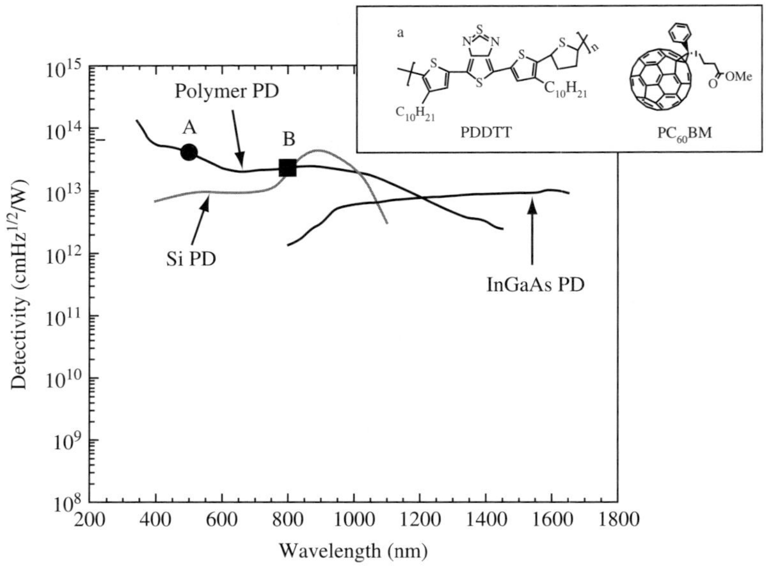

Fig. 13.32 Detectivities of a Si photodetector, InGaAs photodetector and polymer photodetector versus wavelength. The inset shows molecular structures of PDDTT and PC60BM. [Reproduced from X. Gong *et al.*, *Science*, **325**, 1665 (2009) with permission from AAA.]

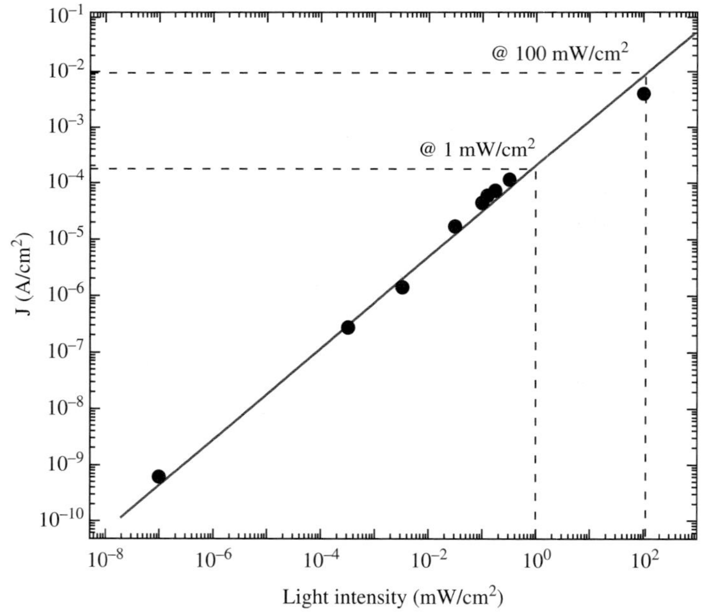

Fig. 13.33 Linear dynamic range of the polymer photodetectors with ITO/PS-TPD-PFCB/PDDTT: $PC_{60}BM/C_{60}/Al$ architecture. [Reproduced from X. Gong *et al.*, *Science*, **325**, 1665 (2009) with permission from AAA.]

The linear dynamic range (LDR) or photosensitivity linearity (typically quoted in dB) is another important figure of merit for photodetectors.

$$\text{LDR} = 20\log(J*_{ph}/J_d), \tag{13.12}$$

where J^*_{ph} is the photocurrent, measured at light intensity of 1 mW/cm^2. Figure 13.33 shows the photocurrent versus the light intensity for the polymer photodiode structure. Under illumination at $\lambda = 800$ nm, the LDR is more than 100 dB which is close to that of a Si photodetector (120 dB) and significantly higher than that obtained from InGaAs photodetectors (66 dB).

The achievement of high-performance polymer photodetectors implies that semiconducting polymers can be used in a variety of possible sensor applications. The photoresponsivity and detectivity results open exciting opportunities for the creation of detectors with unusually wide spectral range and for the fabrication of high-resolution detector arrays for optical communications, chemical/biological sensing and day- and night-time surveillance.

Further reading

C. J. Brabec, N. S. Sariciftci, and J. C. Hummelen, *Adv. Funct. Mater.*, **11**, 15 (2001).

C. Brabec, V. Dyakonov, J. Parisi and N. S. Sariciftci (eds.), *Organic Photovoltaics*, Springer, Berlin, 2003.

C. Brabec, V. Dyakonov and U. Scherf (eds.), *Organic Photovoltaics*, Wiley–VCH, Weinheim, 2008.

G. Dennler, N. S. Sariciftci and C. J. Brabec, in: *Semiconducting Polymers*, edited by G. Hadziioannou and G. G. Mallarias, Wiley-VCH, Weinheim, 2007, Chapter 11.

G. Dennler, M. C. Scharber and C. J. Brabec, *Adv. Mater.*, **21**, 1323 (2009).

A. Moliton, *Optoelectronics of Molecules and Polymers*, Springer, Berlin, 2006.

A. J. Mozer and N. S. Sariciftci, in: *Handbook of Conducting Polymers*, 3rd edition, edited by T. Skotheim and J. R. Reynolds, CRC Press, Taylor and Francis Group, Boca Raton, 2007, Chapter 10.

H. S. Nalwa (ed.), *Handbook of Organic and Conductive Molecules and Polymers*, John Wiley, Chichester, 1997.

T. Soga (ed.), *Nanostructured Materials for Solar Energy Conversion*, Elsevier, Amsterdam, 2006.

S.S. Sun and N. S. Sariciftci (eds.), *Organic Photovoltaics: Mechanisms, Materials and Devices*, Taylor and Francis, Boca Raton, 2005.

S. S. Sun and L. Dalton (eds.), *Introduction to Organic Electronic and Optoelectronic Materials and Devices*, Taylor and Francis, Boca Raton, 2008.

P. Würfel, *Physics of Solar Cells*, Wiley–VCH, Weinheim, 2005.

14
Polymer field effect transistors (FETs)

14.1 Introduction

A field effect transistor (FET) is a three-terminal device, an electronic "relay". The electrical current that flows in the channel between the source and drain contacts, I_{ds}, is a function of the source drain voltage (V_{ds}), but it is controlled (turned on and off) by the gate voltage (see Fig. 14.1). The principle of the field-effect was first proposed by J. E. Lilienfeld in 1930 in a patent.[1] Two decades later, in 1947, the p-n-p transistor was introduced by W. Shockley, J. Bardeen and W. Brattain. In 1956 they were awarded the Nobel Prize for their discovery of the transistor. The FET is the fundamental building block of the circuits in microprocessors, and today's modern devices. The first silicon-based metal-oxide-semiconductor field effect transistor (MOSFET) was fabricated by Kahng and Atalla in 1960.[2]

A FET is basically a capacitor with mobile charges. By applying a voltage between the gate and the source/drain electrodes, the surface of the semiconductor next to the dielectric becomes an equipotential surface, and charges are induced at the semiconductor–dielectric interface. Because the charges in the semiconductor are mobile, current will flow in the channel when a small voltage is applied between the source and drain.

Organic field effect transistors (OFETs) are essentially identical to inorganic thin-film transistors (TFTs) based on the metal–insulator–semiconductor field-effect (MISFET) architecture. The first organic FET was reported by Koezuka and coworkers in 1986[3] using polythiophene as the semiconductor. In the past few decades impressive progress has been made in developing both p-channel and n-channel semiconducting polymers. OTFT based on polymers typically exhibit relatively low hole and electron mobilities, $< 1\ \mathrm{cm^2\ V^{-1}\ s^{-1}}$ and often well below $10^{-2}\ \mathrm{cm^2\ V^{-1}\ s^{-1}}$ at ambient conditions. Despite the low mobilities, organic FETs open up the possibility of compact, low-cost, flexible devices. OTFTs are potentially useful in flexible active matrix OLED display, back-planes for e-paper, wireless smart cards (RFID tags), wearable computers, and artificial nose (sensors). The goal of this chapter is to introduce the OTFT and light-emitting FET device concepts with emphasis on FETs fabricated with semiconducting polymers as the material in the channel. We will focus on the fundamental physics and the device science.

[1] J. E. Lilienfeld, US Patent, 1745 175 (1930).

[2] D. Kahng and M. M. Atalla, IRE Solid-State Devices Research Conference, Carnegie Institute of Technology, Pittsburgh, PA (1960).

[3] A. Tsumura, *et al.*, *Appl. Phys. Lett.*, **49**, 1210 (1986).

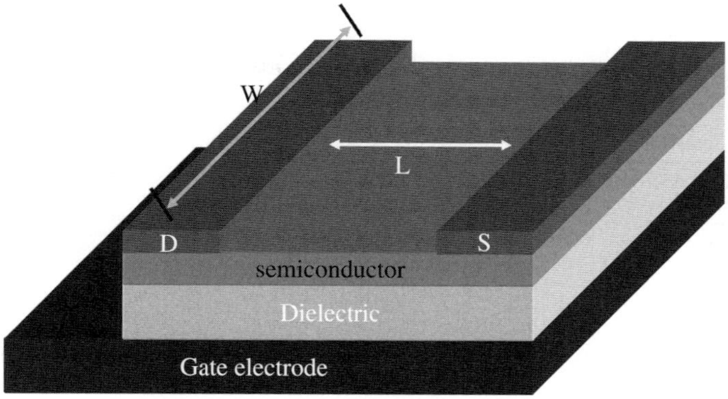

Fig. 14.1 The structure of an organic field effect transistor.

14.2 Operating principles of OFETs

The basic structure of the OFET is shown in Fig. 13.1. It is a three-terminal device with the terminals designated as gate, source and drain. The gate dielectric is made of a thin film of organic or inorganic insulating material sandwiched between the gate electrode and the semiconducting polymer. The important device parameters are the channel length (L), which is the distance between the two metal electrodes (source and drain), the channel width (W), and the capacitance per unit area (C).

OTFTs are primarily operated in the accumulation mode and the organic semiconductor employed in the FET structure is intrinsically in an undoped state as opposed to MOSFET where materials are intentionally doped and operating modes can be inversion (and accumulation).

In principle, both p- and n-type operation is possible in OTFTs depending on the injection barrier at the source/drain contacts, the operating voltage and the stability of the semiconducting polymer. For example, because P3HT is unstable with electrons in the π*-band, P3HT cannot be used as the material in n-channel FETs. An organic semiconductor is said to be n-type when it is electronegative; i.e. the electron affinity is large, implying that the energy of the bottom of the π*-band (the LUMO) is at a relatively low energy. An n-type organic semiconductor is an acceptor material. Similarly an organic semiconductor is said to be p-type when it is electropositive, i.e. the ionization potential is large, implying that the energy of the top of the π-band (the HOMO) is at a relatively high energy. A p-type organic semiconductor is a donor material.

When there is no gate voltage, all Fermi levels are lined up and the bands in both the dielectric and semiconducting polymer are flat (see Fig 14.2a). This is called the flat band condition. When a negative voltage (V) is applied to the gate dielectric, this raises the Fermi level of the gate electrode towards the higher energy and creates an electric field in the dielectric. This electric field bends the HOMO and LUMO energy levels of the semiconductor to self-consistently enable charge injection. Due to band bending, the HOMO energy level comes much closer to the Fermi level (of the semiconductor) at the interface and thus leads to the accumulation of the holes at the interface. The accumulated carrier density of holes

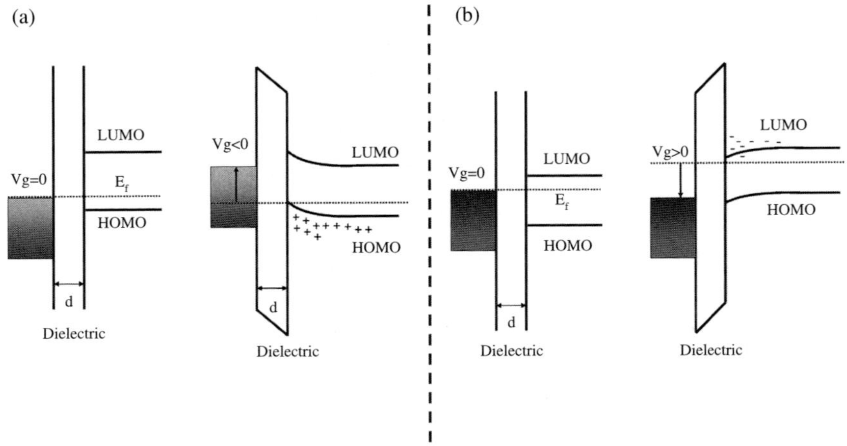

Fig. 14.2 Band diagram and operating modes of OTFT with applied gate voltages: (a) p-type operation; (b) n-type operation. The band bending occurs over a length determined by the residual impurities (see Chapter 8).

depends exponentially on the energy difference between the HOMO and the Fermi energy. If the Fermi level of the source/drain metal is close to the HOMO level of the semiconductor, then holes can be injected and extracted efficiently by applying a voltage between the source and drain electrodes. The amount of the accumulated charge at the semiconductor/dielectric interface can be measured from the gate capacitance and the gate voltage.

When a positive gate voltage is applied to the OTFT, bands bend downwards and the LUMO energy level comes much closer to the Fermi level (of the semiconductor) at the interface. By selecting source drain electrodes with work function closer to the LUMO and applying a second independent bias source to the drain (the source is assumed to be grounded), electrons (or electron polarons) can be injected and transported between the source and drain.

The charge induced in the accumulation layer of the semiconductor by the gate voltage is given by the simple formula

$$Q = CV_g \tag{14.1}$$

where C is the capacitance per unit area of the parallel plate capacitor associated with the gate dielectric insulator, $C = \varepsilon_o\varepsilon(A/d)$, ε_o is the permittivity of free space, d is the thickness of the insulator and $\varepsilon_R$is the dielectric constant (relative permittivity of the insulator). Thus, the gate voltage induces a charge density per unit area in the accumulation layer of the semiconductor,

$$n = \epsilon_o\epsilon(V_g/ed) \tag{14.2}$$

where e is the electron charge. These gate-induced charges are confined by the electric field from the gate voltage to a few monolayers within the semiconductor near the interface between the semiconductor and the gate dielectric.

Typically, however, there is a finite density of electronic traps, n_t, per unit area in the semiconductor near the interface with the gate dielectric. Thus, a minimum gate voltage, the threshold voltage, is required to fill these traps:

$$V_{\text{th}} = (ed/\epsilon_o\epsilon)n_t \tag{14.3}$$

The density of mobile carriers per unit area in the accumulation layer is, therefore,

$$n_{\text{m}} = \epsilon_o\epsilon(V_g - V_{th})/ed \tag{14.4}$$

The mobile charge density given by Eqn 14.4 leads to conductivity in the channel; when a small voltage is applied between the source and drain contacts, current flows from source to drain. To enable current injection and collection, the source and drain contacts should make good "ohmic" electrical contact to the semiconducting material.

The source–drain current as a function of the source drain voltage is sketched in Fig. 14.3 which shows the three regimes of FET operation.

Figure 14.3 assumes that the gate voltage is set to a constant value larger than the threshold voltage ($V_g > V_{Th}$). For $V_{sd} << V_g$, the mobile carrier density is uniform across the channel (see Fig 14.4). In this linear regime, the channel functions like an ohmic resistor as a result of the gate-induced mobile carriers; i.e. I_{ds} is proportional to V_{ds} (e.g. the "linear" regime), and the current is approximately determined from the following equation:

$$I_{DS} = \frac{1}{2}\mu C\frac{W}{L}(V_g - V_T)V_{ds} \tag{14.5}$$

An important parameter in the operation of a transistor is the field-effect mobility, μ [cm^2 V^{-1} s^{-1}] of the semiconductor in the channel. In the linear regime, the mobility can be calculated from the transconductance, *gm*, defined as:

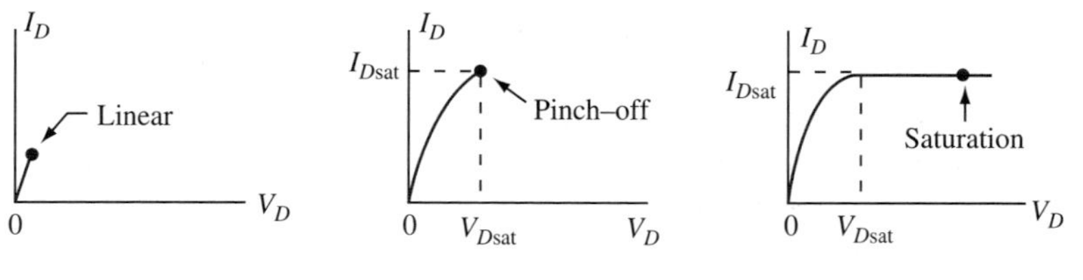

Fig. 14.3 I_{ds} as a function of V_{ds} in the linear regime, at the pinch-off point and in the saturation regime.

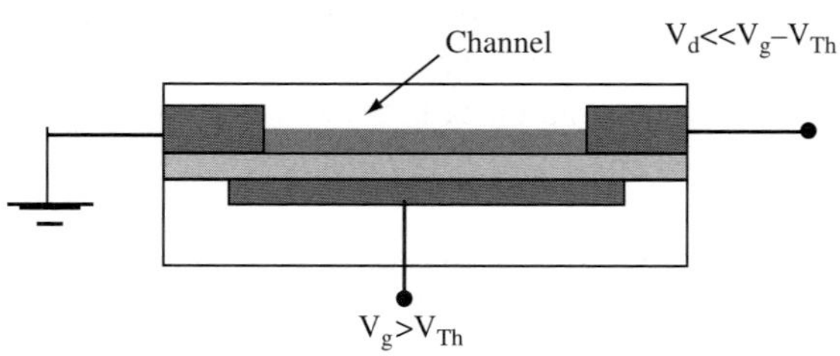

Fig. 14.4 Cross-section of an OFET device in the linear I_{ds} regime. The dark region shows the conductive channel between the source and drain electrode.

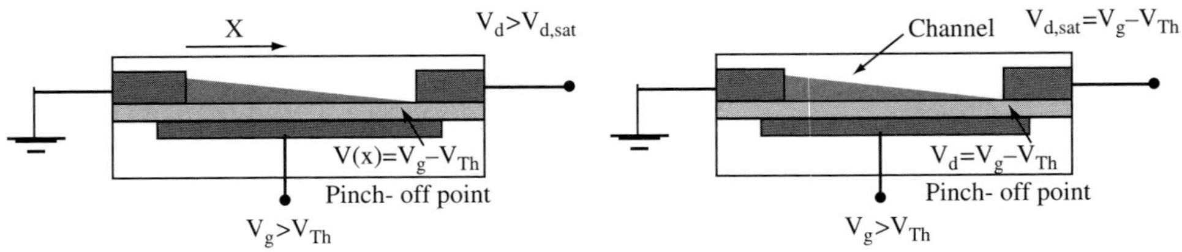

Fig. 14.5 Left: at $V_d = V_g - V_T$ the conductive channel is pinched off by the depletion layer at the drain electrode. Right: at $V_d > V_g - V_{Th}$, the pinch-off point moves towards the source–electrode, and the conductive channel is pinched off over larger distances in the channel.

$$gm = \left(\frac{\delta I_{ds}}{\delta V_g}\right) V_{ds=constant} = \mu C \frac{W}{L} V_{ds} \tag{14.6}$$

When V_{ds} is increased further to values approaching V_g, the carrier density is no longer uniform across the channel. As a result, the channel resistance increases and the current versus V_{ds} becomes sublinear. When the source–drain voltage is increased still further, $V_{ds} > V_g - V_{Th}$, the mobile carrier density within the channel is "pinched off"; i.e. a depleted region is formed near the drain contact where the density of mobile carriers is very small (see Fig 14.5). Because only a space charge limited saturation current $I_{d,sat}$ can flow across the depletion zone, $I_{d,sat}$ remains constant beyond the pinch-off point.

In the saturation regime the current is given as

$$I_{DS} = \frac{1}{2} \mu C \frac{W}{L} (V_{GS} - V_T)^2 \tag{14.7}$$

An example of the current–voltage characteristics of such a poly(2,5-thienylenevinylene) (PTV) OFET are shown in Fig. 14.6. The threshold voltage V_{th} and the carrier mobility can be estimated from the "transfer characteristics" by plotting $\sqrt{I_{ds}}$ versus V_g (see Eqn 14.7).

The current on–off ratio, $I_{ds,on}/I_{ds,off}$, and the field effect mobility μ are the critical device performance parameters. In terms of integrated circuits μ determines the maximum switching frequency f_s of the transistor:

$$f_s = \frac{\mu_0}{L^2} \cdot \frac{1}{(V_g - V_{Th})} \tag{14.8}$$

The on–off ratio $I_{ds,on}/I_{ds,off}$ is a measure of the quality of the FET as an electronic switch; $I_{ds,on}/I_{ds,off}$ is determined by the purity of the semiconductor (residual doping induces mobile carriers even when $V_g = 0$) and the leakage current across the gate insulator, I_{gs}.

14.3 Different architectures used in organic FETs

Advanced organic FET structures use solution-processible polymers or small-molecule systems for the semiconductor layer. A variety of insulating polymers have been utilized as the gate dielectric material. These insulating polymers can also be processed from solution.

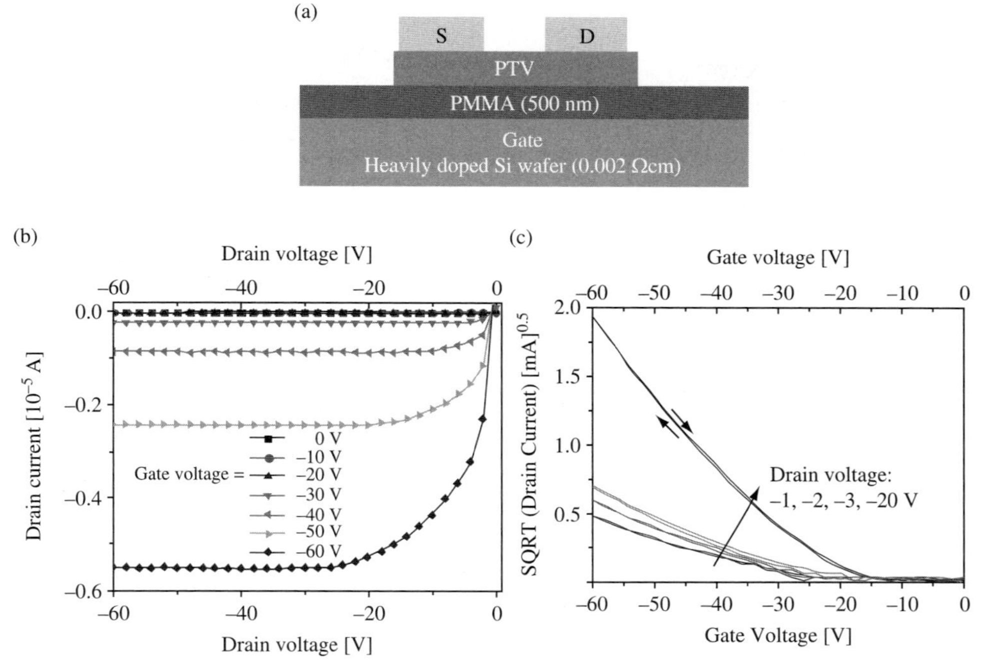

Fig. 14.6 (a) Device structure of an organic FET built on a silicon gate with poly methylmethacrylate (PMMA) as the gate insulator. (b) Output characteristics of (drain–source voltage, V_{ds}, versus drain–source current, I_{ds}) of a p-channel polymer FET. (c) Transfer characteristics ($\sqrt{I_{ds}}$ versus gate voltage, V_g for different V_{ds}) with an arrow indicating the forward and reverse direction of sweeping of V_g. The linear field effect mobility of 7×10^{-4} cm^2/Vs and saturated mobility of 1×10^{-3} cm^2/Vs can be extracted from these devices with channel length, L, of 25 μm and channel width, W, of 1.4 mm ($W/L = 40$).

Although OFETs fabricated with organic semiconductors and organic polymer dielectrics show relatively good performance, in some cases approaching that of amorphous Si, the source–drain current, the on–off ratio and the gate voltage needed to reach saturation are limited by the mobility of the organic semiconductor. For specific product applications, organic FETs offer the promise of cost-effective production by implementing large-scale printing techniques.

The use of organic semiconductors and organic dielectrics enables the fabrication of OFETs in several device geometries (see Fig. 14.7).

The most commonly used geometry is the bottom-gate top-contact configuration (Fig. 14.7a) which resembles the architecture of thin-film silicon transistors comprising a thermally grown SiO_2 insulator on a heavily doped silicon substrate as the gate. However, the ability to use organic dielectrics as gate insulators enables the possibility of bringing the gate insulator polymer on top of the organic semiconductor layer. It may be advantageous to embed the

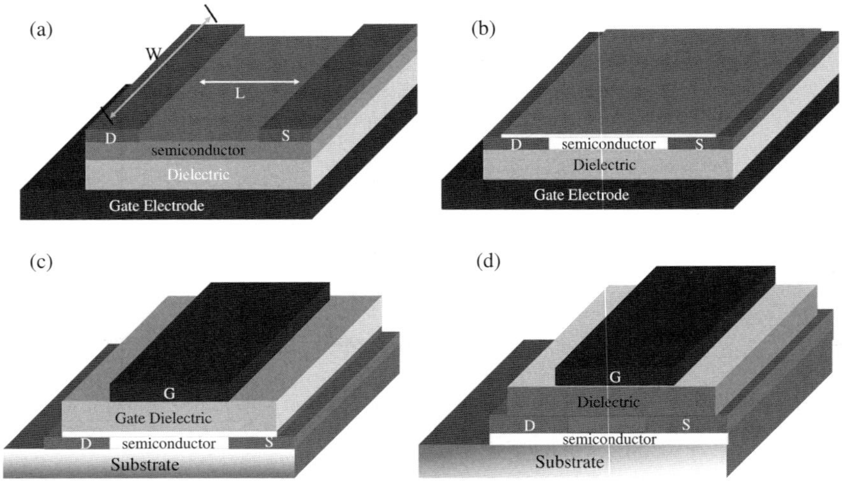

Fig. 14.7 Schematics of p-type semiconductor OFETs with (a) bottom-gate top-contact (BG/TC), (b) bottom-gate bottom-contact (BG/BC), (c) top-gate bottom-contact (TG/BC) and (d) top-gate top-contact (TG/TC) structures.

organic semiconductor and the source–drain contacts underneath the gate insulator which can then function to protect the device against environmental degradation.

14.4 Contact resistance and errors in the mobility measurements

In an ideal FET, the contacts between the semiconductor and electrodes (source/drain) are ohmic; the value of the contact resistance is negligible compared to the total channel resistance. Contact resistance in OTFTs can lead to serious errors in mobility measurements. As an example, we focus on data obtained from n-channel OFETs fabricated with the soluble fullerene, PC61BM and PC71BM, as the semiconductor and with injection of a titanium suboxide (TiO_x) layer. The device structure is like that of Fig. 14.7(a), but with a thin layer of TiO_x on top of the semiconductor; the source and drain are deposited onto the TiO_x layer.

Figure 14.8 shows the total resistance between the source and drain (R_{total}) as a function of the channel length (L). Since the channel resistance is proportional to L, the contact resistance (R_c) can be obtained from extrapolation of R_{total} versus L to zero channel length. The results show that devices with a TiO_x layer have lower R_{total}-values and that R_{total} extrapolates (at L=0) to $R_c \approx 0.3$ MΩ with PC61BM as the semiconductor in the channel and to $R_c \approx 0.4$ MΩ with PC71BM as the semiconductor in the channel. Note that the slope of the R_{total} versus L line did not change by introducing the TiO_x layer in either PC61BM FETs or PC71BM FETs, implying that the true channel resistances are not affected by the addition of the TiO_x layer.

Comparison of device characteristics fabricated with and without a TiO_x injection layer indicates that insertion of a TiO_x layer yields an increase in I_{ds} by approximately an order of magnitude to values approaching those obtained using Ca electrodes. The lower I_{ds} values obtained

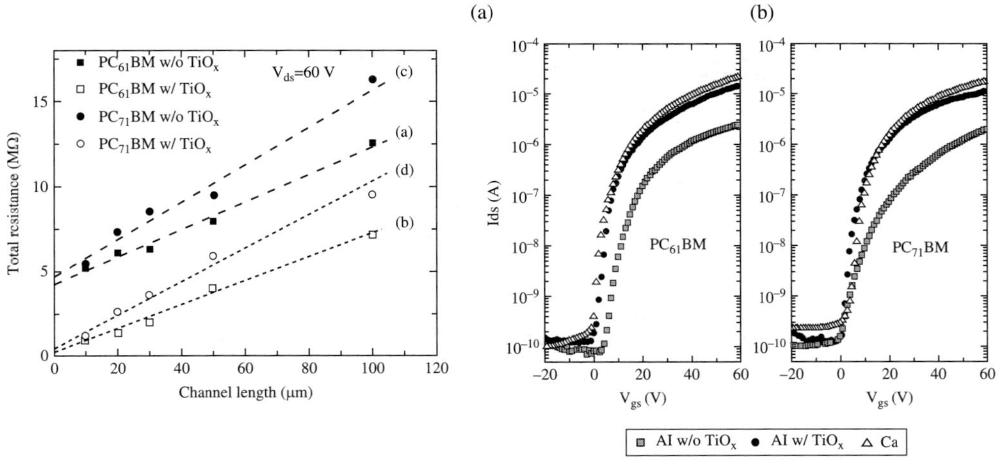

Fig. 14.8 Left: total resistance between the source and drain electrodes ($R_{total}=R_c+R_{ch}$) as a function of the channel length (L). (a) $PC_{61}BM$ FET without TiO_x injection layer (■), (b) $PC_{61}BM$ FET with TiO_x injection layer (□), (c) $PC_{71}BM$ FET without TiO_x injection layer (●), and (d) $PC_{71}BM$ FET with TiO_x injection layer (○). Right: transport characteristics of the n□type (a) $PC_{61}BM$ FET and (b) $PC_{71}BM$ FET with (●) and without (■) TiO_x injection layer. The results obtained from the devices fabricated with Ca electrodes (▲) are given for comparison. [Reproduced from S. Cho *et al.*, *Adv. Funct. Mater.*, **19**, 1459 (2009) with permission from Wiley□VCH Verlag GMBH.]

without the TiO_x layer result from the high contact resistance, rather than from the apparently reduced electron mobility that would be calculated from Eqn 14.7. The origin of the error is in the voltage drop between source and drain; Eqn 14.7 assumes that the contact resistance is zero. If R_c is large, the voltage distribution across the channel is seriously perturbed. As a result, when R_c is large, Eqn 14.7 is not valid and one simply cannot infer a mobility from a plot of $(I_{ds})^{1/2}$ versus V_s.

For the devices without a TiO_x layer, the measured R_c-values increase with increase in the work function of the electrode material, implying an increase in the height of the energy barrier between the fullerene semiconductor and the metal contact (see Fig. 14.9). For FETs with a TiO_x layer, R_c-values are lower, and the sensitivity to the metal work function nearly disappears (Au gives a slightly higher R_c than the other metals). With the TiO_x injection layer, the measured mobilities are relatively insensitive to the work function of the source–drain metals and, again, close to the maximum values obtained from devices fabricated with Ca electrodes. The "apparent mobility" values obtained without TiO_x with different work function electrodes demonstrate the potential for serious error (approximately an order of magnitude).

Although we utilized the TiO_x layer as an example of a method to reduce the contact resistance, the point made here is very general. Any report of mobility values from a new material must demonstrate that the contact resistance is negligible. The potential errors are large, an order of magnitude or more.

14.5 Polarons as the gate induced carriers in OFETs

As would be expected from the discussions presented in Chapters 3, 4 and 8, the charge carriers induced in the accumulation layer of the semiconducting polymer by the field-effect

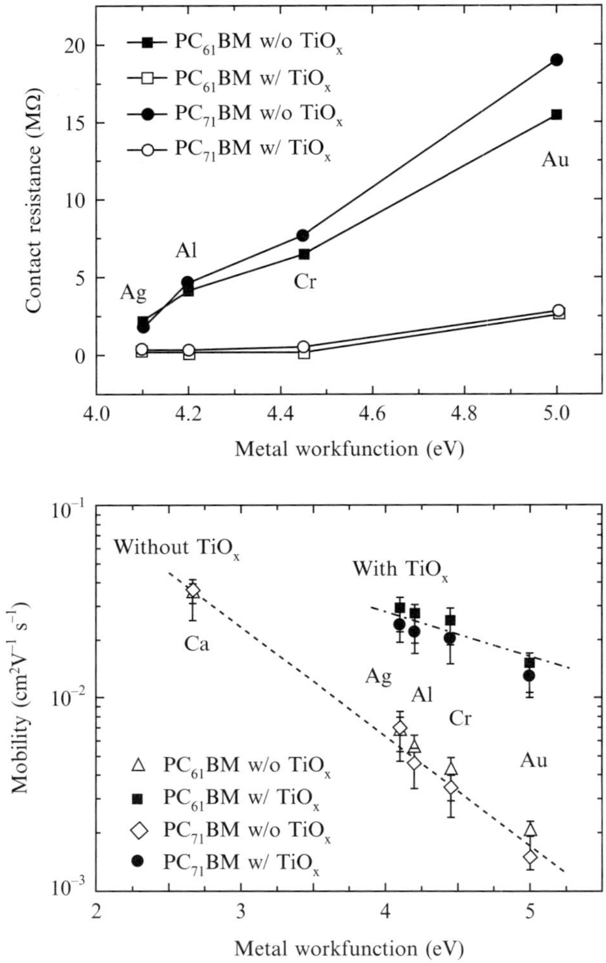

Fig. 14.9 Top: contact resistance of the n-type PC61BM FET and PC71BM FET (with and without TiO_x) fabricated with various source and drain electrodes. Bottom: electron mobilities of the n-type $PC_{61}BM$ FET and $PC_{71}BM$ FET (with and without TiO_x) fabricated with various source and drain electrodes. Because of the high contact resistance, the mobilities obtained from Eqn 14.7 for fabricated FETs are not valid. The apparent values obtained without TiO_x are shown only to demonstrate the potential for serious error (approximately an order of magnitude). [Reproduced from S. Cho *et al.*, *Adv. Funct. Mater.*, **19**, 1459 (2009) with permission from Wiley-VCH Verlag GMBH.]

(i.e. the gate voltage) are polarons. This was proven by direct spectroscopic measurements of the changes in IR transmission through the accumulation layer in OFETs fabricated from regio-regular poly(3-hexyl thiophene).[4]

[4] Z. Q. Li *et al.*, *Nano Lett.*, **6**, 224 (2006).

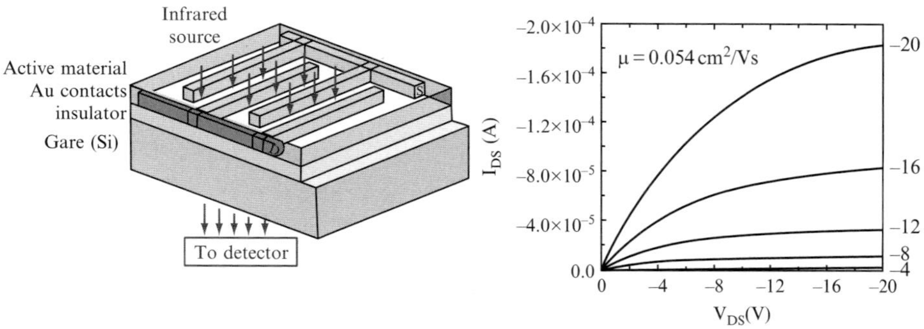

Fig. 14.10 Left: schematic of a FET device in the bottom contact geometry for infrared characterization of the injected charges. Right: the semiconductor used was P3HT; the gate dielectric was either SiO_2 or TiO_2. TiO_2 was used as the gate dielectric. [Reproduced from Z. Q. Li *et al.*, *Nano Lett.*, **6**, 224 (2006) with permission from the American Chemical Society.]

Figure 14.10 shows schematics of the FET device used for the IR measurements (left) and the corresponding data of I_{ds} versus V_{ds} at different gate voltages (right). Both SiO_2 and TiO_2 (sputtered onto the Si substrate) were used in different devices for the gate dielectric.

Figure 14.11 shows the voltage induced absorption spectra, $\Delta\alpha d = 1 - T(V_g)/T(0)$ obtained with the device configuration shown in Fig. 14.10. Here, $\Delta\alpha$ is the change in the absorption coefficient of the P3HT with applied gate voltage and d is the thickness of the accumulation layer.

All spectra show the characteristic fingerprints of polaron formation: IRAV modes in the 1000–1500 cm^{-1} range and the polaron band at 3500 cm^{-1}. The oscillator strength of $\Delta\alpha d$ from the polaron band and the IRAV modes increases with gate voltage. Detailed analysis of the data shows that for the SiO_2 gated devices, the increase in oscillator strength (proportional to the density of carriers in the two-dimensional accumulation layer) increases linearly up to the highest gate voltages (100 V) consistent with the capacitance model as expressed by Eqn 14.4. For the TiO_2 gated devices a tendency toward saturation occurred as a result of leakage through the dielectric.

Typically in the OFET literature, the authors refer to electron or hole injection and to p-type or n-type operation. The data in Fig. 14.11 demonstrate that the transport in OFETs results from the injection of electron-polarons and hole-polarons.

14.6 Charge transport at high carrier density

Charge transport in FETs at high gate voltages is of interest in the general context of the insulator-to-metal transition. Can one achieve a sufficiently high gate-induced carrier density to reach the metallic state? This would be especially interesting because with field-effect injection, there are no counter-ions.

Poly(2,5-bis(3-tetradecylthiophen-2-yl)thieno[3,2-b]thiophene) (PBTTT) is considered one of the most promising p-type semiconducting polymers. Thin films are composed of

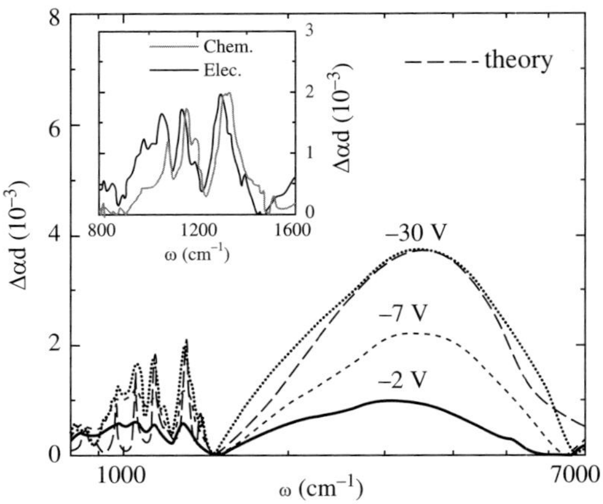

Fig. 14.11 Voltage-induced absorption spectra $\Delta\alpha d$ for the P3HT layer under applied gate voltages in a device with TiO_2 as the gate dielectric. The inset shows the $\Delta\alpha d$ spectrum at $V_G = -30$ V along with the data from chemically doped P3HT (the latter plotted as the difference between the absorption of chemically doped P3HT with 1% PF_6^- and that of pure P3HT). [Reproduced from Z. Q. Li *et al.*, *Nano Lett.*, **6**, 224 (2006) with permission from the American Chemical Society.]

highly ordered crystalline domains ranging in size from hundreds of nanometers to a few microns. The high crystallinity implies high charge mobility. With the goal of reaching charge densities sufficiently high to approach the insulator-to-metal transition, electrical properties of PBTTT-FETs were studied over a wide range of temperatures, magnetic fields, carrier densities (gate voltages) and source–drain voltages.

Figure 14.12 shows typical four-probe conductivity (σ) as a function of inverse temperature from 0.26 K to 200 K.

For high V_G and high V_{SD}, the conductivity decreases monotonically with decreasing temperature with a cross-over to a temperature-independent regime that persists down to 0.26 K indicating a transition from insulator to metal (finite conductivity as $T \rightarrow$ zero) as a function of the gate voltage. Transport in the "metallic" regime at $V_G = -150$ V is highly nonlinear, increasing by two orders of magnitude when V_{SD} is increased by only a factor of 2.

Detailed analysis of data show (see Fig 14.13) that measured *I-V* curves at different temperatures fall in onto a universal curve when $I/T^{\alpha+1}$ is plotted against eV/k_BT, as described by the Luttinger liquid (LL) model[5] of one-dimensional (1D) metals,

$$I = I_0 T^{\alpha+1} \cdot \sinh(\gamma eV/k_B T).\ |\Gamma((\beta+1)/2 + i\gamma eV/\pi k_B T)| \tag{14.9}$$

where k_B is the Boltzmann constant, e is the electron charge, I_0 and γ are constants independent of T and V, and $\Gamma(x)$ is the gamma function. The γ^{-1} parameter has been interpreted as related to the number of tunneling barriers that lie along the transport path in the LL.

[5] J. Voit, *Rep. Prog. Phys.*, 58, 977 (1995).

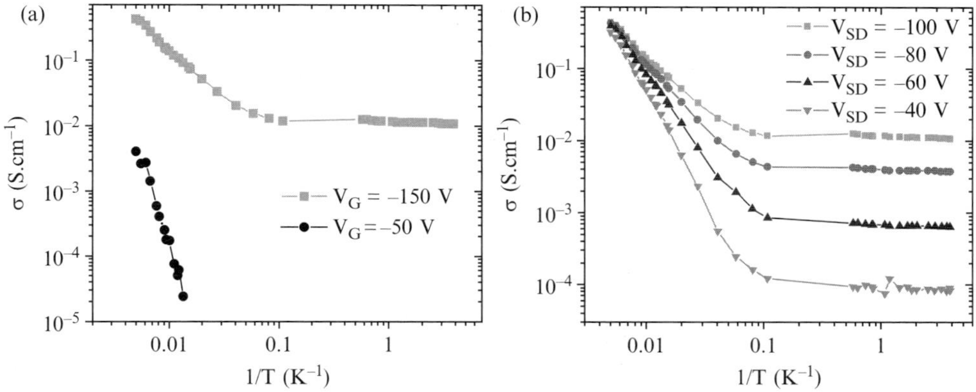

Fig. 14.12 Typical temperature dependence of conductivity (σ) for PBTTT FETs log-log plots of conductivity versus inverse temperature from 200 K to 0.26 K: (a) $V_{SD} = -100$ V for various V_G and (b) $V_G = -150$ V for various V_{SD}. The data were obtained by four-probe measurements on PBTTT in the FET channel. The dashed curves are guides to the eye. For high gate voltages and high source–drain voltages, the conductivity decreases monotonically with decreasing temperature as $\sigma \sim T^{\alpha}$, with a cross-over to a temperature-independent regime. [Reproduced from J. D. Yuen *et al.*, *Nature Materials*, **8**, 572 (2009) with permission from Nature Publishing Group.]

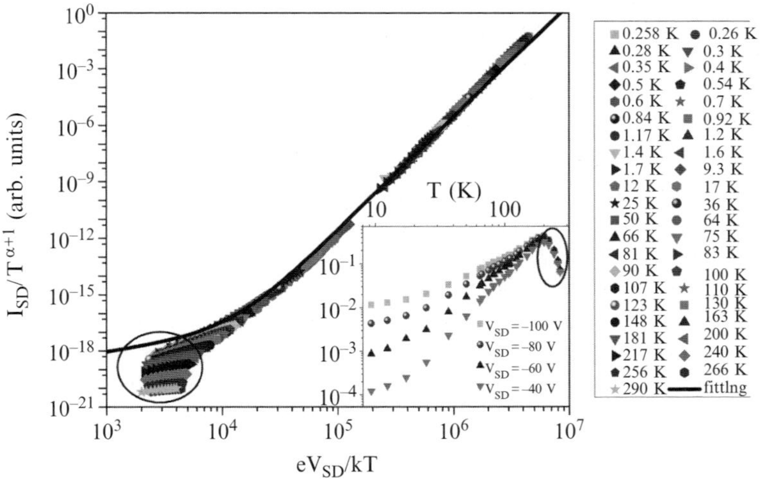

Fig. 14.13 FET data scale onto a "universal" curve for PBTTT FET (SiO_2 as the gate dielectric) measured from 0.26 K to 260 K for $V_G = -150$ V ($\alpha = 5.43$, $\beta = 6.46$). The σ versus T plot in the inset shows the high temperature data in greater detail, indicating that the deviation from the universal LL transport corresponds to the change in slope of σ versus T above 200 K, as shown by the circled regions. [Reproduced from J. D. Yuen *et al.*, *Nature Materials*, **8**, 572 (2009) with permission from Nature Publishing Group.]

In the LL model of the 1D metal, charge and spin are decoupled. The elementary excitations along the chain are collective oscillations of charge density and spin density, propagating coherently, but in general at different velocities. The transport properties of the correlated LL electronic state are described in terms of interaction-dependent power law correlations, $I \sim V^{\beta}$ at low temperatures and $\sigma \sim T^{\alpha}$ at low bias (I and V are the current and voltage, respectively, and σ and T are the temperature and conductivity, respectively), where $\alpha+1 = \beta$. Equation 14.9 assumes that tunnelling is the dominant conduction mechanism, through defects or impurities along a single LL chain and between LL chains. The exponents α and β are related to the tunnelling density of states.

The nonlinear transport data obtained from PBTTT at relatively high carrier densities are in agreement with the predictions of the Luttinger liquid theory. The data obtained using a wide range of parameters (temperature, gate-induced carrier density, source–drain voltage and doping level) are consistent with the prediction that $\alpha+1 = \beta$ and scale onto the universal curve predicted for transport in systems with electronic structure described by the Luttinger liquid model. Variation of the charge density in the channel via gating confirmed the existence of the LL electronic state (α^2 is proportional to $1/V_G$).

The observation of a “new state of matter” at high carrier densities, a correlated 1D metal as described by Luttinger, offers the promise of fundamental studies of broad theoretical interest using organic FETs, rather than restricting the use of such devices for simple characterization of the mobility of a new material.

14.7 Organic dielectrics as gate insulators in OFETs

Organic dielectrics have many advantages over inorganic dielectrics: (i) organic dielectrics can be solution-processed, (ii) they provide smooth films on transparent glass and plastic substrates, (iii) they are suitable for optoelectronics like photoresponsive OFETs due to their high optical transparency, (iv) they can be thermally stable up to 200 °C with a relatively small thermal expansion coefficient, and (v) they have dielectric constants as high as $\varepsilon \sim 10$. The performance of OFETs with organic gate dielectrics critically depends on the choice of the gate dielectric. The use of special gate electrets, for example, allows for the production of nonvolatile all-organic field-effect transistor memory elements due to the large in-built hysteresis in the transfer characteristics. Furthermore, the fabrication of organic FETs that can operate as n-channel and p-channel devices enables their use in integrated circuits.

A selection of polymer dielectrics used in OFETs is displayed in Fig. 14.14.

Figure 14.15(a) shows an example of an OFET using BCB as the gate dielectric and C_{60} as the organic semiconductor.

In Fig. 14.15(b) the currents obtained by sweeping the gate voltage in the forward and backward directions are slightly different causing a hysteresis in the transfer characteristics. This hysteresis can be very large for some organic dielectric/organic semiconductor combinations. This hysteresis can have many different origins such as:

1. Ionic impurities in the organic dielectric layer (Na^+, etc.) can move to the interface between the organic dielectric and organic semiconductor, causing an interfacial charge accumulation.

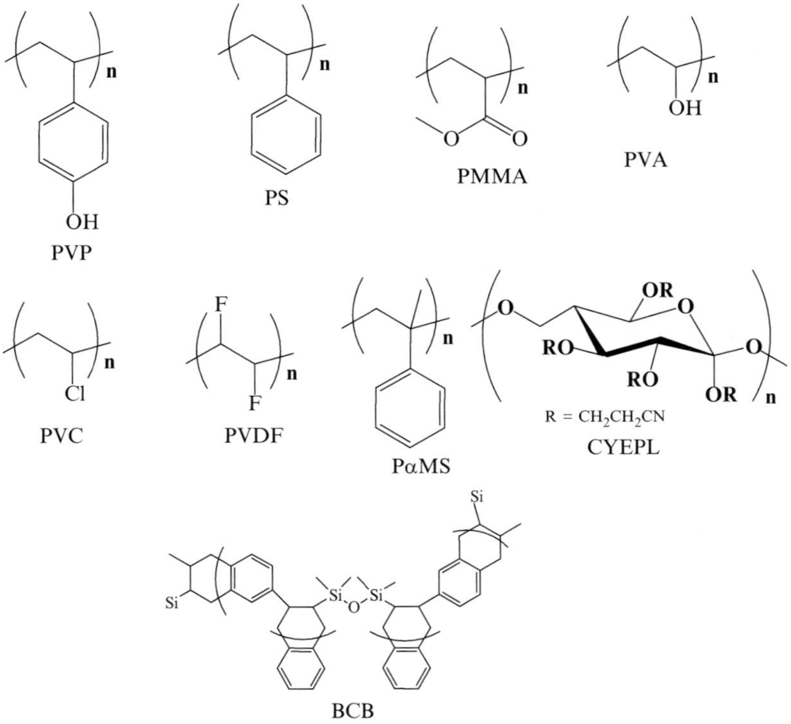

Fig. 14.14 Molecular structure of some commonly used organic dielectrics. PVP: Poly(4-vinyl phenol); PS: Polystyrene; PMMA: polymethyl-methacrylate; PVA: polyvinyl alcohol; PVC: polyvinylchloride; PVDF: polyvinylidenfluoride; PαMS: poly[α-methylstyrene], CYEPL: cyano-ethylpullulan and BCB: divinyltetramethyldisiloxane-bis(benzocyclobutene).

2. The field induced charges can polarize the dielectric creating persistent dipoles. The disordered pattern of pre-existent dipoles in the dielectric can be polarized by the charges in the channel which can in turn affect the charge transport in the channel as well as being responsible for the hysteresis; see Fig. 14.16.
3. There can be charge exchange and charge transfer interactions between the organic semiconductor and organic dielectric layers which create permanent charging of the organic dielectric and organic semiconductor interface. These permanent charges can act either to amplify or shield the electric field from the gate voltage.
4. Large hystheresis can also be demonstrated using ferroelectric polymers (electrets) as gate insulators; the polarization of the gate dielectric is metastable in ferroelectric materials.

The metastability can be useful, for example, in memory type devices where the source–drain current can be turned on and off by applying a write or erase voltage pulse to the gate; see Fig 14.17.

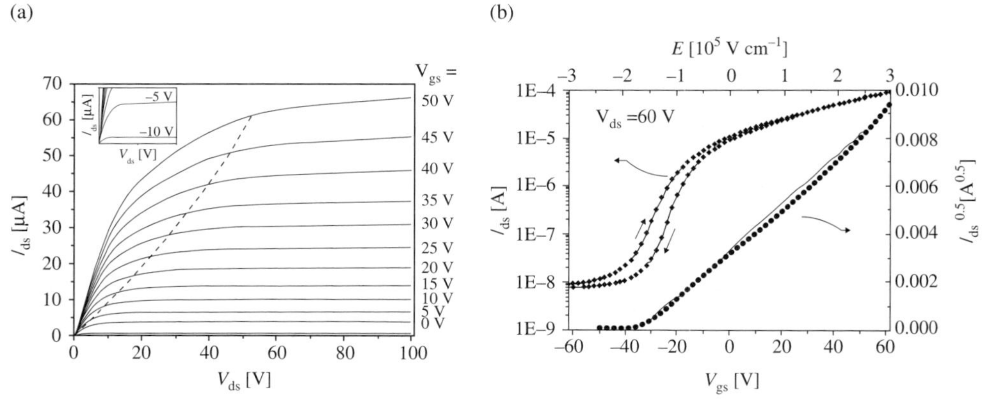

Fig. 14.15 (a) $I_{ds}(V_{ds})$ curves with increasing V_g. The dotted line is a guide to the eye for V_{ds}(sat). Inset: $I_{ds}(V_{ds})$ in the depletion mode. (b) Semilogarithmic plot of I_{ds} versus V_{gs} (left hand scale) and plot of (I_{ds} versus V_{gs} (right hand side scale) for the same C_{60} OFET. The full line shows a fit to Eqn 14.7 from which the field effect mobility $\mu = 0.63$ cm^2/Vs is obtained. [Data reproduced from Th. B. Singh *et al.*, *Organic Electronics*, **6**, 105 (2005) with permission from Elsevier B.V.]

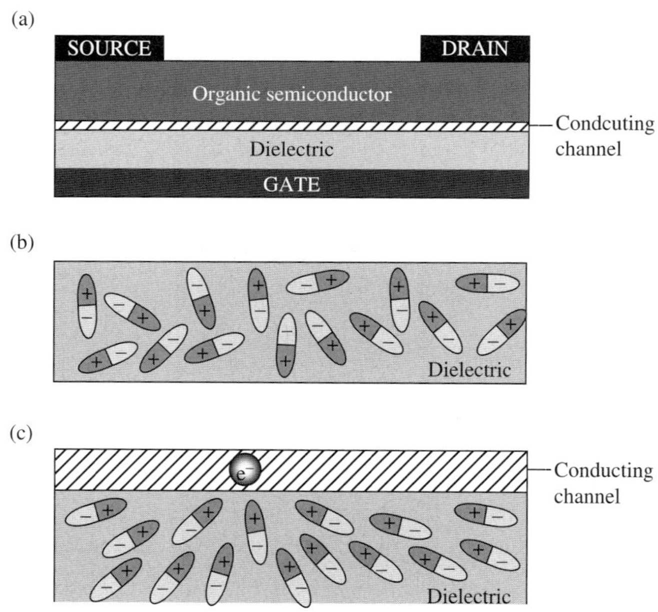

Fig. 14.16 The effect of charges in the channel polarizing permanent dipoles in the dielectric insulator. [Reproduced from J.-L. Bredas *et al.*, *Nature Materials*, **5**, 929 (2006) with permission from Nature Publishing Group.]

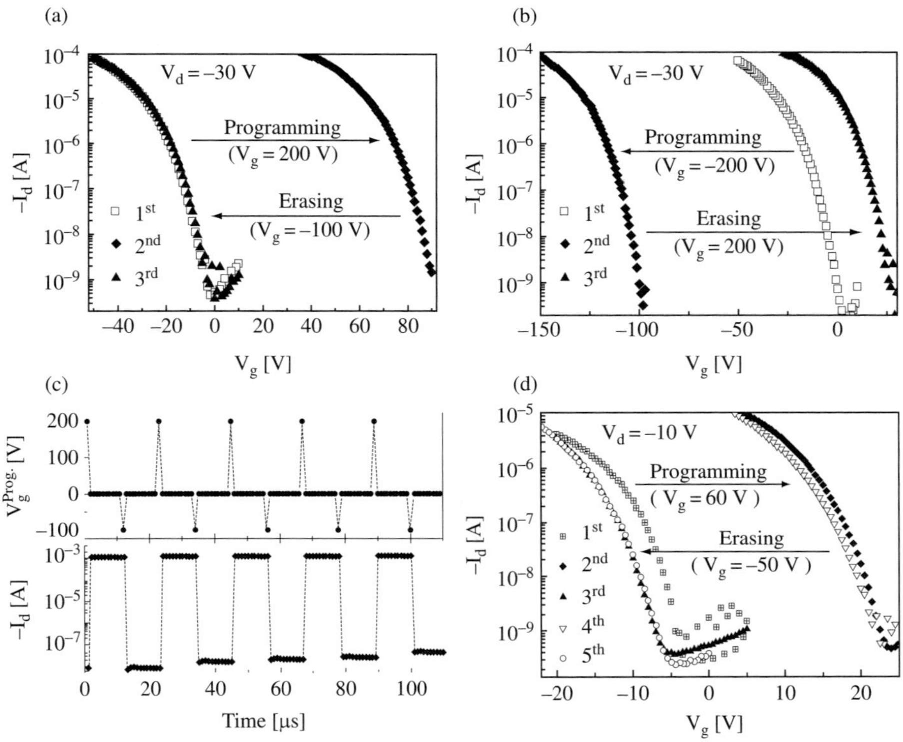

Fig. 14.17 Memory type devices based on Si/SiO$_x$/poly(a-methylstyrene)/pentacene type OFETs. [Reproduced from D.-Y. Kim *et al.*, *Adv. Mater.*, **18**, 3179 (2006) with permission from Wiley-VCH.]

14.8 OFETs for both n-channel and p-channel circuit operation

Many of the known semiconducting polymers are charge symmetric; they are stable in the presence of either electron injection or hole injection. This is quite clear for polymer LEDs where electrons are injected into the π*-band at the cathode and hole are injected into the π-band at the anode. The band structure diagrams shown in Fig. 14.2 imply that with appropriate choice of the semiconducting polymer and the gate dielectric, it should be possible to make OFETs that can be operated either in the p-channel mode or the n-channel mode.

Unipolar OFETs are of course common. Unipolar operation can be caused by instability of the semiconducting polymer to one sign of carrier (as for the polythiophenes which are unstable with electrons in the π*-band) or by selective trapping of carriers with one sign by traps at the semiconductor–dielectric interface. Showing electron transport in polymer FETs had proven difficult. The Cambridge group[6] determined that hydroxyl groups serve as traps for

[6] L.L. Chua *et al.*, *Nature*, **434**, 194 (2005).

electrons at the polymer–silicon dioxide (SiO_2) interface in FETs fabricated on Si with SiO_2 as the gate dielectric. Therefore, by using nonpolar polymers as the gate dielectric materials or by passivating SiO_2, with pure, nonpolar polymer dielectrics, electron transport was demonstrated in a variety of conjugated polymers.

Figure 14.18 compares the output characteristics of the two pentacene OFETs demonstrating strikingly different transport properties. Devices using a dielectric polymer PVP in a p-channel

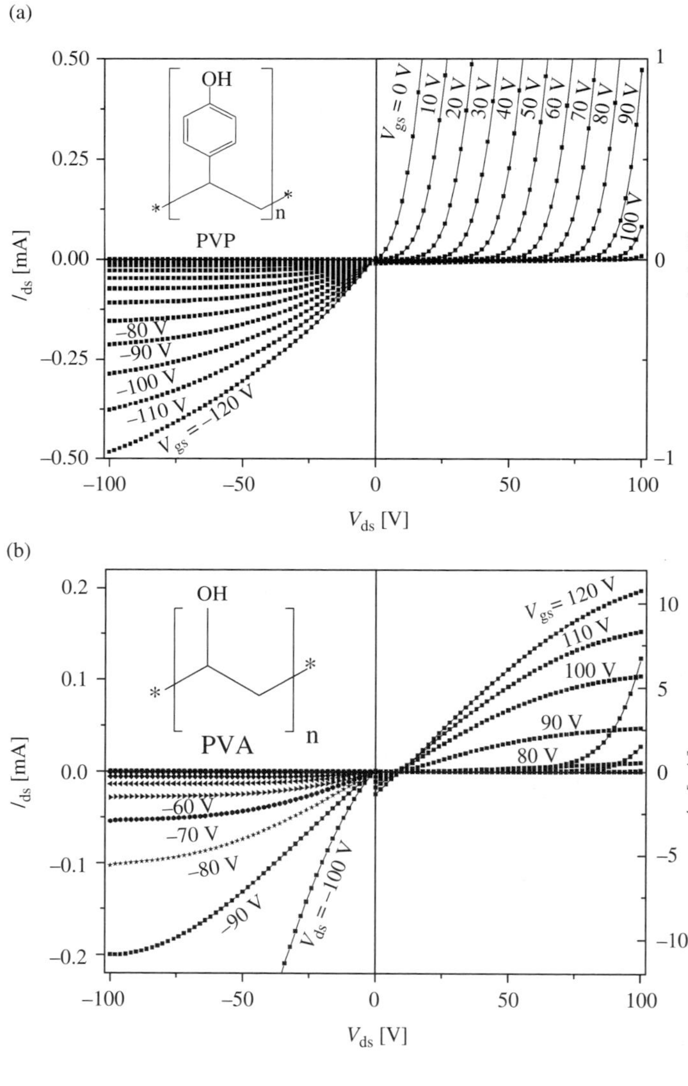

Fig. 14.18 (a) Output characteristics of unipolar type pentacene OFETs using PVP as gate insulator; (b) ambipolar pentacene OFET using PVA as insulator. [Data reproduced from Th.B. Singh *et al.*, *Adv. Mater.*, **17**, 2315 (2005) with permission from Wiley-VCH.]

mode show large on-current. In the n-channel mode, the hole current is depleted by applying a larger gate voltage and there is no evidence for an electron current. For the devices using a dielectric polymer PVA, both p-channel and n-channel formation is observed when the direction of the gate voltage is reversed. The hysteresis observed in PVA based devices is discussed above and attributed to the presence of mobile ionic charges.

14.9 Light-emitting field effect transistors (LEFETs)

Light-emitting field effect transistors (LEFETs) combine the electrical properties of a transistor with light emission in a single device architecture. Due to theire unique properties, LEFETs are attractive for new applications such as simplified pixels in flat panel displays, as optoelectronic devices in communications, sensors, and potentially as electrically driven lasers. The architecture and fabrication step of the LEFET is shown in Fig. 14.19.

The gate dielectric is made of thin film SiN_x (or SiO_2) coated with a thin layer of saturated polymer (e.g. PPCB, PVDF, PMMA, etc.). A thin layer of saturated polymer removes the electrons traps and facilitates electron transport.

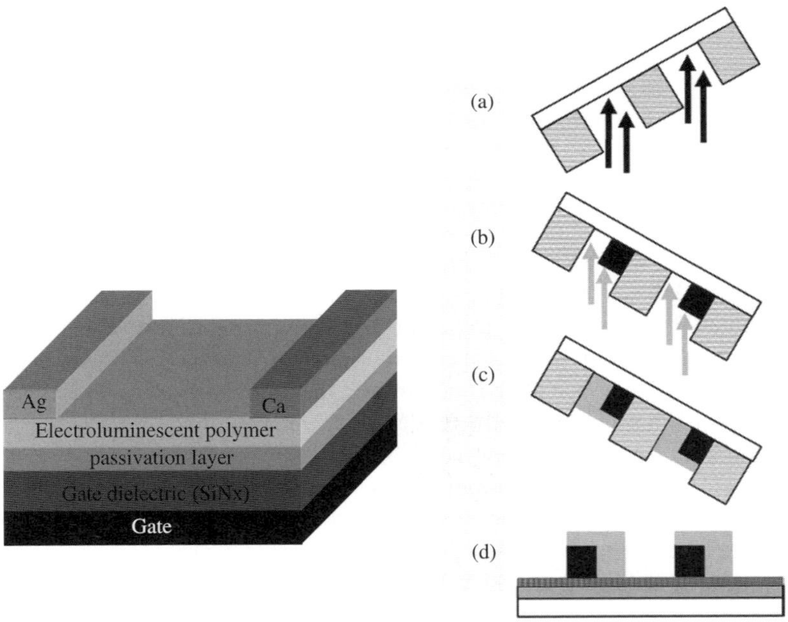

Fig. 14.19 LEFET architecture and schematic of the fabrication of the top-contact two-color electrode geometry by the angled evaporation technique. (a) The substrate is mounted on a silicon shadow mask at an angle to the metal sources, and the first metal, Ca, is evaporated. (b) The angle of the mask is changed and the second metal, Ag, is evaporated. (c) The deposition of the two-color electrodes is complete. (d) Final structure of the two color LEFET after the removal of the silicon shadow mask.

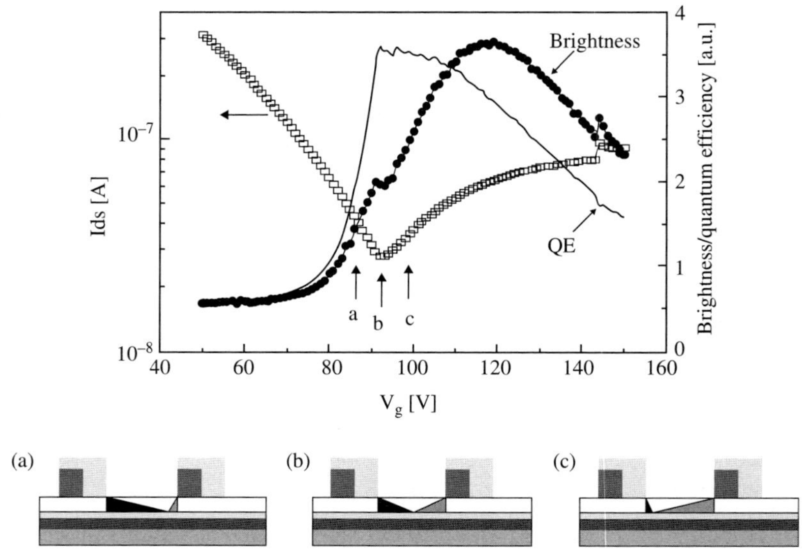

Fig. 14.20 Transfer scan for the ambipolar LEFET along with the corresponding emitted light intensity. (a) Hole channel is well developed and hole current dominates; (b) shows a situation where ambipolar current is present; and (c) the electron channel is well developed and electron current dominates. [Data reproduced from James Sherman Swensen, Ph.D. thesis, *Light emission from an ambipolar semiconducting polymer field-effect transistor*, UCSB 2007.]

The electrons and holes are injected into the semiconducting polymer from source and drain electrodes made from high and low work function metals. The high work function electrode serves as the source for holes into the HOMO and the drain for electrons from the LUMO; the low work function electrode serves as the source for electrons into the LUMO and the drain for holes from the HOMO. Depending on the electron and hole mobility, injected electrons and holes recombine within the channel leading to the electroluminescent emission. Both ambipolar and unipolar operation is possible. In the following sections we will discuss in more detail the operating principles of LEFETs. As an example, we focus on data obtained from superyellow as the light-emitting semiconducting polymer. The electron passivation layer in ambipolar LEFET was polypropylene-co-1-butene (PPCB).

Figure 14.20 shows a transfer scan together with the gate dependence of the light emission of ambipolar LEFETs. The low work function source electrode (Ca) was grounded and the gate voltage (V_g) was swept from 0 to 200 V by keeping V_{ds}=200V. When $V_g = 0$, there is no voltage drop between the source and gate. There is, however, a 200 V drop between the drain and gate which polarizes the gate dielectric and induces a hole channel in the vicinity of the high-work-function drain electrode (Ag). The current in this region (at lower V_g values) of the transfer scan is hole dominated (Fig. 14.20a).

As V_g increases, the voltage drop between the drain and gate decreases, causing the magnitude of the hole current to decrease. Simultaneously, the voltage drop between the source and the gate increases, but is oppositely charged, causing the gradual buildup of an electron channel near the low-work-function source electrode (Ca). At about $Vg = 90$ V, I_{ds} reaches a minimum and then begins to increase again. This is the cross-over point from hole dominated current to electron dominated current (Fig. 14.20b). Cross-over is expected to occur at $V_g = V_d$. At higher gate voltages, there is very little voltage drop between the gate and drain. There is, however, a large voltage drop between the gate and source, causing an electron channel to develop, extending across the channel from the low work function calcium electrode (Fig 14.20c).

The light intensity data corresponding to the transfer scan is also shown in Fig. 14.20. The light intensity data begins increasing around 80 V while the hole current still dominates, reaching a maximum around 120 V, well into the electron dominated current regime. At the cross-over point the hole and electron currents are nearly equal to each other, and the quantum efficiency for light emission is maximum.

The emission zone is not stationary within the channel, but in fact moves from the source to the drain as the gate voltage is swept from 88 V (a) to 93 V (b) to 98 V (c). The images shown in Fig. 14.21 correspond to data points shown Fig. 14.20 with arrows a, b, and c.

In the limit of infinite recombination rate, the recombination zone (and thus the emission zone) would be infinitely narrow. In reality, however, the width of the emission zone is determined by the recombination rate; the smaller the recombination rate, the wider the emission zone. The emission zone was spatially resolved using confocal microscopy. The full width at half maximum of the emission zone is consistently 2 μm as it moves across the channel.

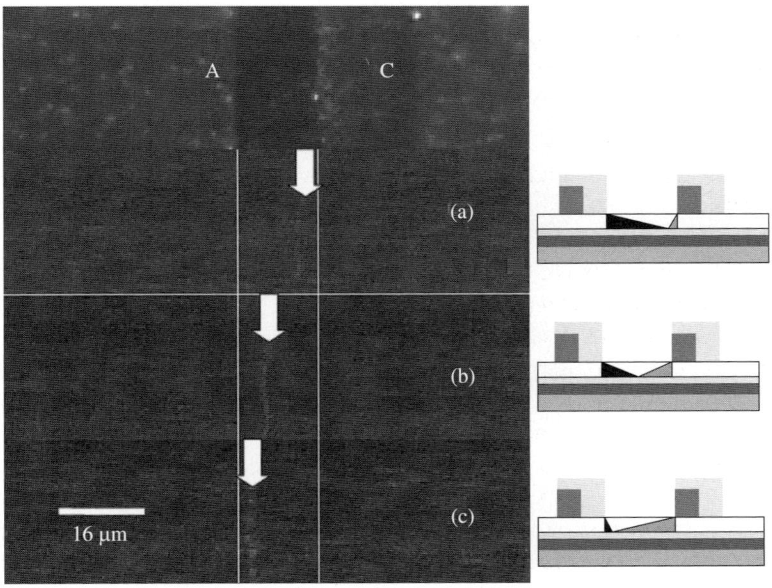

Fig. 14.21 The top photograph shows the Ag and Ca electrodes and the channel of the two-color LEFET. Photographs (a), (b), and (c) are taken in the dark while the device is operating. The emission zone moves from near the Ca electrode, (a) $V_g = 88$ V, to mid channel, (b) $V_g = 93$ V, and then approaches the Ag electrode, (c) $V_g = 98$ V. [Data reproduced from J. S. Swensen *et al.*, *Appl. Phys. Lett.*,**87**, 253511 (2005) with permission from the American Physical Society.]

14.10 High brightness LEFETs

The brightness and light-emission efficiency of LEFETs is poor compared to the OLEDs. This is mainly due to the challenges in developing LEFET materials. To achieve higher performance, it is essential to obtain materials (or material combinations) with both high photoluminescence quantum efficiency and high carrier mobilities. A number of conjugated polymers and small molecules single crystals have been investigated. Conjugated polymers possess moderate photoluminescence, but the amorphous nature of the thin film usually results in low carrier mobility. Single crystals offer very high carrier mobility due to the well-ordered packing of molecules. However, concentration quenching leads to a significant reduction of photoluminescence quantum efficiency.

There are two approaches to the strategy for improving LEFET performance:

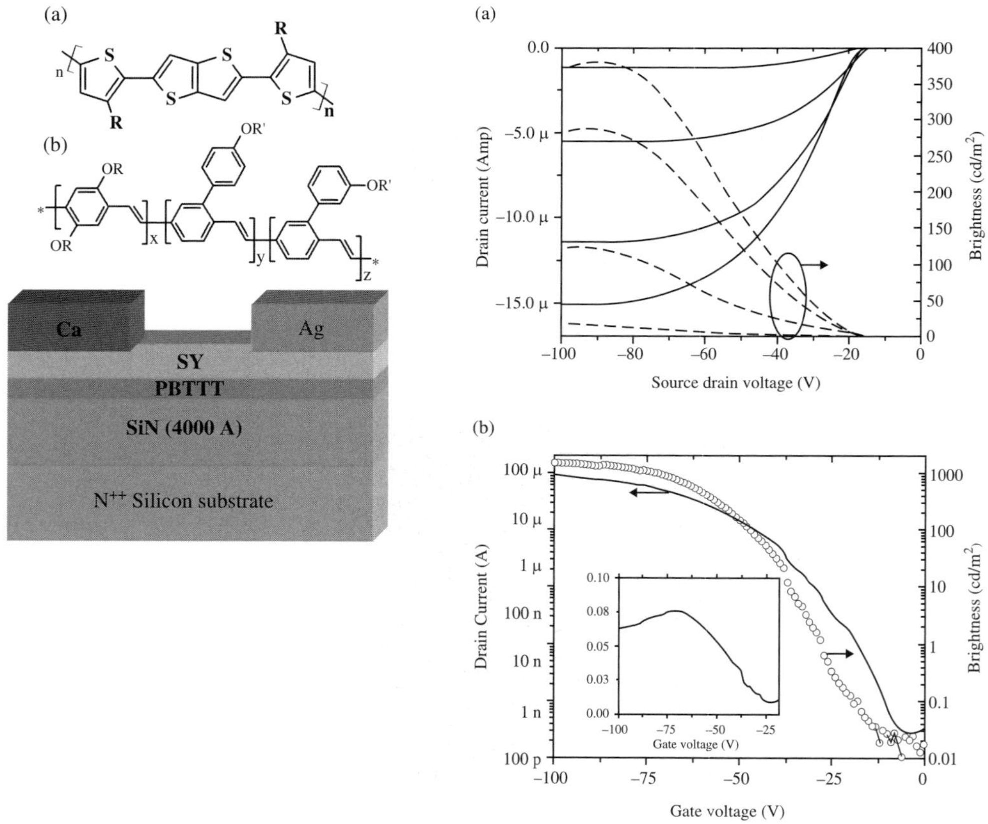

Fig. 14.22 Left: (a) molecular structures of PBTTT ($R=C_{14}H_{29}$) and (b) Superyellow and schematic diagram of the device architecture. Right: electrical and optical characteristics of LEFET. [Data reproduced from E. B. Namdas *et al.*, *Appl. Phys. Lett.*, **92**, 183304 (2008) with permission from the American Physical Society.]

(a) bilayer films comprising a charge transport layer and a light-emitting layer;
(b) phosphorescent materials in a suitable charge transport host.

14.10.1 Bilayer LEFETs

Figure 14.22 shows bilayer LEFETs comprising a hole transporting polymer, poly(2,5-bis(3-alkylthiophen-2-yl)thieno[3,2-*b*]thiophene) (PBTTT) and a light-emitting polymer, Superyellow, a polyphenylenevinylene (PPV) derivative. The brightness increases with V_{DS} and V_{GS} and then reaches saturation. The maximum brightness reached approximately 1500 cd/m^2 at V_{GS}, $V_{DS} = -100$ V. In this device recombination occurs near the interface of the PBTTT/SY and close to the edge of Ca electrode and the emission zone does not move with the gate voltage.

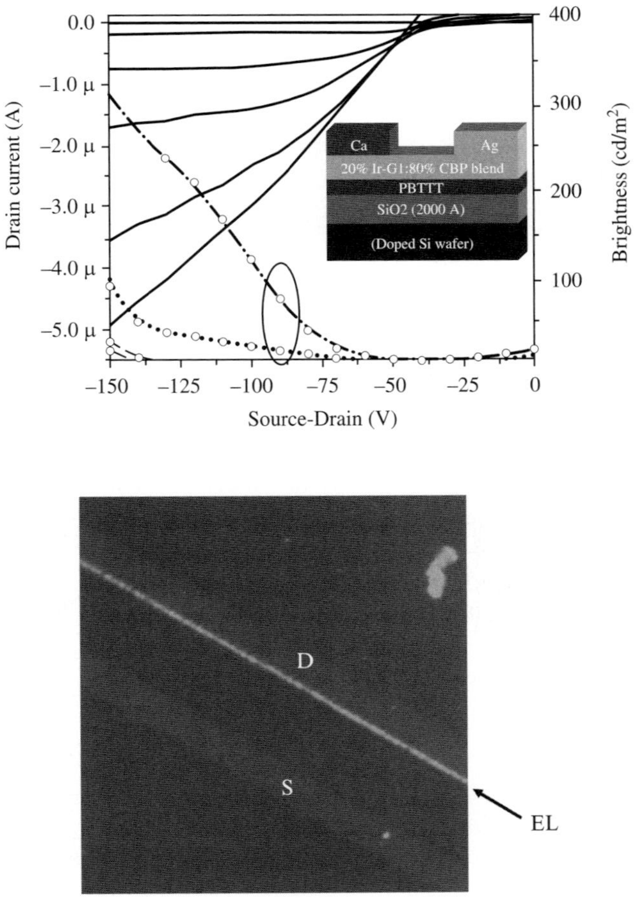

Fig. 14.23 Top: electrical and optical output characteristic of 20 wt% Ir-G1:CBP LEFET at various gate voltages; the inset shows the device architecture. Bottom: microscope image of LEFET device in operation. [Data reproduced from E. B. Namdas *et al.*, *Adv. Mater.*, **21**, 4951 (2009) with permission from Wiley-VCH.]

14.10.2 Phosphorescent LEFETs

Phosphorescent materials have been successfully used to improve the efficiency of OLEDs. The incorporation of heavy atoms into organometallic complexes induces spin–orbit coupling which enables efficient intersystem crossing from the singlet to the triplet states and therefore leads to efficient OLED devices that are able to harvest the emission from both singlet and triplet excitations. However, the channel length in LEFETs is three orders of magnitude larger than the film thickness in OLEDs, meaning that the mobility of phosphorescent OLED materials is insufficient for their direct application in LEFETs. In order to overcome this challenge, it is necessary to (i) blend the phosphorescent materials in a suitable charge transport host; and (ii) fabricate bilayered films comprising a charge transport layer and a light-emitting layer. Figure 14.23 shows an example of a phosphorescent LEFET. The highest EQE was in the range of 0.28–0.45% at 310–740 cd/m^2. These relatively high efficiency and brightness values for LEFETs result from utilizing phosphorescent dendrimers in the bilayer structure and blending in a suitable charge transport host.

Further reading

Z. Bao and J. Locklin (eds.), *Organic Field Effect Transistors*, Taylor and Francis, Boca Raton, 2007.

G. Hadziioannou and G. G. Mallarias (eds.), *Semiconducting Polymers*, Wiley-VCH, Weinheim, 2007.

H. Klauk (ed.), *Organic Electronics*, Wiley–VCH, Weinheim, 2006.

A. Moliton, *Optoelectronics of Molecules and Polymers*, Springer, Berlin, 2006.

H. S. Nalwa (ed.), *Handbook of Organic Electronics and Photonics*, American Scientific Publishers, 2008.

R. Shinar and J. Shinar (eds.), *Organic Electronics in Sensors and Biotechnology*, Mc Graw-Hill, New York, 2009.

S. S. Sun and L. Dalton (eds.), *Introduction to Organic Electronic and Optoelectronic Materials and Devices*, Taylor and Francis, Boca Raton, 2008.

J. Zaumseil, R. H. Friend, H. Sirringhaus, *Nature Materials*, **5**, 69 (2006).

Index

absorption
doping induced 91
nondegenerate ground state polymers 101–3
photoinduced 97–101
as a loss mechanisms 181–4
nondegenerate ground state polymers 105–7
time resolved 101, 106–7
trans-$(CH)_x$ 101
trans-polyacetylene 97–100
structural disorder effects 122–3
acceptors see electron–acceptors; hole–acceptors
acetylene 8
amplified spontaneous emission (ASE) 198–9
amplified stimulated emission (ASE) 184–6
gain influence 188–9
spectral narrowing 187
thresholds 188
Anderson localization 46–8
Anderson transition 46
aniline 36
see also polyaniline (PANI)
autocorrelation function (ACR) 226–7

band gap engineering 133–5
band structure 5–12
dimerized *trans*-$(CH)_x$ chain in tight-banding approximation 8–10
experimental verification of 10–12
uniform *trans*-$(CH)_x$ chain in tight-banding approximation 5–8
bimolecular recombination of charge carriers 128
bipolarons 88–9, 101–2
neutral 86
Bose-Einstein condensation 89
bulk heterojunction (BHJ) materials 219
nanomorphology of BHJ solar cells 219–21
characterization 222–7
photodetectors 246–9
self-assembly 219
thermal annealing 221–2
see also solar cells

charge carriers
organic field effect transistors 257–9
charge transport at high carrier density 259–62
photogeneration 125–6
polymer based LEDs 139–48
improved carrier injection approaches 146–8
single carrier devices 140–5
recombination 127–8
bimolecular recombination 128
monomolecular recombination 127–8
transport in disordered semiconducting polymers 128–30
charge injection 18–21
at metal–semiconducting polymer (MS) interface 29
electrochemical 24–5
see also doping
charge transfer 21–2
organic field effect transistors 259–62
see also conductivity; photoinduced electron transfer
chemical doping 21–2
see also doping
cis-polyacetylene 81
see also polyacetylene
conducting polymers 3–4
band structure 5–12
experimental verification 10–12
blends with insulating polymers 37–9
electronic structure 4–5
molecular structures 12, 13
processing approaches 31–2
counter-ion induced processing 35–6
dispersions of metallic polymers in water 36–7
precursor route 34–5
side-chain functionalization 33–4
see also metallic properties of conducting polymers
conductivity 19, 76
critical regime 52–3
estimating 78–80

conductivity (*cont.*)
higher conductivity achievement 75–8
PANI/CSA 40, 52–3, 71–5
see also charge carriers; metallic properties of conducting polymers; photoconductivity
contrast transfer function 223

dangling bonds 111
Davidov splitting 179–80
DFB polymer laser 193–5
DFB structure 196–8
dielectric function 58
disorder 12–14, 47
consequences of strong disorder 44–6
in semiconducting polymers 122–3
charge carrier transport 128–30
dispersions of metallic polymers in water 36–7
donor–acceptor (D–A) bilayer heterojunctions 217–19
doping 18–21
chemical 21–2
electrochemical 22–4
induced absorption 91
nondegenerate ground state polymers 101–3
interfacial 29
methods 20
n-type 23–6
p-type 23–6
photoinduced 28–9
protonic-acid 27–8
Drude behavior 60–1, 75
DuPont Gen 3 156, 157

electrical pump lasing 195–8
electroabsorption (EA) spectroscopy 111–12, 117
electrochemical charge injection 24–5
electrochemical doping 22–4
see also doping
electroluminescence efficiency
limits 149–51
fluorescent LED 150–1
phosphorescent LEDs 149–50
thickness dependence 148
electron spin resonance (ESR)
photoinduced, *trans*-polyacetylene 100–1
spin susceptibility measurements 91–4
nondegenerate ground state polymers 104–5
electron transfer see conductivity; photoinduced electron transfer
electron–acceptors 146–8
donor–acceptor (D–A) bilayer heterojunctions 217–19
photoinduced electron transfer to 201–4
ultrafast 206–11
photophysics of polymer/acceptor composites 202–4
electron–electron repulsive interaction 14–16
emeraldine 27–8, 35
ethylene-dioxythiphene (EDOT) 36
exciplex 114
excitons 113–15
binding energy 151
measurement 126–7

Fermi energy control 111–13
Fermi glass 45–6, 47, 49
field effect transistors (FETs) 250
light-emitting (LEFETs) 267–9
high brightness 270–2
see also organic field effect transistors (OFETs)
fill factor 215
flexible LED displays 153
fluorescent LED 149–50
Fowler–Nordheim tunneling in polymer based LEDs 138–48
improved carrier injection approaches 146–8
MIM model limits 145
operating voltage and efficiency 145
single carrier devices 140–5
Frenkel exciton 114
frozen junction LECs 169–72

Hagen–Rubens (H–R) relation 59–60
hexagonal symmetry 2
high brightness light-emitting field effect transistors (LEFETs) 270–2
bilayer LEFETs 271
phosphorescent LEFETs 272
hole-blocking layers 146–8
hole–acceptors 146–8
Huang–Rhys factors 119

infrared (IR) measurements, metal–insulator transition 56–65
infrared active vibration (IRAV) modes 73–5, 94–7, 101, 125–6
nondegenerate ground state polymers 105
interfacial doping 29
interpenetrating networks 38
Ir metal–organic complex, Ir(HFP)3 153, 156

Jablonski diagram 117–19

Knight shift 68–9
Kramers–Kromig (K–K) relations 58

Lambertian source 136
laser materials 177
amplified stimulated emission (ASE) 184–6
gain influence 188–9
spectral narrowing 187
thresholds 188
electrical pump lasing 195–8
laser resonant structures 191–5
DFB polymer laser 193–5
microcavity polymer laser 192–3
lowering the lasing threshold with energy transfer 189–91
optical gain 177–81
photoinduced absorption as a loss mechanism 181–4
light induced ESR (LESR) 205–6
trans-polyacetylene 100–1
light-emitting diodes (LEDs) 132, 214
see also polymer based light-emitting diodes (PLEDs)
light-emitting electrochemical cells see polymer based light-emitting electrochemical cells (LECs)
light-emitting field effect transistors (LEFETs) 196, 197, 267–9
high brightness 270–2
bilayer LEFETs 271
phosphorescent LEFETs 272
localization-modified Drude model (LMD) 61–2
luminescent microcavity 199
luminous efficiency 137, 138
see also electroluminescence efficiency
luminous intensity (luminance) 136
measurement 136–8
luminous power efficiency 138

magnetic field, metal–insulator transition and 54–6
magnetic field induced cross-over 53–4
magnetic susceptibility see spin susceptibility
MBL-PPV see poly(2-methoxy,5-(2'-butoxy solfonate)-p-phenylene vinylene) (MBL-PPV)
MEH-PPV see poly(2-methoxy,5-(2'-ethylhexoxy)-*p*-phenylene vinylene) (MEH-PPV)
metal-oxide semiconductor field effect transistor (MOSFET) 250
metallic properties of conducting polymers 66–7
magnetic susceptibility 67–9
reflection 70, 72
specific heat 69
thermopower 69–70, 71
transport measurements 71–5
metal–insulator (M–I) transition 14–16, 18, 46
critical regime 51–2
electrical conductivity 52–3
infrared measurements 56–65
magnetic field and 54–6
magnetic field induced cross-over 53–4
metal–semiconductor–metal (MSM) diodes 108–11, 216–17
Fowler–Nordheim tunneling 138–48
improved carrier injection approaches 146–8
MIM model limits 145
operating voltage and efficiency 145
single carrier devices 140–5
methyltrioctylammonium trifluoromethanesulfonate (MATS) 174–5
microcavity polymer laser 192–3
midgap electronic state 84–7, 91
midgap transition 98–9
monomolecular recombination of charge carriers 127–8
Mott transition 14–16, 46

naphthalene 21
neutral bipolaron 86
nondegenerate ground state polymers 85–7
bipolarons 88–9
doping induced absorption 101–3
IRAV modes 105
photoinduced absorption 105–6
time resolved 106–7
polarons 87, 101–2
spin susceptibility, ESR measurements 104–5

open circuit voltage 215
optical beam induced current (OBIC) microscopy 164–5
optical gain 177–81
optical properties
conducting polymers 67
semiconducting polymers 113–15
oriented and aligned polymer chains 115–22
vibronic structure 117–22
organic dielectrics 262–4
organic field effect transistors (OFETs) 250
architectures 254–6
charge transport at high carrier density 259–62
contact resistance 256
for both n-channel and p-channel circuit operation 265–7
mobility measurement errors 256–7
operating principles 251–4
organic dielectrics as gate insulators 262–4
polarons as gate induced carriers 257–9
see also field effect transistors (FETs)

organic light-emitting diodes (OELDs) 132
see also polymer based light-emitting diodes (PLEDs)

Pauli spin susceptibility see spin susceptibility
PCPDTBT 233, 239, 240–3
recombination in 244–6
Peierls gap 10
Peierls instability 10
percolation 37–8
phosphorescent dendrimers 149–50
photoconductivity
charge carrier photogeneration 125–6
onset measurement 123–4
sensitization of 211–13
see also photoinduced electron transfer
photodetectors 214–17, 246–9
photoinduced absorption
as a loss mechanism 181–4
nondegenerate ground state polymer 105–7
time resolved 101, 106–7
trans-$(CH)_x$ 101
trans-polyacetylene 97–100
photoinduced doping 28–9
see also doping
photoinduced electron transfer 201–4
from semiconducting polymers to C60 204–5
sensitization of 211–13
ultrafast 206–11
see also photoconductivity
photoinduced ESR, *trans*-polyacetylene 100–1
photoluminescence 203–4
decay 206–9
photopic luminosity function 136
photovoltaic cells 214–17
see also solar cells
π-electron 4
π-orbital 4
pixelated displays 151–5
flexible displays 155
polarons
as carriers in organic field effect transistors 257–9
nondegenerate ground state polymers 87, 101–2
photogeneration 125–6
semiconducting polymers 114–15
poly-para-phenylene (PPP) 85–6
poly(2,5-bis(3-tetradecylthiophen-2-yl)thieno [3,2-b] thiophene) (PBTTT) 259–62, 271
poly(2,5-thienylenevinylene) (PTV) 254
poly(2-methoxy,5-(2'-butoxy solfonate)-*p*-phenylene vinylene) (MBL-PPV) 33
poly(2-methoxy,5-(2'-ethylhexoxy)-*p*-phenylene vinylene) (MEH-PPV) 33
in polymer based LEC 161–2
in polymer based LEDs 133, 140–5
optical properties 115–22, 179, 180
photoinduced absorption 183–4
photoinduced conductivity 212
structural disorder effects 122–3
poly(3-alkylthiophenes) (P3ATs) 33
poly(3-hexylthiophene) (P3HT) 33, 34, 232
P3HT:PCBM BHJ solar cells 224–6
thermal annealing 221–2
poly(3-octylthiophene) 208–9
polyacetylene 3–4, 8, 81–3
band structure 5–12
experimental verification 10–12
doping 18, 19, 21, 22
induced absorption 91
electronic structure 4–5
higher conductivity achievement 75–6
molecular structure 81–2
photoinduced absorption 97–100
photoinduced ESR 100–1
reflectance 70, 72
solitons in 83–5
spin susceptibility, ESR measurements 91–4
thermopower 69, 70
polyaniline (PANI) 27–8
counter-ion induced processing 35–6
PANI/CSA 35, 38, 40
conductivity 40, 52–3, 71–5
metal–insulator transition 54–7
thermopower 69–70, 71
self-stabilized dispersion polymerization (SSDP) 71–3
transport measurements 71–5
polyethylene 3
poly(ethylenedioxythiophene)-polystyrene sulfonate (PEDOT-PSS) 36
polyfluorene (PFO) 154, 156
polymer based light-emitting diodes (PLEDs) 132–5, 214
band gap engineering 133–5
dual function polymer diodes 214–17
electroluminescence efficiency 148–51
limits 149–51
thickness dependence 148
electronic structure 138–9
Fowler–Nordheim tunneling 138–48
improved carrier injection approaches 146–8
MIM model limits 145
PLED operating voltage and efficiency 145
single carrier devices 140–5

measurement of parameters 135–8
operating lifetime 157
pixelated displays 151–5
flexible displays 153
versus LECs 167–8
white light emission 154, 156–7
see also metal–semiconductor–metal (MSM) diodes
polymer based light-emitting electrochemical cells (LECs) 158
device architecture 158–9
frozen junction LECs 169–72
junction formation
confirmation 164–6
in "sandwich" configuration 169
operating lifetimes 174–6
operating mechanism 159–63
phase separation control 166–7
surface geometry configuration 163–4
versus LEDs 167–8
with ionic liquid additives 174–6
with ionic luminescent polymers 172–4
polymer battery 26
poly(phenylene vinylene) (PPV) 12, 33, 34
polythiophene
doping induced absorption 101–3
IRAV modes 105
photoinduced absorption 105–6
spin susceptibility 104
power spectral density (PSD) 223–5
printed plastic electronics 32
protonation 27–8

recombination of charge carriers 127–8
bimolecular recombination 128
monomolecular recombination 127–8
reflectance 70, 72
measurements 56–65
resonators 191–5

self-stabilized dispersion polymerization (SSDP) 71–3
semiconducting polymers 1–2, 130–1
as laser materials *see* laser materials
charge carrier photogeneration 125–6
charge carrier recombination 127–8
charge carrier transport 128–30
exciton binding energy 126–7
Fermi energy control 111–13
metal–semiconductor–metal (MSM) diodes 108–11, 216–17
Fowler–Nordheim tunneling 138–48
optical gain in 177–81
optical properties 113–15
oriented and aligned polymer chains 115–22
vibronic structure 117–22
photoconductivity onset 123–4
photoinduced electron transfer from 201–5
structural disorder in 122–3
see also conducting polymers; doping
short circuit current 215
side-chain functionalization 33–4
solar cells 214–17
bulk heterojunction (BHJ) materials 219
characterization 222–7
nanomorphology 219–21, 222–7
self-assembly 219
thermal annealing 221–2
design rules 235
donor–acceptor (D–A) bilayer heterojunctions 217–19
dynamics 230–2
high internal quantum efficiency 240–4
open circuit voltage origin 227–30
processing additives 232–4
tandem solar cells 238–40
titanium oxide as multifunctional components 235–8
solitons
distortion 84
photogeneration of spinless solitons 100–1
spin-charge relations 85
trans-polyacetylene 83–5
sp^2 hybridization 2
sp_2p_z hybridization 1–2
conducting polymers based upon 3–4
sp_3 hybridization 1–2
traditional polymers based upon 3
specific heat 69
spin susceptibility 66–9
electron spin resonance (ESR) experiments 91–4
nondegenerate ground state polymers 104–5
Stark effect 111
stimulated emission (SE) 178–9
amplified stimulated emission (ASE) 184–6
gain influence 188–9
spectral narrowing 187
thresholds 188
Su–Schrieffer–Heeger (SSH) Hamiltonian 5–6, 8

tandem solar cells 238–40
tetrahedral symmetry 2
thermopower 67, 69–70, 71
thin film diode 108–11
thin film transistor (TFT) 153
titanium oxide 233–8

traditional polymers 3
trans-$(CH)_x$ chain
band structure 5–12
dimerized chain in tight-banding approximation 8–10
experimental verification of 10–12
uniform chain in tight-banding approximation 5–8
electronic structure 4–5
time resolved photoinduced absorption 101
trans-polyacetylene 81–3
photoinduced absorption 97–100
photoinduced ESR 100–1
solitons in 83–5
see also polyacetylene
transparent metallic conductors 40–3

vandalized grid 37, 38
variable range hopping (VRH) 48–50, 129–30
vibronic structure 117–22

Wannier exciton 114
waveguiding 199
white light emission 154, 156–7

Zabrodskii plots 54–6, 57